Approaches to Plant Evolutionary Ecology

APPROACHES TO PLANT EVOLUTIONARY ECOLOGY

G.P. Cheplick

OXFORD
UNIVERSITY PRESS

OXFORD
UNIVERSITY PRESS

Oxford University Press is a department of the University of Oxford.
It furthers the University's objective of excellence in research, scholarship,
and education by publishing worldwide. Oxford is a registered trade mark
of Oxford University Press in the UK and in certain other countries

Published in the United States of America by
Oxford University Press
198 Madison Avenue, New York, NY 10016,
United States of America

Library of Congress Cataloging-in-Publication Data
Cheplick, G. P. (Gregory Paul), 1957–
Approaches to plant evolutionary ecology / Gregory P. Cheplick.
p. cm.
Includes bibliographical references and index.
ISBN 978-0-19-998832-7
1. Plants—Evolution. 2. Plant breeding. 3. Selection (Plant breeding) I. Title.
QK980.C44 2015
581.7—dc23
2014042333

Contents

Preface

In this age of electronic media and information overload, one may well ask whether any new scientific books are needed anymore. Although a book in print form may be increasingly irrelevant, the information contained therein is not. Reference books can become outdated as time goes on, yet they invariably function as a snapshot of what is known about a particular topic at a particular time. A scientific work should synthesize what is known, and not yet known, about the subject at hand.

Every active researcher perhaps at some midpoint in his or her career probably secretly harbors a desire to halt the enormous flow of information, the steady stream of publications that never ceases, and to sit back, take it all in, and ponder the current state of the subject of interest. Would that we all had the time necessary to do just that! Writing a book allows, indeed *forces*, one to do just that (although the information tap continues to flow with no way to turn it off). It forces one to "catch up" on all those studies over the years that may have been missed and to, hopefully, put it all together in a way that creates a coherent, integrative picture. It is my hope that this book provides graduate and undergraduate students who have an interest in ecology, evolution, and/or plants an opportunity to learn of the diverse approaches used by biologists to understand the evolutionary ecology of natural populations. In addition, because no researcher necessarily is aware of all pertinent aspects of his or her subject of interest, and the many related topics that likely overlap with it, it is my hope that professional ecologists find this book to be useful as well. "Useful" in the sense of summarizing the relevant literature and the enormous range of approaches investigators have used in the broad field of plant evolutionary ecology. Note, moreover, that many of the approaches considered in *Approaches to Plant Evolutionary Ecology* have also been widely used to investigate *animal* ecology and evolution.

To gain a clearer picture of what this book is about, let's first begin with two points regarding what this book is *not*. First, it is not a compendium of ecological or evolutionary theory, and it is assumed that the reader has some basic background in ecology and evolutionary biology. Many excellent texts already exist that provide the theoretical foundations of the vast subjects of evolution and ecology. Rather than providing much redundant theory or mathematical models, this book takes an empirical approach, emphasizing experimental studies of specific plant species and what they reveal about microevolution and adaptation in relation to the many agents of natural selection. The focus is at the population and genotype levels, because this is where microevolutionary dynamics are most evident.

Second, this text is not a methods manual or handbook of statistical techniques used in evolutionary ecology. Again, there are many review papers, edited volumes, and texts available that contain details on methodology and data analyses appropriate to the approaches covered (as in molecular ecology in Chapter 5). I have tried to cite a good amount of the literature to guide interested readers to potentially useful sources.

What, then, *is* included in this book? It contains an overview of the diversity of approaches that have been, and are being, used to investigate different aspects of the evolutionary ecology of plant populations. This is a vast topic to pursue, as I soon discovered while compiling and reading the published literature. Each chapter topic is broad enough to be a book unto itself. Thus, I had to be selective regarding which studies to include and use as key examples to illustrate specific topics, and which types of data to present. My emphasis is on empirical research reporting original data, although review articles are cited when useful to summarize earlier literature. I hope I have included a fair sampling of plant species with a wide range of growth forms, breeding systems, and habitat distributions. I have not ignored the earlier literature; much of what we know today, as well as the types of approaches adopted in modern research, exists as an offshoot of earlier studies and approaches. I also have the uneasy feeling that a good fraction of the junior research scientists coming along are not well versed in the findings of the vast pool of ecologists who came before (and also largely developed the field as we currently know it). Unsure if this is true? If you are a well-seasoned, senior ecologist in academia, try asking a graduate student to name the five top ecologists, alive or not, who have had the greatest effect on plant ecology. Or, try pulling a "big name" out of the proverbial hat, one of the most influential plant ecologists you can fathom, and ask the student to describe his or her research accomplishments.

When attempting to write a book on such a vast subject, it may be best to proceed by asking, in this order, three key questions:

1. What *should* be included?
2. What *can* be included?
3. What *will* be included?

As one proceeds through these questions, one can narrow the topic increasingly and focus on the primary points and the studies that illustrate them. This way of thinking has been behind the writing of this book. Many more published studies (and reviews) are out there than can ever be included in a text of reasonable size. Despite this, the reference list is quite extensive (final update: June 2014). Hopefully, this book and the studies described herein provide a snapshot of the field of plant evolutionary ecology at this time and serve as a springboard for future studies.

Acknowledgments

I thank the College of Staten Island of the City University of New York for granting me a sabbatical for the 2013/2014 academic year, which provided the time needed to compile information, organize it, develop figures and tables, and (finally) actually write the book. I am indebted to the following for reading and commenting on one or more chapters: Diane Byers, Jeffrey K. Conner, Susan A. Dudley, Stanley H. Faeth, Jose Fornoni, Joanna R. Freeland, David J. Gibson, Nancy C. Johnson, Matthew A. Parker, James A. Quinn, Nishanta Rajakaruna, and Jacob Weiner.

The following patiently answered my queries or provided pertinent literature and/or photographs: Ragan M. Callaway, Joanna R. Freeland, James L. Hamrick, Nancy C. Johnson, Adrian Leuchtmann, Pierre Liancourt, John T. Lovell, David B. Lowry, Shaibal Mitra, and Nishanta Rajakaruna.

Finally, I'd like to extend thanks to the highly professional editorial and production personnel at Oxford University Press who made this book possible: Erik Hane, Manikandan Kuppan, Jeremy Lewis, and Hallie Stebbins.

Approaches to Plant Evolutionary Ecology

1 The Domain of Evolutionary Ecology

1.1 INTRODUCTION: THE INDIVIDUAL IN ECOLOGY AND EVOLUTION

The scientific discipline of ecology is concerned with the interactions of organisms in nature and the environmental factors that affect them. These environmental factors include physical (abiotic) features, such as temperature or precipitation, and other living organisms (biotic factors), such as competitors, predators, and pathogens. Because no individual exists in isolation from abiotic and biotic features, all organisms are affected by their local environment. Microenvironmental factors at a local scale affect the growth and reproduction of the individual, as well as its probability of survival. Within a natural population, the myriad effects of microenvironmental factors on individuals ultimately determine the distribution and abundance of species within and between variable habitats. Thus, one aspect of modern ecological research is to determine how the environment affects spatial and temporal patterns in the distribution and abundance of organisms in nature (Begon et al. 2006; Scheiner & Willig 2011).

Key to understanding short-term evolutionary change within population gene pools (microevolution) is the individual, defined as a single genotype. The differential responses to the environment of genotypes within a genetically variable population result in evolutionary change. Similar to ecological research at the population level, these responses typically involve the "big three" components of life history: survival, growth, and reproduction. The explicit tracking of the genetic element, whether genotypes or gene pools, separates evolutionary studies from those of traditional ecology. The tracking may involve specific genotypes or genetically related groups (such as siblings), or the identification of patterns of genetic variation within and among populations through the use of molecular and quantitative genetic techniques.

In evolutionary ecology, the emphasis is on genotypes within population gene pools, and their interaction with abiotic and biotic features of the environment. This interaction determines the Darwinian fitness of individuals and the genetic changes that characterize evolving populations.

A good sampling of the broad domain of evolutionary ecology research is provided in the volume edited by Fox et al. (2001) titled *Evolutionary Ecology: Concepts and Case Studies*. In the preface, the editors state that "evolutionary ecologists consider both historical and contemporary influences on patterns of variation and study variation at all levels" (Fox et al. 2001, p. *v*). The evolutionary consequences of phenotypic variation (Mazer & Damuth 2001), natural selection (Fairbairn & Reeve 2001), adaptation (Reznick & Travis 2001), phenotypic plasticity (Pigliucci 2001), and breeding systems (Waser & Williams 2001) are several of the "recurring themes" explored in the first part of the book. Recognizing that populations of a particular species do not exist in isolation, six other contributions examine interspecific interactions, including plant–herbivore relations (Berenbaum 2001), mutualism (Bronstein 2001), and coevolutionary dynamics (Thompson 2001). Although interspecific interactions have, traditionally, been considered a major part of the domain of community ecology (Morin 2011), the population of one species influences the dynamics of the other species, and vice versa, providing the opportunity for changes in the gene pools of both species' populations—in other words, coevolution (Thompson 2005).

A sample of the types of questions addressed by research in evolutionary ecology includes the following:

- What are the patterns of genetic variation found within and between populations, and what ecological factors or aspects of species' life history account for them?
- How is the variation in a population of diverse genotypes filtered by natural selection, resulting in adaptive evolution?
- Do populations show special adaptations to local habitat conditions?
- How does genetic variation in molecular traits relate to phenotypic variation in ecologically relevant quantitative traits?
- In what ways do biotic interactions, such as competition or symbiosis among two or more species, affect their coevolutionary dynamics?

Of course, no single study could ever hope to answer completely any of these conceptually broad questions that might be posed by basic researchers working on any particular species or system. Lest one surmise that only basic research is done by evolutionary ecologists, rest assured that many applied questions can also be posed:

- How can information on population genetic structure and demographic processes be used to improve the management of rare or endangered species?
- How can the available genetic variation in wild populations of agricultural crop species be used to enhance desirable quantitative traits, directing evolution with artificial selection techniques?
- How can knowledge of the breeding system of an invasive plant species help in understanding its demography and population genetics, and suggest ways to minimize its potential spread?

The applied field of conservation biology has especially close ties to evolutionary ecology (Kinnison & Hairston, Jr. 2007), as reflected in calls to integrate demographic and genetic approaches to plant conservation (Oostermeijer et al. 2003). Articles in the journal *Conservation Genetics* highlight many issues important to evolutionary ecology. In an editorial of the inaugural issue in 2000, research topics suggested included population genetic structure of natural and managed populations, including identification of "evolutionary significant units"; assessments of the level of genetic variation in small or endangered populations; and investigations of the effect of inbreeding and outbreeding (Hoelzel 2000). All of these topics fall at the intersection of evolution and ecology. Common research themes in evolutionary ecology can also shed light on the evolution of weeds in agricultural ecosystems (Baucom & Holt 2009).

While developing a general theory of ecology, Scheiner and Willig (2011) defined the domain as the range in space and time, and the phenomena addressed by a theory or model. The domain of evolutionary ecology includes all the recognized characteristics of microevolution relevant to population biology: natural selection and adaptation, genetic variation, gene flow, genetic drift, and nonrandom mating (Roughgarden 1979). Because microevolutionary processes act within ecological communities, the dynamics of the component populations are an integral part of the interplay between evolution and ecology. Indeed, Roughgarden (1979, p. 296) maintained that "evolutionary ecology should be called evolutionary *population* ecology." As the editor-in-chief of *Evolutionary Ecology* pointed out in an editorial that served as his introduction to the journal, articles in the field should be concerned with the evolutionary influences on ecological processes, the ecological influences on evolutionary processes, or both, if possible (Endler 2010). The same sentiment is echoed by recent explorations of "ecoevolutionary dynamics" in a theme issue of the *Philosophical Transactions of the Royal Society B* (Pelletier et al. 2009). Clearly, ecology is important to understanding evolutionary mechanisms: "everything the ecologist looks at is the result of evolution" (Bradshaw 1984, p. 20).

1.2 PLANT EVOLUTIONARY ECOLOGY

Although a recognizable discipline of its own, plant evolutionary ecology emphasizes many concepts and techniques common to ecology, genetics, botany, and conservation (Fig. 1.1). Much of modern population ecology is rooted in evolutionary theory. As noted earlier, the genetic element is critical to the definition of evolutionary ecology. Both quantitative and molecular genetic approaches are useful, depending on the questions and hypotheses addressed. Basic information provided by many fields of plant science, including botany and plant systematics, contribute much to

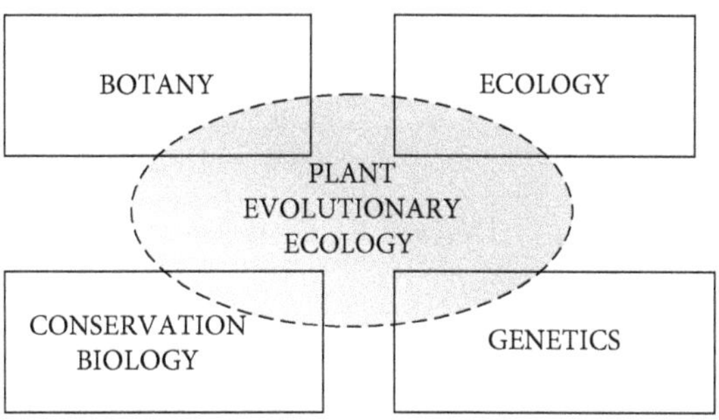

FIGURE. 1.1
Plant evolutionary ecology overlaps with other biological disciplines.

the understanding of phenotypic evolution and the traits examined by evolutionary ecologists. Information on genetic and phenotypic variation in natural populations can inform decisions regarding the conservation and management of plant species.

Many of the questions posed in evolutionary ecology research apply to all types of organisms. However, certain universal features of plant life history are notable in regard to their influence on studies that focus specifically on the evolutionary ecology of plant populations.

Plants are sessile, so most of their ecological interaction is with the prevailing local environment, including their immediate neighbors (Bradshaw 1972; Silvertown & Charlesworth 2001). This means that phenotypic plasticity is expected as individuals alter their growth and development in relation to changes in their abiotic and biotic environments (Bradshaw 1965; Sultan 1987). Because they show indeterminate growth throughout their life span, the ability to change morphology and physiology over time is well developed in plants (compared with animals, with determinate growth patterns). Thus, the range of phenotypic variation in a population able to respond to the agents of natural selection is especially great in a spatially or temporally heterogeneous environment. Genotype-by-environment interactions are especially relevant to research on phenotypic variation and selection in plant populations (Des Marais et al. 2013).

Another consequence of the sessile nature of plants is that gene flow occurs mostly via pollen and seed dispersal. Because the distances moved by pollen or seeds can be quite short, alleles, genotypes, and phenotypes may be aggregated spatially in nature. These considerations are clearly important to the interpretation of genetic structure and population differentiation in plants (Heywood 1991; Linhart & Grant 1996; Ennos 2001; Epperson 2007). Inbreeding, with its attendant reduction of heterozygosity and allelic diversity, can be common if neighbors are related genetically, having emerged from poorly dispersed seeds from one or a few parents. In addition, as a consequence of indeterminate growth and vegetative propagation, intermingled shoots (ramets) from the same individual (genet) may flower and self-fertilize effectively.

In contrast to animals, most plants are hermaphroditic (de Jong & Klinkhamer 2005); they have both male and female organs within individual flowers. This characteristic makes it possible for self-fertilization and seed set to occur in some species or under specific conditions, such as when conspecific mates are rare or pollinating insects are not available. The breeding system typical for a particular species influences population features important in evolutionary ecology, such as genetic structure and variation (Chapter 5).

The spatial aggregation of genetically related individuals that results from limited seed dispersal and the production of ramets by clonal growth can result in competitive or facilitative interactions among similar or identical genotypes. These interactions are probably more common in plant populations than is generally recognized. As becomes apparent in Chapter 7, the genetic relatedness of interacting individuals can affect the outcome of competition and its effectiveness as an agent of natural selection.

1.3 THE TIMESCALE OF EVOLUTIONARY ECOLOGY

Considerable evolutionary change can occur within and between populations in just a few generations (Hairston et al. 2005; Carroll et al. 2007), and the strength of selection has been shown to vary greatly from year to year (Siepielski et al. 2009).

Contemporary evolutionary processes clearly affect population dynamics and ecological interactions (Carroll et al. 2007); thus, the timescale for most evolutionary ecology research is typically one to several generations of a short-lived species. This is true whether a study examines genetic variation within and between natural populations, how genetic and phenotypic variation among individuals changes over time (microevolution), or whether specific populations along a geographic cline show local adaptation to the prevailing environment. Ecological experiments, such as population manipulations or mesocosm studies, to address evolutionary hypotheses can and have been conducted over one to several years (Travis & Reznick 1998). Note that some studies explore ecological and evolutionary processes *as they occur* (such as gene pool dynamics), whereas others explore the *end result* of these processes (adaptation, for example). The point to note is that most deal with phenomena occurring over the relatively short timescale of microevolution, at or below the species level. This recent realization that evolution can occur over ecological timescales is part of an emerging synthesis (Schoener 2011) that attempts to bridge the gap between ecology and evolution through the examination of ecoevolutionary dynamics (Pelletier et al. 2009).

1.4 PRINCIPLES AND GENERAL THEMES OF EVOLUTIONARY ECOLOGY

This text is organized around the premise that there are well-established principles central to evolutionary ecology. These principles act as guiding concepts underlying the varied approaches used in the study of natural populations and their dynamics:

- Populations are composed of groups of genetically variable individuals. Both molecular methods (Chapter 5) and quantitative genetic techniques (Chapter 2) have been used to reveal extensive genetic differences among individuals and populations in many species.
- Individuals show variation at the molecular (genotypic) and organismal (phenotypic) levels, providing the raw material for microevolution. Temporal change in allele frequencies or the distribution of phenotypic trait values is evidence for microevolution (Chapter 2).
- Population gene pools can change and become differentiated as a result of the action of natural selection, genetic drift, gene flow, and nonrandom mating, but only natural selection can result in adaptation. Common gardens have been used extensively to reveal genetic differentiation among populations for phenotypic traits (Chapter 3), whereas reciprocal transplant experiments have been used to document adaptation to local habitat conditions (Chapter 4).
- Phenotypic variation within populations arises from the interaction of different genotypes with their local environment. The phenotypic responses of different genotypes to abiotic (Chapter 6) and biotic environmental factors are manifested in changes in development, physiology, and morphology.
- The ubiquity of interspecific interactions (biotic factors) in nature is likely to affect the microevolution of populations (Chapters 7, 8, and 9), which can result in reciprocal evolutionary changes (coevolution) in the interacting species' populations.

2 Natural Selection in the Plant Population

Natural selection is ubiquitous enough to be found in a wide variety of organisms and . . . strong selection is by no means uncommon in natural populations (Endler 1986, p. 224).

2.1 NATURAL SELECTION AS A POPULATION ATTRIBUTE

Some time ago, in his book *Stages in the Evolution of Plant Species*, Clausen (1951, p. 11) recognized the primacy of the local population by referring to it as the "basic evolutionary unit." Within the ecological hierarchy of life, populations have recognizable attributes (emergent properties) that individuals do not possess. Age and size structure, density, gene pool, mortality and natality rates, and sex ratio are typical examples found in introductory ecology texts (e.g., Smith & Smith 2012). Natural selection, of course, often gets its own chapter in such texts. Mayr (1982, p. 479) stated that the idea of natural selection logically followed from specific "facts derived in part from population ecology." Because selection can be expected to occur in all populations, Reed (1981) maintained that it was a law of nature. Pianka (2011, p. 7) concurred, stating "the single concept closest to deserving the status of 'law' in ecology and one that is shared with all of biology, is natural selection." Thus, natural selection should be considered another of the attributes shown by a population.

The Darwinian postulates that provide the basis for natural selection can be found in many textbooks of evolutionary biology. Herron and Freeman (2014) list them as follows:

- Individuals within populations are variable (with regard to phenotypic traits).
- This variation is, at least in part, heritable (i.e., able to be passed from parents to offspring).

- Some individuals are more successful at surviving and reproducing than others (i.e., there are differences in fitness among individuals).
- Survival and reproduction of individuals is tied to variation in phenotypic traits among individuals. In other words, there is a consistent relationship between phenotypic traits and relative fitness (= phenotypic selection [Endler 1986]).

Therefore, those individuals with phenotypic trait values that are better able to survive and reproduce (i.e., that have greater Darwinian fitness) pass on more of their genetic variants to future generations (compared with individuals with phenotypic trait values that result in poor survival and/or low reproductive output). This results in the genetic and phenotypic characteristics of the population changing with each generation—that is, microevolution (Section 2.1.3).

As a thought exercise to emphasize why selection should be considered a usual population attribute, it is instructive to consider the circumstances under which selection would *not* occur. In relation to the previous postulates, selection in a population would not be expected if

- There was no genetically based phenotypic variation among individuals. In other words, all genotypes were the same, or different genotypes all expressed similar phenotypes that could not be sorted differentially by selection. Examples are a rhizomatous species producing many ramets by clonal growth that have become separated and are now distinct individuals, or an apomictic species (e.g., dandelion) that has made many genetically identical seeds and offspring.
- None of the phenotypic variation important to fitness was heritable—that is, phenotypic variants of successful parents with high fitness would not be passed on to their offspring. Note that, in this instance, strictly speaking selection could still be occurring, but it would not have evolutionary consequences; adaptation to environmental conditions cannot occur if there is no effective transmission of genetic information across generations.
- There is no differential survival or reproduction among genotypes. All individuals show equivalent chances of survival and all individuals produce the same number of viable offspring (i.e., there are no differences in fitness).
- There is no relationship between variation in phenotypic traits and variation in fitness (survival and reproduction). In other words, the quantitative value for a phenotypic trait of an individual is completely independent of the ability of that individual to survive and produce offspring. Although this may be possible for some specific trait (e.g., variation in leaf pubescence that does not affect fitness), many traits are expected to show some relationship to fitness.

When considering these circumstances, the possibility that selection would *not* occur in a population appears extraordinarily remote, and a solid understanding of natural selection is essential to evolutionary ecology (Pianka 2011). Natural selection is indeed ubiquitous in nature (Endler 1986; Hoekstra et al. 2001; Bell 2008) and, unless shown otherwise, is expected to be a natural attribute of populations everywhere.

2.1.1 Classifying the Agents of Selection

Any factor in nature that can influence survival and reproduction within a population can function as an agent of natural selection. Identifying selection agents and the way they interact to produce their effects on populations is an important area of evolutionary ecology (Wade & Kalisz 1990; MacColl 2011). Anthropological changes in the environment such as air pollution, climate warming, insecticide and herbicide application, and heavy metals in soils can all function as potent agents of selection (Chapter 6). They provide superb examples of adaptive evolution in plants (Antonovics et al. 1971; Macnair 1987; Linhart & Grant 1996; Reznick & Ghalambor 2001; Jump & Peñuelas 2005; Dechamps et al. 2008). The rates of microevolutionary change in plants exposed to human-generated agents of selection can be quite rapid and involve only a few generations of exposure (Bone & Farres 2001).

Other agents of selection that are not anthropogenic in origin can result in genetic differentiation and adaptive evolution (Linhart & Grant 1996; Mazer & LeBuhn 1999). An attempt is made to classify these agents in Table 2.1. An obvious first separation is to delimit abiotic from biotic components of the environment. Most abiotic agents can be grouped into those aspects of climate and soil that influence plant survival, growth, and reproduction. Precipitation, temperature and photoperiod are agents that vary on a geographic scale, but temperature and light availability can also vary on a much smaller (local) spatial scale.

Table 2.1 Classification of the agents of natural selection.

Category	*Agent*	*Examples*
Abiotic	Climate	Precipitation
		Temperature regimes, growing season length
		Photoperiod, light availability
	Edaphic	Soil minerals and fertility, metals, salinity
		Soil moisture
		Soil pH
Biotic (macroscopic)	Competitors	Neighbors of the same and/or different species
	Neighbors	Allelopathy, nurse plants, litter
	Herbivores	Insects, nematodes, mammals
	Seed predators	Insects, birds, mammals
	Pollinators	Insects, bats, birds
	Seed dispersers	Birds, mammals
Biotic (microscopic)	Endosymbionts	Bacteria, fungi
	Pathogens	Bacteria, fungi, viruses
Indirect[a]	Disturbance, habitat fragmentation, anthropogenic impacts	
	Elevation, latitude	
	Landscape topography, hydrology	

Within a population, any agent of natural selection can lead to the evolution of local adaptation. Examples of genetic differentiation in relation to some of these agents can be found in Linhart and Grant (1996). [a]Act through effects on the abiotic and biotic agents listed in the table.

Heterogeneity in environmental factors is pervasive in plant communities (Wilson 2000). Soil properties along with the associated levels of minerals and water often show small-scale spatial variability (Cain et al. 1999; Fitter et al. 2000; James et al. 2003). Consistent exposure to soils of a particular type (e.g., serpentine [Kruckeberg 1954; Brady et al. 2005; O'Dell & Rajakaruna 2011]) will select for those genotypes best able to survive and produce viable offspring under the prevailing edaphic (i.e., soil-related) conditions. Over time, this leads to adaptation of the population to the local soil environment (Pregitzer et al. 2010). Thus, climatic and edaphic factors are major abiotic agents of selection and have been much studied by evolutionary ecologists investigating population differentiation and local adaptation (Chapter 6).

Any living component of the environment inhabited by a plant population may function as a biotic agent of selection, including macroscopic and microscopic organisms in diverse taxonomic groups (Table 2.1). Both intraspecific and interspecific competition may act by reducing the availability of some of the same abiotic factors noted earlier (Keddy 2001), or by producing allelochemicals (Chapter 7). Herbivores, seed predators, and pollinators are other notable macroscopic selection agents (Chapter 9). Symbiotic microbes on and within plants are extremely common in nature and function as internal selection agents in specific ways that depend on whether interactions are mostly antagonistic or mutualistic (Chapter 8).

Several other environmental features studied by ecologists such as disturbance, elevation, latitude, and landscape topography and hydrology do not generally act as direct agents of selection because they affect the abiotic and biotic agents listed in Table 2.1, which in turn affect plant growth and reproduction. Thus, in the classic studies of plants adapted to different portions of an elevation gradient (Clausen et al. 1948; Clausen & Hiesey 1958), the selection agents were undoubtedly the climatic and edaphic factors that varied along the gradient.

2.1.2 Natural Selection as Cause Versus Effect

When considering the operation of natural selection, confusion sometimes arises regarding the process and its effect. An analogy with artificial selection, in which humans deliberately choose which individuals (each a unique phenotype and genotype) will breed, can illustrate these points. The *act* of choosing which individuals will breed is artificial selection, whereas the *result* is that only specifically selected individuals will be reproducing. This is, in essence, "differential reproduction"—a result or *effect* of artificial selection. In other words, the act of choosing which individuals breed *causes* differential reproduction.

In an analogous way, attributes of the natural environment act as agents of selection and "choose" which individuals will survive, mate, and produce viable offspring successfully. In this way, natural selection is a sorting or filtering process during which some individuals are eliminated or reproduce less than other individuals. Like in artificial selection, the end result or *effect* is differential reproduction. In other words, the selection of individuals that survive and reproduce by nature *causes* differential reproduction. In short, to state emphatically that natural selection is nothing more than differential reproduction—that is, variation in reproductive success among individual genotypes—confuses the cause with the effect.

As MacColl (2011, p. 514) asserted, "the cause of natural selection and, therefore, of adaptive evolution, is any environmental factor (agent of selection) that results in differential fitness among phenotypes." However, stating that the agents of selection *cause* natural selection seems a bit circular. Perhaps we might clarify these relationships as follows:

$$\textbf{Agents of selection} \xrightarrow{\textit{Natural selection}} \textbf{Differential survival and reproduction}$$

$$[\text{CAUSE}] \qquad\qquad\qquad\qquad [\text{EFFECT}]$$

Here I have placed natural selection in a somewhat murky zone between cause and effect. Selection is a process of differential sorting of genotypes (i.e., some genotypes survive and/or reproduce more than other genotypes) caused by interacting abiotic and biotic variables in a complex habitat. Note that selection differs from other biological processes in that it is not readily decomposable into a series of progressive steps. For example, the *process* of photosynthesis is often rendered as a series of steps (photo-oxidation of pigments, electron flow, photolysis of water, adenosine triphosphate synthesis by chemiosmosis, carbon fixation by Rubisco, and so on), with the end result being the production of carbohydrate. Likewise, the *processes* of mitosis and meiosis are taught as a continuum of steps that ultimately results in a specific number of diploid or haploid cells. With the ecological process of natural selection, we know roughly what to expect as an end result, but there is no readily definable sequence of causal steps.

2.1.3 How Natural Selection Causes Microevolution

Population microevolution through natural selection can be viewed as changes over time in (1) the frequencies of alleles at variable gene loci (the gene pool) or (2) the frequency distribution of phenotypes (the phenotype pool). However, it is worth noting that selection acts on phenotypes in nature, even when we use the shorthand phrase "differential survival and reproduction of genotypes."

Allele frequency changes in a gene pool may occur as a within-generation response to selection as a result of differential survival among genotypes (Fig. 2.1, upper). Suppose the initial population in year *t* consists of 12 individuals, each carrying two of three possible alleles at gene locus A. Sometime during the same year, two individuals die (A_1A_1 and A_1A_2), indicated by an X through those genotypes in the "year *t*, late" population in Figure 2.1. The new allele frequencies in the remaining population of 10 individuals are clearly different.

Allele frequency changes should also be expected between generations because selection causes differential reproduction among genotypes (Fig. 2.1, lower). In year $t + 1$ in our example, the population has grown to 15 individuals with new allele frequencies different from that of the prior population in "year *t*, early." Note that it is assumed that all 12 individuals from the prior year survived, reproduced, and then died (e.g., an annual plant species with discrete generations).

Although microevolutionary changes are depicted separately for those caused by differential survival and differential reproduction, in most natural populations both responses to selection are likely. In addition, for continuously breeding, perennial populations, the distinction between within- versus between-generation responses is expected to blur. For example, genotypes from year *t* may not reproduce, yet some proportion will survive to year $t + 1$, rendering within-generation responses to selection operational between years. Allele frequencies in year $t + 1$ will be a function of the genotypic composition of survivors from year *t* plus that of new recruits in year $t + 1$.

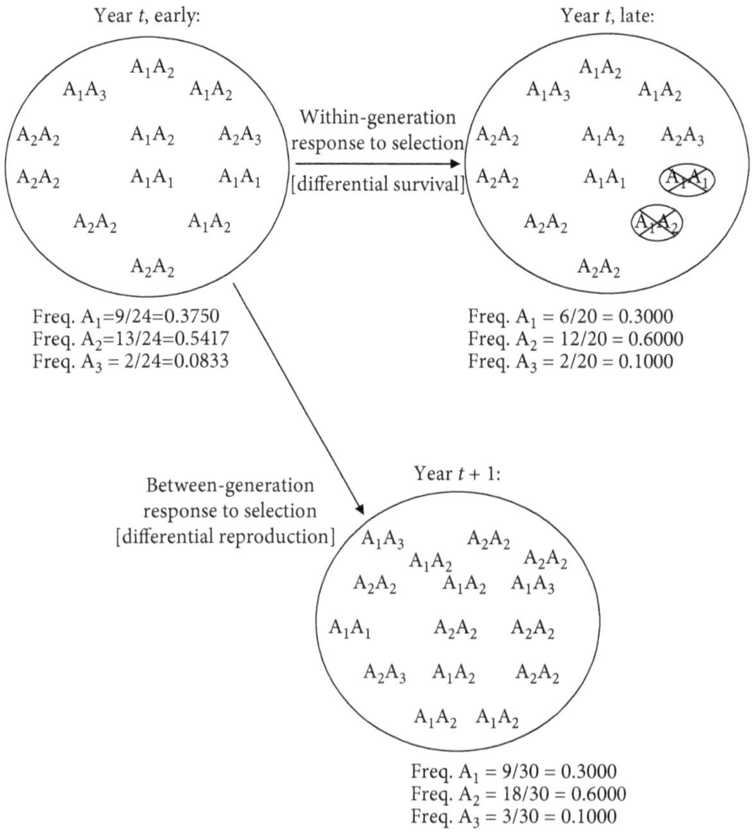

FIGURE 2.1
Hypothetical within- and between-generation changes in a population gene pool. One gene locus (A) with three alternative alleles (subscripts) is used as the example.

Analogous to changes in a gene pool, the collection of phenotypes in a population (i.e., the phenotype pool) can be changed because selection affects the relative frequencies of phenotypic categories. This occurs because phenotypes with specific ranges of trait values survive better, or reproduce more, than other phenotypes with a different range of values for the same trait. This change in the frequency distribution of phenotypes is illustrated in Figure 2.2 for arbitrary values of a quantitative trait shown by a phenotype pool before and after selection. Note that for a trait showing continuous variation, this categorization of phenotypic measurements into seemingly discrete classes is somewhat contrived. However, such categorization can highlight the way selection can cause microevolution of quantitative traits by altering the distribution of phenotypic trait values.

In the hypothetical scenario devised in Figure 2.2, selection has occurred within or between generations by causing differential survival and reproduction among phenotypic classes. By time $t + 1$, the frequencies of phenotypes with the greatest trait values have declined, whereas those of phenotypes with the lowest trait values have increased. This implies that directional selection has reduced the survival and/or reproductive fitness of phenotypes with larger values of the trait.

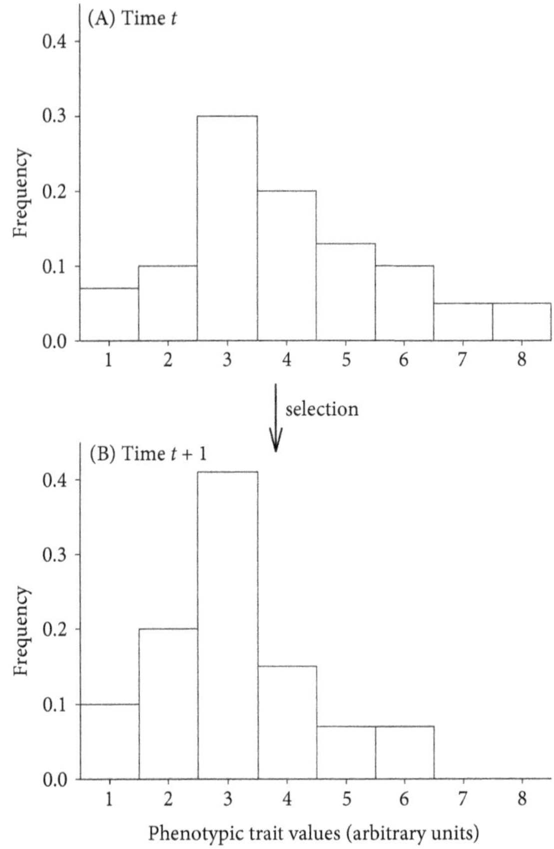

FIGURE 2.2
(A, B) Hypothetical change in the distribution of values for a phenotypic trait under selection from one time (t) (A) to a later time ($t + 1$) (B).

The close relationship of the selection processes in Figures 2.1 and 2.2 may not be apparent initially. Allele frequency changes at a specific gene locus must be associated with changes in the proportion of phenotypes that carry the alleles, which means that the gene locus must affect survival or reproductive fitness of phenotypes. So, as selection changes the distribution of phenotypes that carry specific alleles at this locus, it simultaneously changes the frequencies of those alleles. Furthermore, if most phenotypic traits are quantitative traits, their expression is governed by some unknown number of quantitative trait loci (QTLs). Clearly, the frequencies of alleles at different QTLs responsible for the expression of a specific trait will change as selection alters the frequency distribution of phenotypes in the phenotype pool.

2.1.4 The Importance of Genotype-by-Environment Interactions

When the phenotypes of different genotypes depend on environmental conditions, then genotype-by-environment (G × E) interactions are apparent (Stearns 1992). Different selection regimes can operate in different environments, often changing the rank order of fitness shown by the same set of genotypes (see Figs. 3.7, 6.9, 7.3, and 8.6 for examples). This can be shown experimentally by planting cloned replicates of the same genotypes into different environments—a technique exemplified in the transplant studies by Clausen and coworkers (Clausen et al. 1940, 1948; Clausen & Hiesey 1958; reviewed by Núñez-Farfán and Schlichting [2001]). Quantitative traits associated with fitness (i.e., functional traits [*sensu* Violle et al. 2007]) depend on both genotype and environment. In a two-way analysis of variance, this is manifested as a statistically significant interaction term (Stratton 1994; Conner & Hartl 2004).

As a consequence of G × E interactions, divergent selection pressures favor different genotypes in different environments and help maintain genetic variation (Heywood & Levin 1984; Stratton 1994; Byers 2005). For the common dandelion *Taraxacum officinale* growing in Vancouver, Canada, a different mix of the predominant genotypes (identified using microsatellite markers) was found in regularly mown versus unmown microhabitats (Fig. 2.3 [McLeod et al. 2012]). Experimental clipping treatments of the same genotypes revealed that genotypes with the greatest growth under clipping (e.g., genotype 24) were most common in mown fields, whereas those that grew best under unmown control conditions (e.g., genotypes 2 and 9) were more common in unmown microhabitats (Fig. 2.3). Thus, small-scale G × E interactions appear to have established localized adaptation to specific microhabitats, each with markedly divergent selection pressures (McLeod et al. 2012).

One aspect of G × E interaction that differs from the traditional definition concerns the process *within* a particular homogeneous environment that determines the fitness of distinct individuals. This individual-by-environment interaction (Nussey et al. 2007) is similar to a G × E interaction, except the environment (and selective regime) can be presumed to be the same for all genotypes. Differences between genotypes in growth or fitness traits indicate genetic variation in responses to a given set of conditions and selective agents. Intrinsic developmental processes, perhaps related to differential gene expression, hormonal activity, regulatory proteins, and so on, are primarily responsible for the phenotype exhibited by a particular genotype in a given environment (West-Eberhard 2003; Griffiths & Gray 2004; Larsen 2005). The resulting natural selection resulting from differential reproduction between genotypes occurs in the absence of identifiable variation in environmental conditions or selection pressures within the population. This is an example of "internal selection," in which selection pressures "derive from the internal dynamics of a functioning organism" (Schwenk & Wagner 2004, p. 395). Several of the molecular mechanisms for developmental variation are reviewed by Willmore and Hallgrimsson (2005).

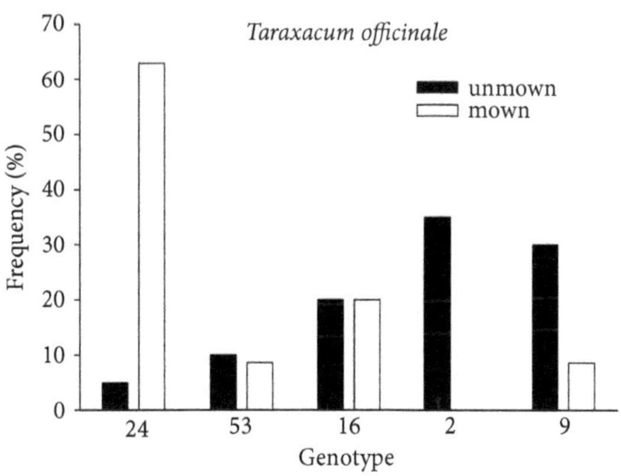

FIGURE 2.3
Frequency distribution of five dandelion genotypes in mown and unmown microhabitats in Vancouver, Canada. Drawn from data in McLeod et al. (2012).

Although the internal molecular milieu determines developmental trajectories, it should be recognized that external environmental conditions such as precipitation and temperature can also affect gene expression, as demonstrated in the perennial grass *Andropogon gerardii* (Travers et al. 2010). For a review of the use of genomics tools in the analysis of G × E interactions in relation to abiotic conditions, see Des Marais et al. (2013).

2.1.5 Can Selection Occur without an External Agent?

To address this question it is helpful to perform a simple thought exercise. Imagine three distinct genotypes of an annual plant species in a common garden where they experience identical growing conditions and there are no biotic interactions. Assume these conditions are not limiting and there is adequate soil moisture, minerals, and light for optimal growth and reproduction. Genotype A produces 10 seeds, genotype B produces 8 seeds, and genotype C produces 6 seeds. This difference in reproductive output among the genotypes indicates that natural selection has occurred. Yet, there is no discernible external agent at work. The phenotypic variation in reproduction can only be the result of inherent properties that differ between genotypes. These properties at the physiological level might include growth rate, photosynthetic efficiency, the ability to take up soil water and minerals, or any number of other attributes that vary among genotypes and are correlated positively with reproductive output.

The variation in a population of genotypes in our hypothetical homogeneous, unlimited environment translates into a population of variable phenotypes. This phenotypic variation is a natural consequence of genetic interactions with both the external environment and the internal physiological and biochemical environment within individuals (Kalisz & Kramer 2008). Any internal factors that can influence phenotypic variation affect the process of natural selection. For example, variation in gene regulation and expression can be important to phenotypic evolution (Fay & Wittkopp 2008). The evolutionary consequences are fitness differences among genotypes and their associated phenotypes. This is natural selection without a defined external agent. It is the result of an individualized developmental process that represents the combined, interactive effect of a specific genotype growing in a specific environment.

2.1.6 Internal Agents and the Evolutionary Role of Development

Developmental processes are critical determinants of plant growth (Körner 1991) and, by extension, reproductive fitness. Developmental trajectories can vary among genotypes and are highly responsive to environmental conditions (Schlichting & Pigliucci 1998; Cheplick 2003; Frank 2011). It is the many genotypes developing under a potentially diverse array of environmental conditions that leads to a variable population of phenotypes. Some phenotypes show greater Darwinian fitness than others; hence, the population shows natural selection. In essence, development has produced the "phenotypic variation that is screened by selection" (West-Eberhard 2003, p. 89). Selection occurs because only a subset of all alleles in the gene pool is sampled to constitute the next generation, and the selected alleles do not represent a random sample (Price 1995).

Epigenetic modification of the genotype during an individual's development can result in variable expression of ecologically important traits (Bossdorf et al. 2008). In general, epigenetics refers to the interaction of genes and their products during development (West-Eberhard 2003; Bird 2007). Gottschling (2007) defines an epigenetic phenomenon as a change in phenotype that is heritable but does not involve DNA mutation. There are many ways in which epigenetic effects can occur (see papers in Hallgrimsson and Hall [2011]), but they all involve internal biochemical processes occurring within the organism. Here, only experimental manipulation of DNA methylation is considered to illustrate the consequences of one type of epigenetic modification to phenotypic expression in plants.

Bossdorf et al. (2010) used genotypes of the model plant *Arabidopsis thaliana* to determine possible consequences of DNA methylation for phenotypic traits and their plasticity in response to mineral nutrient levels. Demethylation was accomplished for a subset of plants of each genotype using the chemical agent 5-azacytidine. They found significant genotypic differences in the extent to which demethylation affected flowering time, size (rosette diameter) at the time of flowering, and biomass (Bossdorf et al. 2010). In addition, for several phenotypic traits, demethylation reduced overall plasticity. They concluded that epigenetic variation among individuals, independent of DNA variation, can cause substantial changes in phenotypes.

Environmental stress can also induce DNA methylation alterations and show heritable transmission to offspring, as shown for the common dandelion (*Taraxacum officinale* [Verhoeven et al. 2010]). Epigenetic phenomena can act as internal agents of phenotypic trait expression and are increasingly recognized as having important implications for plant evolution and ecology (Kalisz & Purugganan 2004; Bossdorf et al. 2008; Herrera & Bazaga 2011; papers edited by Donohue [2014]).

In addition to the many physiological and biochemical features that comprise the internal environment inside individual plants, there may be endosymbiotic microorganisms with the potential to influence many aspects of growth and development (see Section 8.5). Endosymbionts, which include bacteria and fungi, are not limited to specialized groups, but are widespread and probably found in all plant taxa (Arnold 2007; Rodriguez et al. 2009a). Endophytic fungi of grasses have been investigated extensively in terms of their effects on their hosts (reviewed in Cheplick and Faeth [2009]). The hyphae of clavicipitaceous endophytes are obligate biotrophic symbionts, living off organic nutrients within the intercellular spaces of the host's leaves (Bacon et al. 2009). Some of those in the well-studied genus *Epichloë* (formerly included in *Neotyphodium*) are clandestine and do not reproduce sexually or produce any symptoms of infection.

For endosymbionts to act as internal agents of selection there must be genetic variation among hosts in their phenotypic responses to infection. Given an appropriate experimental design in which endophyte-infected (E+) and uninfected (E−) clones of the same host genotype are exposed to a specific set of environmental conditions, phenotypic trait measurements may be related to host genotype and infection status. The source of phenotypic variation of interest in an analysis of variance is therefore the two-way interaction of host genotype by infection status. As Table 2.2 attests, such interactions have been well documented for a variety of host traits in the much-studied grass *Schedonorus arundinaceus* (tall fescue; = *Lolium arundinaceum*) infected by the endophyte *Epichloë coenophiala* (formerly *Neotyphodium coenophialum*). If genotypes are replicated sufficiently so that E+ and E− plants can be

Table 2.2 Significant two-way interactions of host genotype by endophyte infection status (i.e., present or absent) for phenotypic traits of tall fescue.

Trait	*Number of genotypes*	*Reference*
Morphological		
Number of tillers	5	Belesky et al. (1987)
	4	Faeth and Sullivan (2003)
Plant volume	4	Faeth and Sullivan (2003)
Vegetative dry mass	2	Belesky and Fedders (1996)
Root-to-shoot ratio	4	Morse et al. (2007)
Physiological		
Carbon exchange rates	13	Marks and Clay (1996)
Leaf conductance	4	Morse et al. (2007)
Leaf net photosynthesis	4	Morse et al. (2007)
Nonstructural carbohydrate content	2	Belesky and Fedders (1996)
Relative growth rates	2	Belesky and Fedders (1996)
	4	Morse et al. (2007)
Reproductive		
Number of panicles	29	Rice et al. (1990)
Number of seeds	29	Rice et al. (1990)
Total seed dry mass	29	Rice et al. (1990)
	4	Faeth (2009)

exposed to two or more environments, then the three-way interaction of host genotype × infection status × environment may be of interest (Cheplick 1997; Cheplick et al. 2000; Morse et al. 2007).

2.2 ALLELIC, GENOTYPIC, AND PHENOTYPIC SELECTION

Because selection causes microevolution at the molecular level by changing allele frequencies in the gene pool (Fig. 2.1) and by changing the frequency distribution of quantitative trait values in the phenotype pool (Fig. 2.2), analyses of selection may be focused at the molecular genetic or organismal level. Because ecologists traditionally have measured traits at the organismal level, they have mainly researched quantitative phenotypic traits. However, with the advent of many molecular techniques now available to the ecologist (Chapter 5), there are more opportunities for researchers (with the appropriate training or help) to bridge molecular and quantitative genetic analyses of life history traits (e.g., Steinger et al. 2002; Brock et al. 2010). Of course, the distinction between molecular and quantitative genetic analyses is somewhat arbitrary because phenotypic traits always have some genetic basis. In model plants such as *Arabidopsis thaliana* and some crop species, the genes underlying quantitative variation (QTLs) in ecologically relevant phenotypic traits are continually being identified (Borevitz 2004; Murren & Kover 2004; Brock et al. 2010; Grillo et al. 2013). In the yellow monkeyflower *Mimulus guttatus*, two populations that were readily distinguished based on life history features such as flower size, number, and phenology could also be characterized by specific QTLs that contributed significantly to population divergence in phenotypes (Hall et al. 2006). Sometimes only a few genes are implicated for ecologically important life history traits. For example, Moritz and Kadereit (2001) reported that relatively few QTLs may be involved in determining the primary phenotypic traits that distinguish the cosmopolitan weed *Senecio vulgaris* var. *vulgaris* from its European progenitor subspecies.

Sometimes, the allelic variation at specific gene loci can be related to variation in quantitative traits (e.g., Hamrick & Allard 1975). When the variation shows correlation with one or more environmental variables, selection is often invoked as the causative agent. For example, in cork oak (*Quercus suber*), individuals homozygous for one allele (QpZAG46–188) at a particular locus made larger leaves than individuals homozygous for the alternative allele or heterozygous at that locus (Ramírez-Valiente et al. 2010). Going a step further, these researchers also found that the frequency of this allele for 13 populations in a common garden showed a significant, positive correlation with mean annual temperature (MAT) in the original habitats ($r^2 = 0.48$, $p < 0.01$; Fig. 2.4A). A second allele at the same locus showed a negative correlation with annual temperature (Fig. 2.4B). Although this type of methodology that involves correlation of a genetic trait with an environmental factor does not demonstrate natural selection directly (Endler 1986), for cork oak it does suggest that an increase in the frequency of one allele and an associated increase in large-leaf phenotypes are both favored under high annual temperatures (Ramírez-Valiente et al. 2010). Other examples of the relationship of various molecular markers to adaptive phenotypic traits are considered in Chapter 5.

2.2.1 The Classic Case of *Avena barbata*

Earlier work on allele and genotype frequencies inferred from electrophoretic analysis of allozymes (i.e., different forms of enzymes coded by different alleles) has shown that plant species can be differentiated genetically with respect to microenvironmental factors (Clegg & Allard 1972; Hamrick & Allard 1972; Hamrick & Holden 1979; Nevo et al. 1986, 1988, 1994; Gram & Sork 2001). In the slender wild oat *Avena barbata*, the frequencies of genotypes at five enzyme-coding gene loci appeared to be related closely to a moisture gradient from mesic to xeric conditions in California (Hamrick & Allard 1972). A later study of ribosomal DNA (rDNA) also revealed concordance between allozyme and rDNA genotypes across 48 sites, suggesting that selection favored particular multilocus combinations of alleles in different habitats (Cluster & Allard 1995). Furthermore, four of five quantitative traits differed significantly between plants homozygous for the genotype found in mesic sites and those homozygous for the genotype found in xeric sites when grown in a common garden (Hamrick & Allard 1975). Plants from the mesic sites flowered and matured their seed earlier, were shorter, and produced more tillers than plants from xeric sites. These differences were interpreted as having been driven by selection for competitive ability in the dense plant communities found at the mesic sites (Hamrick & Allard 1975). Indeed, a later study did show that the mesic genotype was competitively superior to the xeric genotype in a replacement series experiment (Latta et al. 2004). This classic example of ecotypic divergence (Latta 2009) showing a nonrandom association between genotypes and specific microhabitats, whether involving differences in

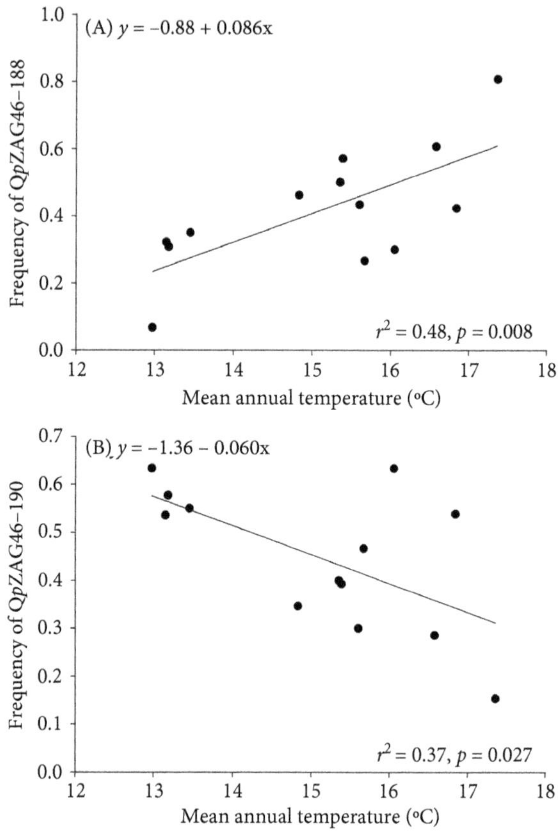

FIGURE 2.4

(A, B) Significant correlations between two alleles in 13 populations of cork oak (*Quecus suber*) and annual mean temperature.

Reprinted with kind permission from Springer Science and Business Media, and the author. *Source:* Ramírez-Valiente, J.A., Lorenzo, Z., Soto, A., Valladares, F., Gil, L. & Aranda, I. 2010. Natural selection on cork oak: allele frequency reveals divergent selection in cork oak populations along a temperature cline. *Evol. Ecol.* 24: 1031–1044.

allozymes, DNA, or quantitative traits, illustrates a population genetic approach to inferring the action of selection (Ennos 1989).

However, nothing is ever as simple as it seems and the story doesn't end here. Using artificially founded "colonies" of *Avena barbata* at several field sites in California, Jain and Rai (1980) presented some experimental evidence showing the possible role of selection in population divergence; however, contemporary analyses have called into question the inference that *A. barbata* shows genetic adaptation to specific habitat conditions. Using QTL mapping and reciprocal transplant techniques, Latta (2009) found that the performance of the mesic genotypes of *A. barbata*, in terms of lifetime reproductive success, were consistently the greatest across all sites and years. Also, for the QTLs that affect reproductive success, the same alleles were favored in all environments examined. Latta (2009) concluded that the favored genotype was currently spreading and increasing in frequency and that local adaptation had not yet been achieved.

2.2.2 Deviations from Hardy-Weinberg Expectations

Deviation of population genetic data from null models, such as the Hardy-Weinberg formulation that predicts genotype frequencies in the absence of selection (and other processes), is often used to infer that natural selection has been responsible for the deviation (Endler 1986). However, there are several reasons other than selection that allele or genotype frequencies observed in a population may not match the Hardy-Weinberg expectations: nonrandom mating (e.g., selfing) or genetic drift can also change population genetic parameters (Hartl & Clark 2006; Hedrick 2011). Knowledge of selfing rates and the examination of large populations unlikely to have been exposed to the sampling errors associated with drift can help in ruling out nonrandom mating or drift as reasons for deviations from Hardy-Weinberg expectations.

The phenomenon of heterozygote superiority has been described in a few plant species (e.g., Jain & Allard 1960; Farris & Mitton 1984; Stilwell et al. 2003) and is manifested as a significantly greater frequency of heterozygosity observed in a population relative to what is predicted by the Hardy-Weinberg principle. One study of the self-fertilizing annual *Hordeum vulgare* (barley) took advantage of a composite cross among 31 varieties that was then propagated, without conscious selection, for 18 generations (Jain & Allard 1960). The propagated populations were very large (10,000–15,000 individuals per generation) and closed—that is, genetic drift and migration likely had only minor or no effects at all on allele or genotype frequencies. Seven gene loci were studied as markers, each affecting alternatives of a specific phenotypic trait; for example, the B locus controlled pericarp color (B = black, b = white). Observed genotype frequencies were reported for the 3rd and 5th generations, for several other intermediate generations, and for the final, 18th generation (Jain & Allard 1960). Because allele frequencies do not change with repeated episodes of selfing, and the authors reported estimates of outcrossing rates for several of the gene loci, I was able to calculate expected frequencies of genotypes using equations provided by Roughgarden (1979) for situations in which selfing and random mating can co-occur.

The observed and expected frequencies of the three possible genotypes at four loci are shown for the 5th and 18th generations in Table 2.3. With the extremely high level of selfing (0.98–1.00) in this species, one would expect an increase in homozygosity with a concomitant decrease in heterozygosity over time. For two loci (B and S), the observed and expected frequencies of heterozygous genotypes after 18 generations were similar, suggesting that selection was not favoring the particular "chromosome segments" carrying these marker genes (Jain & Allard 1960). In contrast, by the 18th generation, the observed frequency of heterozygotes at the E locus was more than four times greater, and that of the R locus was 11 times greater, than expected (Table 2.3). Thus, there may have been an advantage to individuals containing both alternative alleles for these marker genes. Additional analyses of allele frequencies at four esterase loci were presented by Clegg et al. (1978a) for the 8th, 19th, and 28th generations from this same original cross and revealed that allele frequency change could be substantial even between different life cycle stages (zygote to adult) within the same generation.

Heterozygote superiority was again suggested by data collected from four populations of the American chestnut (*Castanea dentata*) (Stilwell et al. 2003). Leaf samples were used for starch gel electrophoresis to obtain allele frequencies at five polymorphic gene loci. Tree growth data were also recorded during a three-year period in one

Table 2.3 Observed and expected frequencies of genotypes for four gene loci in barley after 5 and 18 generations of propagation.

Gene locus	*Frequency*	*Generation 5*			*Generation 18*		
		AA	*Aa*	*aa*	*AA*	*Aa*	*aa*
B	Observed	0.074	0.020	0.906	0	0	1.000
	Expected	0.106	0.035	0.858	0.122	0.004	0.873
E	Observed	0.270	0.114	0.616	0.248	0.059	0.693
	Expected	0.387	0.039	0.559	0.400	0.014	0.586
R	Observed	0.370	0.073	0.557	0.489	0.100	0.411
	Expected	0.311	0.148	0.659	0.322	0.009	0.670
S	Observed	0.148	0.107	0.745	0.261	0.008	0.731
	Expected	0.193	0.091	0.715	0.232	0.014	0.754

Expected frequencies were calculated based on measured rates of selfing (0.98–1.00). Data from Jain and Allard (1960).

population. Comparisons of the observed and expected frequencies of heterozygotes (based on Hardy-Weinberg expectations) revealed that the observed heterozygosity averaged over the five loci was 0.158, which was significantly greater than expected (= 0.149, $\chi^2 = 244.0$, $p < 0.0001$). Individuals heterozygous for at least one locus also had a greater growth rate, measured as the relative increase in basal diameter during the three-year period relative to homozygous individuals. In the absence of sexual reproduction in chestnut trees for more than 70 years as a result of the continued attack by the Chestnut blight fungus (*Cryphonectria parasitica*), selection may be acting slowly to favor heterozygous individuals (Stilwell et al. 2003).

2.2.3 Selection Analysis of Quantitative Traits

2.2.3.1 Correlation with Environmental Variables

The method illuminated earlier, in which an attempt is made to correlate geographic variation in phenotypic traits with environmental variables that could be the relevant agents of selection, is a well-established, indirect way to infer the operation of natural selection (Endler 1986). Analysis and identification of environmental variables in the natural habitat that explain a substantial fraction of the phenotypic (or genotypic) variation among populations can provide clues regarding which agents of selection function as ecological causes of evolution (MacColl 2011). Common garden approaches, detailed in Chapter 3, may entail growing plants collected from populations distributed along a climatic, geographic gradient in a homogeneous environment to reveal genetically based phenotypic variation that is correlated with the gradient (Maron et al. 2004; Stinchcombe et al. 2004; Rutter & Fenster 2007; Oyarzabal et al. 2008; Monty et al. 2009; Scheepens et al. 2010; Novy et al. 2013).

An example of this approach is provided by Monty et al. (2009), in a study in which seeds were collected from the invasive perennial *Senecio inaequidens* distributed along an altitudinal gradient from 2 to 1,695 m in southern France (see Section 3.2 for additional examples). Plants in a common garden were reared either directly from the field-collected seeds or, to control for potential maternal environmental effects (Section 3.6), reared from seeds of maternal plants that had grown for a generation in a climate-controlled incubator. Regardless of the seed collection conditions, plants from populations of higher elevations were shorter and had reduced biomass compared with those from populations of lower elevations (Fig. 2.5). Overall, more than 38% of the variation in biomass of plants in the common garden could be explained by the altitude of the source population. The authors speculated that the smaller stature of plants from high elevations reflected adaptation to harsher conditions and shorter growing seasons (Monty et al. 2009). However, the specific environmental variables responsible for clinal differentiation can be difficult to identify, revealing one limitation of this approach. Nonetheless, Monty and Mahy (2009) used principal components analysis to regress a composite axis of phenotypic traits of *S. inaequidens* onto a climate axis that primarily depicted a gradient in temperature and summer drought conditions. The climate axis explained 62.5% of the variance in plant traits. Mean height and biomass of populations showed highly significant correlations with the climate axis. Thus, the relevant selection agents may be predominantly abiotic along the altitudinal gradient. MacColl (2011) suggested a useful refinement whereby data are collected on several environmental variables and then are used in a multiple regression analysis to identify which variables were the most critical selective agents.

2.2.3.2 Univariate Methods and Selection Differentials

Statistical regressions have been generally used to estimate the strength of selection on specific quantitative traits, especially since the early 1980s with the development of multivariate methodology for analyzing effects of selection on several traits simultaneously (Lande & Arnold 1983). Because these approaches were developed primarily at the University of Chicago, they are sometimes called *Chicago School approaches* (Brodie et al. 1995; Conner & Hartl 2004). These methods allow one to determine both the type (directional, stabilizing, disruptive) and intensity of phenotypic selection (*sensu* Endler 1986) or natural selection (*sensu* Lande & Arnold 1983) on a quantitative trait based on its relationship to the fitness of individuals. These determinations are made within one generation by measuring the relation between trait values and fitness.

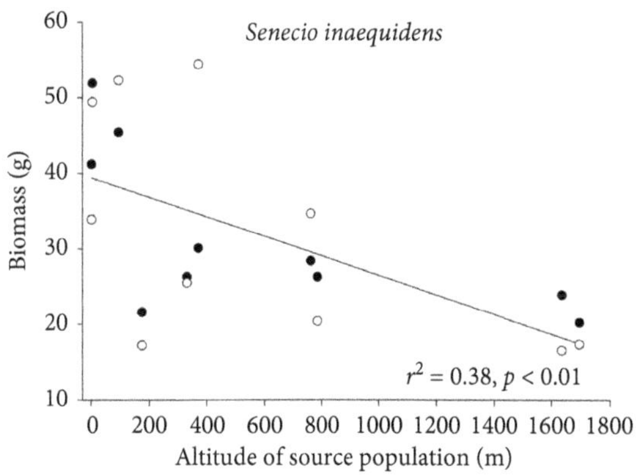

FIGURE 2.5
Biomass of *Senecio inaequidens* originating from sites along an altitudinal gradient in southern France and grown in a common garden. •, plants from seeds collected directly from plants in the field; ○, plants from seeds of plants grown first in a controlled growth-chamber environment. Data from Monty et al. (2009).

Although fitness has been defined and characterized in several ways (Endler 1986; Stearns 1992; de Jong 1994; Metcalf & Pavard 2007), and refinements may be necessary for some plants (Pan & Price 2001), measures of survival, growth (especially size), and/or reproduction are most commonly used as fitness estimates. Primack and Kang (1989) noted that traditional studies of plant fitness have mostly examined seed production (fecundity), although this can only apply to the female component of fitness, and it is assumed that more seeds translate into more offspring recruited to the next generation. Harper (1977, p. 650) cautioned that "the number of progeny produced by an organism is only one component of success in natural selection. . . ." Surrogate measures of fitness such as size, estimated as dry mass or total leaf area, for example, often show high correlation with seed numbers or total seed mass and can provide sensible metrics of absolute fitness. However, as every student of evolution learns, it is relative fitness, which expresses fecundity (or some other measure of absolute fitness) compared with that of others in the population, that matters most to evolution by natural selection (Latta 2010; Herron & Freeman 2014). In many studies, relative fitness of an individual is calculated as its absolute fitness divided by the mean absolute fitness of the population (Stratton 1992a, b; Fairbairn & Reeve 2001; Etterson 2004a).

The covariance between relative fitness (w) and a single phenotypic trait (z) is easily depicted as a fitness function (Schluter 1988; Brodie et al. 1995; Conner & Hartl 2004), which can show three general shapes (Fig. 2.6). A simple linear function implies directional selection because phenotypes with larger (or smaller) values of the trait show greater relative fitness (Fig. 2.6A) and thus are favored by selection. Nonlinear fitness functions are also possible. Figure 2.6B shows a case in which phenotypes with intermediate trait values (nearest the average) show the greatest relative fitness, indicative of stabilizing selection. Another nonlinear fitness function occurs where phenotypes with the intermediate trait values show the *lowest* relative fitness (Fig. 2.6C). This pattern, when selection favors phenotypes with the more extreme values, indicates disruptive selection.

An example of a linear fitness function using tillers collected from a field population of the invasive annual grass *Microstegium vimineum* is shown in Figure 2.7. Mature tillers, each from a single individual bearing a terminal raceme with seeds intact, were sampled from a large population in a woodland in central New Jersey (Cheplick 2005). To incorporate a range of light conditions, tillers were sampled from both sunny, edge, and shady interior regions. The total dry mass of mature seeds per tiller was defined as absolute fitness. Relative fitness for an individual tiller was its dry mass of seeds divided by the mean dry mass of seeds for the 35 tillers collected. For illustrative purposes, the fitness function is used here to quantify the intensity of selection—that is, the standardized selection differential (S)—on the average leaf dry mass of a tiller. Before plotting, the phenotypic values of leaf mass were standardized by subtracting the population mean from each value and dividing by the population's standard deviation (Conner & Hartl 2004). Linear regression revealed a significant positive relationship between leaf mass per tiller and relative fitness ($F = 66.69$, $p < 0.0001$; Fig. 2.7A). The standardized selection differential (0.706 ± 0.087 [standard deviation {SE}]) is the slope of the function line and indicates covariance between fitness and the trait of interest. The specific fitness function in Figure 2.7A therefore shows relatively robust directional selection on the mean mass of leaves produced by a tiller in this *M. vimineum* population.

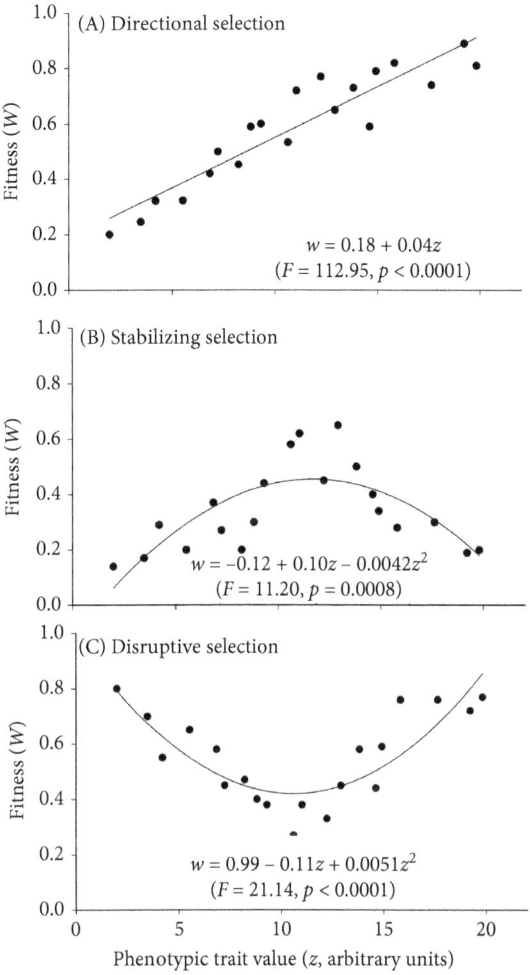

FIGURE 2.6
(A–C) Three possible distributions of fitness in relation to a range of values for a quantitative phenotypic trait. Each fitness function implies a specific mode of selection.

Because it is based on univariate statistics, S is really a composite measure of total phenotypic selection, including direct linear selection on the trait and indirect linear selection resulting from phenotypic correlations with other traits (Lande & Arnold 1983; Pfennig & Kingsolver 2009). It must be recognized that evolutionary ecologists mostly use the phrase *selection differential* to denote the magnitude of the total selection for or against a phenotypic trait (Conner & Hartl 2004). This use differs from the "selection differential" used in the familiar breeder's equation to predict population response (R) to selection as a function of trait heritability (h^2) ($R = h^2S$ [Falconer & Mackay 1996]). In this case, S represents the difference between the mean phenotype of a population before selection and the mean phenotype of the selected individuals (Bell 2008; Herron & Freeman 2014).

Families representing seed sibships collected from individuals in the same population of *Microstegium vimineum* were grown under greenhouse conditions, and data were again recorded on seed dry mass and leaf mass per tiller. Linear regression of the family means for these data, calculated as before (Fig. 2.7B), was statistically significant ($F = 5.13$, $p < 0.05$) but generated a much lower standardized selection differential (= 0.056 ± 0.025 [SE]) compared with the field-collected tillers. Thus, under

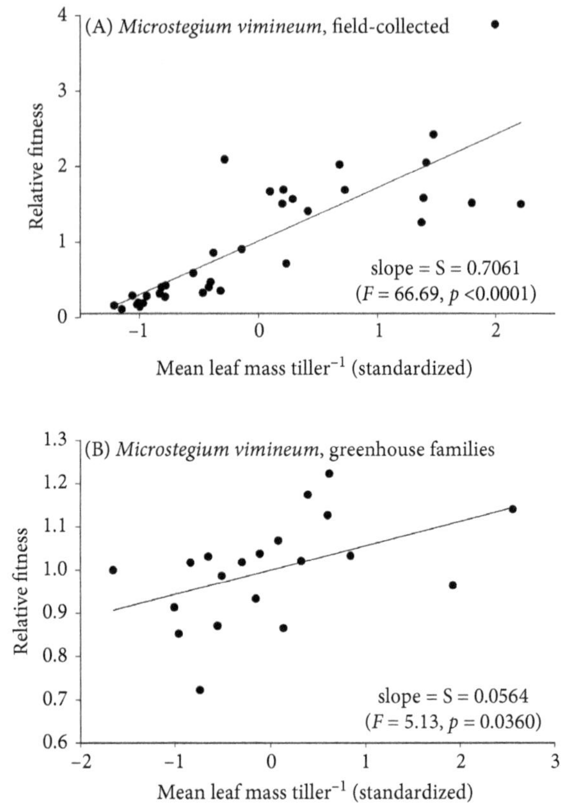

FIGURE 2.7

(A, B) Example of linear fitness functions (based on seed dry mass per tiller) in the invasive grass *Microstegium viminuem*. Relative fitness was regressed on standardized mean leaf mass per tiller for tillers collected directly from a field population in central New Jersey (A) (each point is an individual) and from tillers from families (i.e., sibships) originally collected as seeds from individuals in the same population and grown to adulthood in a greenhouse (B). The slope of the fitness function is the standardized selection differential (S). Data from Cheplick (2005).

the benign conditions of the greenhouse, genotypic selection favored families with a higher average leaf mass per tiller, suggesting that quantitative genetic changes in the population could occur in a sunny environment across generations as some family lineages propagated, mostly by self-fertilization, the genetic combinations that favored greater relative leaf mass. In the field, where light conditions are much more variable (Cheplick 2005), there is a premium on making larger, heavier leaves with a greater surface, which results in the highest relative fitness (Fig. 2.7A), presumably through enhancement of photosynthetic production.

In other plant species, standardized selection differentials have been shown, perhaps unsurprisingly, to vary greatly among traits, populations, and environmental conditions (van Tienderen & van der Toorn 1991b; Stratton 1992a; Stanton et al. 2000; Etterson 2004a; Núñez-Farfán & Schlichting 2005; Donovan et al. 2007; Dudley et al. 2012). Etterson's (2004a) detailed study of *Chamaecrista fasciculata*, an annual legume, across an aridity gradient in the central United States provides an example. Seeds were collected from three populations: Minnesota, Kansas, and

Oklahoma. After rearing and then crossing plants within each population in a greenhouse, offspring from each population were planted into gardens at each of the three collection sites (a reciprocal transplant design [see Chapter 4]). One of several traits recorded was leaf number, which was standardized and used for selection analyses done separately for each population at each site. Relative fitness was fecundity (i.e., per capita seed output) divided by site or population mean fecundity (Etterson 2004a). Standardized selection differentials of leaf number are shown for each population and site in Figure 2.8A. With one exception, all selection differentials were significant and positive. It is perhaps no surprise that, in most populations, selection favors producing more leaves because leaves are the source of photosynthates needed to make seeds and thereby increase fecundity.

2.2.3.3 Multivariate Methods and Directional Selection Gradients

Because the univariate method used to estimate standardized selection differentials does not permit the separation of direct selection on a particular trait from indirect selection on other traits that may be correlated with it (Conner & Hartl 2004), multivariate methods are preferred whenever several phenotypic traits are being measured (Lande & Arnold 1983; Brodie et al. 1995; Lande 2000). Multiple regression is used to regress fitness on several traits during the same analysis, thereby affording a measure of selection—the selection gradient—that quantifies direct selection on a trait after controlling statistically for the effects of correlation with the other measured traits (i.e., indirect selection). The basic model for multiple regression of this sort is

$$w = \alpha + \beta_1 z_1 + \beta_2 z_2 + \beta_3 z_3 + \ldots \beta_i z_i,$$

where w is relative fitness, α is the y-intercept, and β_i is the partial regression slope for each trait z_i (Pfennig & Kingsolver 2009). This partial regression slope (or coefficient) is the selection gradient (or selection coefficient) and it provides a measure of the association between fitness and a particular trait, independent of other measured, correlated traits (Endler 1986; Brodie et al. 1995).

In the study of *Chamaecrista fasciculata* by Etterson (2004a), directional selection gradients (β) were obtained from multiple regressions of relative fitness on three

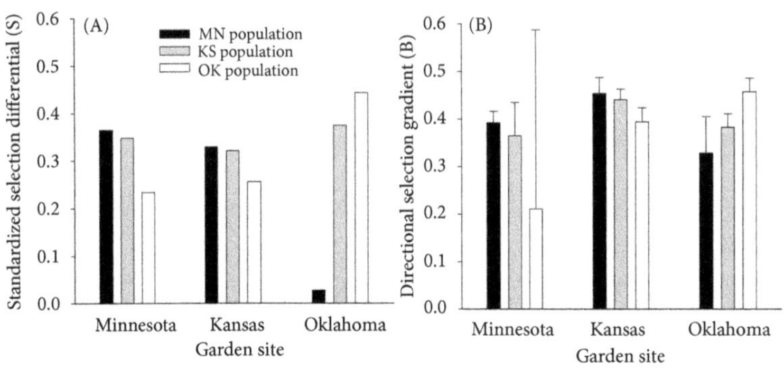

FIGURE 2.8

(A, B) Standardized selection differentials (S) (A) and directional selection gradients (β) (B) for three populations of the annual legume *Chamaecrista fasciculata* planted into three garden sites in the midwestern United States. Data from Etterson (2004a).

traits: leaf number, reproductive stage, and specific leaf area. The values of β for leaf number were somewhat greater than the selection differentials (S) for the three populations and garden sites, but showed broadly similar patterns (Fig. 2.8B). This might be a generalization, as Donovan et al (2009) showed significant, positive relationships between β and S for several leaf traits in populations and species of sunflowers (*Helianthus*). Greater values for β relative to S in *C. fasciculata* (Etterson 2004a) indicate that indirect, correlated selection reduced total direct selection on leaf number, which can be considered a measure of plant size. This result agrees with an analysis of numerous estimates of selection across a wide range of taxa that showed that indirect selection on correlated traits may often reduce total selection on size, although possibly not selection on other types of traits (Kingsolver & Diamond 2011).

Other estimates of standardized selection gradients for size-related traits, all of which were statistically significant, include the following:

- Biomass in the biennial *Oenothera biennis* (β = 0.58 ± 0.11 [Johnson et al. 2009])
- Height in different populations of the perennial *Mimulus guttatus* (range of β values, 0.27–1.42 [Murren et al. 2009]) and in six populations of the perennial grass *Arrhenatherum elatius* in a common garden (range of β values, 0.10–0.31 [Petit & Thompson 1998])
- Leaf length in the perennial *Potentilla glandulosa* (β = 0.16; data of Clausen and Hiesey [1958] reanalyzed by Núñez-Farfán & Schlichting [2005]) and in the annual *Diodia teres* in two habitats (β = –0.035 ± 0.010 and –0.090 ± 0.015 [Jordan 1991])
- Number of leaves in the annual *Impatiens capensis* in two field plots (β = 0.45 ± 0.26 [95% confidence interval] and 0.46 ± 0.41 [Brassard & Schoen 1990], different populations of two subspecies of the perennial *Aquilegia vulgaris* (range of β values, 0.18–0.49 [Alcántara et al. 2010], and a composite index based on leaf number, width, and length in *Plantago lanceolata* at three sites (range of β values, 0.29–0.41 [van Tienderen & van der Toorn 1991b])
- Leaf area in different populations of the annual *Impatiens pallida* (range of β values, –0.33 to 0.51 [Bennington & McGraw 1995], in the fern *Blechnum chilense* in a forest understory in Chile (β = 0.37 ± 0.04 [Saldaña et al. 2007], and in different populations of two species of annual sunflowers, genus *Helianthus* (range of β values, 0.42–0.73 [Donovan et al. 2007])
- Rosette diameter in the annual *Erigeron annuus* (β = 0.54 [Stratton 1992a])
- Stem thickness in *Diodia teres* in two habitats (β = 0.085 ± 0.013 and 0.149 ± 0.020 [Jordan 1991] and in *I. capensis* in two field plots (β = 0.53 ± 0.27 and 0.51 ± 0.17 [Brassard & Schoen 1990])

Often, directional selection on size in many species is positive (Kingsolver et al. 2012), indicating a fitness advantage for larger size, but this is not invariably the case, as noted in the negative gradients for leaf length in *Diodia teres* (Jordan 1991) and leaf area in some populations of *Impatiens pallida* (Bennington & McGraw 1995).

Phenological traits for which significant directional selection gradients have been reported include the timing of seedling emergence (Kalisz 1986; Stratton 1992a; Bennington & McGraw 1995) and flowering (O'Neil 1997; Petit & Thompson 1998; Stanton et al. 2000; Griffith & Watson 2005; Munguía-Rosas et al. 2011), where typically negative values for β suggest selection favoring earlier emergence and/or flowering (Geber & Griffen 2003). Although vegetative morphological traits have been used most commonly to estimate selection gradients (e.g., Jordan 1991; Etterson 2004a; Núñez-Farfán & Schlichting 2005), selection has also been estimated for floral traits (see Section 9.3). In addition, a few studies have begun to report selection for biochemical and physiological traits. For example, selection gradients have been reported for specific components of plant chemistry such as phenolics (including flavonoids) and tannins in *Oenothera biennis* (Johnson et al. 2009). In their review, Geber and Griffen (2003) noted the dearth of selection estimates on physiological traits; however, since then, there have been more studies providing estimates of directional selection on gas exchange rates, transpiration rates, and water use efficiencies (WUEs) (Donovan et al. 2007, Saldaña et al. 2007; Agrawal et al. 2008; Donovan et al. 2009; Dudley et al. 2012).

WUE, a measure of carbon gain by photosynthesis relative to water loss through transpiration, will be used as an example of how greatly directional selection gradients can vary between species and among populations within species for an ecologically important physiological parameter (Table 2.4). The expectation is for selection to favor greater WUE wherever soil moisture is limited, provided that WUE correlates positively with some measure of fitness (Dudley 1996; Saldaña et al. 2007). In several species, selection gradient estimates varied greatly with population and the time of sampling (Table 2.4), and were often negative and insignificant. Both significant positive selection for WUE in dry conditions (Dudley 1996; Saldaña et al. 2007), consistent with predictions, and significant negative selection for WUE (Donovan et al. 2007; Dudley et al. 2012) have been detected. The ecological reasons for selection for lower WUE have not been clear. It was speculated that in *Helianthus anomalus*, selection for lower WUE may be the result of a relationship with increased transpirational water loss that improves the uptake of mobile mineral nutrients from the soil (Donovan et al. 2007), thereby improving individual fitness. In any event, the substantial variation in the magnitude and sign of selection for WUE reveals the difficulty inherent in measuring and interpreting selective pressures with regard to ecophysiological traits (Agrawal et al. 2008).

2.2.3.4 Stabilizing and Disruptive Selection

So far, we have only considered linear selection gradients that indicate a continuous, directional change in fitness as phenotypic trait values change (Fig. 2.6A). Nonlinear selection gradients are also possible, however, indicating stabilizing (Fig. 2.6B) or disruptive selection (Fig. 2.6C), depending on the shape of the fitness function. Nonlinear selection gradients (γ) estimate the curvature of the fitness function, measuring the rate of change in the slope as phenotypic trait values increase (Conner & Hartl 2004). The basic model for quadratic multiple regression is:

$$w = \alpha + \beta z + \left(\frac{\gamma}{2}\right) z^2,$$

where w is relative fitness, α is the y-intercept, β is the linear selection gradient (regression slope) described earlier, γ is the nonlinear (quadratic) selection gradient (i.e., the partial regression slope or coefficient) that describes convexity ($\gamma < 0$) or concavity ($\gamma > 0$) of the fitness function (Brodie et al. 1995), and z is the phenotypic trait (Conner & Hartl 2004; Pfennig & Kingsolver 2009). Thus, γ represents the covariance of fitness with the squared term z^2 (Schluter 1988). The partial regression

Table 2.4 Directional selection gradient (β) estimates (±standard error when available) of water use efficiency for multiple populations of two annual species of *Clarkia* (Dudley et al. 2012), an annual species of sunflower (*Helianthus anomalus* [Donovan et al. 2007]), the fern *Blechum chilense* (Saldaña et al. 2007), and the annual *Cakile edentula* (Dudley 1996).

Clarkia exilis			*Clarkia unguicalata*			*Helianthus anomalus*	
Population	*Before flowering (β)*	*After flowering (β)*	*Population*	*Before flowering (β)*	*After flowering (β)*	*Population, month*	*(β)*
GS	**−0.61 ± 0.36***	−0.19 ± 0.40	CFR	0.06 ± 0.16	—	Mesic, June	**−0.27****
SC	0.09 ± 0.18	−0.08 ± 0.14	GR	0.23 ± 0.17	—	Mesic, July	−0.11
WS	−0.11 ± 0.20	0.04 ± 0.17	JS	0.14 ± 0.29	−0.12 ± 0.24	Dry, June	−0.02
WR	—	**−0.32 ± 0.14***	LO	−0.08 ± 0.16	0.06 ± 0.19	Dry, July	−0.14
			SC	0.10 ± 0.16	0.20 ± 0.17		

Blechum chilense		*Cakile edentula*	
Population	*(β)*	*Population*	*(β)*
Gap	**0.29 ± 0.06*****	Dry	**0.51****
Understory	−0.01 ± 0.04	Wet	0.01

Populations of *Clarkia* were sampled before or after flowering, whereas those of *Helianthus anomalus* were sampled in June and July. Significant selection gradients are in bold type. *$p < 0.05$, **$p < 0.01$, ***$p < 0.001$.

coefficient of z^2 in this quadratic regression analysis may be negative, indicating a constantly decreasing slope with increasing phenotypic trait values, interpreted as stabilizing selection acting on the trait of interest. Alternatively, when the partial regression coefficient is positive, the slope of the fitness function is steadily increasing with increasing phenotypic trait values, interpreted as disruptive selection (Conner & Hartl 2004; Pfennig & Kingsolver 2009).

Negative nonlinear selection gradients appear to be relatively uncommon in populations of many species, calling into question the importance of stabilizing selection in nature (Kingsolver et al. 2001; Kingsolver & Diamond 2011). Furthermore, when estimates of the standard errors are provided by researchers, they typically are quite large, rendering values that are not significantly different from zero (Table 2.5). Nevertheless, in several plant species, significant negative values for γ have been reported; however, the sign and magnitude of γ estimates typically differ among traits examined and among populations (Table 2.5). For example, Dudley (1996) detected significant stabilizing selection on leaf size in *Cakile edentula* growing in a dry environment, suggesting selection against very small or large leaves. At two of the three sites where populations of *Chamaecrista fasciculata* were grown, there was significant stabilizing selection on specific leaf area (Etterson 2004a). Using data from the classic studies of Clausen and Hiesey (1958), Núñez-Farfán and Schlichting (2005) found evidence for stabilizing selection on leaf length and number of days to flowering in *Potentilla glandulosa* (Table 2.5). The flowering date of *Mimulus guttatus* at a coastal site in Oregon also provided a highly significant negative γ value (–0.76, $p < 0.001$), indicating stabilizing selection on flowering time (Hall & Willis 2006).

Positive γ values, indicative of disruptive selection, have also been reported for several traits in a few plant species (Table 2.5). Photosynthetic rate, measured before plants flowered, showed significant positive γ values in the two species of *Clarkia* investigated by Dudley et al. (2012). The interpretation of these estimates is a challenge because it means that low or high photosynthetic rates, but not intermediate rates, are favored selectively early during the growing season. It is difficult to envision a scenario of why this might occur. Furthermore, after flowering, *Clarkia unguiculata* showed a significant *negative* γ value for photosynthetic rate (the γ estimate for *Clarkia exilis* was also negative after flowering, but not statistically significant; Table 2.5). Thus, the form of nonlinear selection varied from one life cycle stage (in one part of the growing season) to another.

Estimates of nonlinear selection can also vary greatly from one population or site to another. For example, stabilizing selection on reproductive stage in *Chamaecrista fasciculata* was detected in populations at a site in Oklahoma, but at sites in Minnesota and Kansas, disruptive selection was detected for this same trait (Table 2.5 [Etterson 2004a]). In addition, significant positive γ estimates were found for leaf number at two sites, but not the third. Again, it is a challenge to devise an explanation for why having a relatively low or high number of leaves would improve fitness in this species and be favored by selection, but having an intermediate number of leaves would not!

2.2.3.5 Correlational Selection

Our previous discussion of nonlinear selection gradients was concerned with univariate multiple regression in which only *one* phenotypic trait was examined in each

Table 2.5 Examples of nonlinear selection gradient (γ) estimates (±standard error when available) for phenotypic traits of several herbaceous species.

Species	*Trait*	*γ Estimates*	*Comments*	*Reference*
Cakile edentula	Leaf size	0.02, **−1.13*****	Wet and dry conditions, respectively	Dudley (1996)
	Water use efficiency	**−0.22***, 0.12		
Chamaecrista fasciculata	Reproductive stage	**1.000 ± 0.118*****, **0.062 ± 0.016*****, **−0.090 ± 0.029*****	Minnesota, Kansas, and Oklahoma sites, respectively	Etterson (2004a)
	Leaf number	0.088 ± 0.023, **0.173 ± 0.012*****, **0.195 ± 0.020*****		
	Specific leaf area	**−0.030 ± 0.020***, −0.003 ± 0.006, **−0.004 ± 0.003***		
Clarkia exilis	Photosynthetic rate	**0.90 ± 0.32***, −0.20 ± 0.44	Before and after flowering, respectively	Dudley et al. (2012)
	Transpiration rate	−0.02 ± 0.30, −0.20 ± 0.32		
	Index of maturity	Not applicable, **0.41 ± 0.21***		
	Water use efficiency	−0.21 ± 0.16, −0.29 ± 0.16		
Clarkia unguiculata	Photosynthetic rate	**0.65 ± 0.29***, **−1.58 ± 1.09***	Before and after flowering, respectively	Dudley et al. (2012)
	Transpiration rate	0.16 ± 0.31, −0.67 ± 1.15		
	Index of maturity	Not applicable, −0.15 ± 0.25		
	Water use efficiency	−0.12 ± 0.12, −0.39 ± 0.28		

Helianthus anomalus	Leaf size	−0.15, −0.28	Mesic and dry sites, respectively (in July)	Donovan et al. (2007)
	Leaf succulence	−0.02, −0.01		
	Nitrogen concentration	−0.05, 0.02		
	Water use efficiency	**−0.16***, −0.01		
Mimulus guttatus	Flowering date	**−0.76*****		Hall and Willis (2006)
Oenothera biennis	Biomass	**−0.37 ± 0.12****		
	Quercetin glucuronide concentration	**0.15 ± 0.07***	A flavonoid associated negatively with herbivory	Johnson et al. (2009)
Potentilla glandulosa	Rosette width	0.039	Reanalysis of data of Clausen and Heisey (1958)	Núñez-Farfán and Schlichting (2005)
	Leaf length	**−0.111***		
	Stem number	−0.076		
	Stem length	−0.030		
	Days to flower	**−0.092***		

For each *Clarkia* species, data were pooled across three to five populations (Dudley et al. 2012). Significant selection gradients are in bold type. $^{*}p < 0.05$, $^{**}p < 0.01$, $^{***}p < 0.001$.

analysis. However, the special case of *two* traits interacting simultaneously to determine fitness may lead to correlational selection (γ) in which specific combinations of phenotypic trait values may (or may not) be favored selectively (Conner & Hartl 2004). A bivariate nonlinear selection analysis, in which partial regression coefficients of cross-product terms are extracted, can be used to estimate selection on correlated traits (Lande & Arnold 1983; Endler 1986; Brodie et al. 1995). Bivariate fitness surfaces, with two horizontal axes for the two phenotypic traits and a vertical axis showing fitness, are a useful way to represent graphically the relationship of the phenotypic traits to fitness (Brodie et al. 1995; Conner & Hartl 2004). Examples of the use of fitness surfaces to visualize the interactive nature of correlated traits in determining fitness can be found in Jordan (1991), Dudley (1996), and Etterson (2004a).

In an approximate manner, bivariate gradients indicate selection is favoring similar combinations of the traits when positive, but opposite combinations of traits when negative (Brodie et al. 1995). In Dudley's (1996) study of *Cakile edentula*, positive correlational selection was indicated between leaf size and WUE in a dry environment. Although univariate stabilizing selection had been noted for leaf size alone (Table 2.5), correlational selection analysis revealed the optimal leaf size increased with increasing WUE. In Etterson's (2004a) study of *Chamaecrista fasciculata*, there was evidence for correlational selection favoring specific combinations of traits (Table 2.6). At the site in Minnesota, "early-developing plants with more and thinner leaves were particularly favored" (i.e., positive γ values for reproductive stage × leaf number, negative γ values for reproductive stage × specific leaf area) (Etterson 2004a, p. 1452). However, at the sites in Kansas and Oklahoma, later-developing plants with more and thicker leaves were favored.

As with univariate nonlinear selection estimates, SEs of the estimates of bivariate correlational selection are typically large, often resulting in statistical insignificance (Table 2.6). Correlational selection, when significant, likely indicates that the pair of phenotypic traits does not act independently with regard to the impact on fitness (Bell 2008). The methods used for bivariate nonlinear selection analysis, however, deal with phenotypic trait correlations, but it is the underlying genetic correlations that ultimately reduce or facilitate evolutionary responses of multiple traits to selection (Stearns 1992). If nothing more, correlational selection analysis emphasizes the important point that, often, no *one* phenotypic trait can be designated easily as *the* specific target of selection. Instead, a number of phenotypically correlated traits are all likely to contribute to fitness to some extent (Bell 2008). This idea is not new, of course, and has long been recognized by researchers interested in the evolution of phenotypic integration, life history trade-offs, and cohesiveness of the phenotype (e.g., Clausen & Heisey 1958; Schlichting & Pigliucci 1998; Merilä & Björklund 2004; Wolf et al. 2004).

2.2.3.6 Path Analysis of Selection

The much-used, multivariate regression techniques of Lande and Arnold (1983) have contributed greatly to our understanding of natural selection, yet they have been modified or criticized for a variety of reasons. For example, logistic regression has been proposed as a useful alternative to multiple linear regression when the fitness measure consists simply of two alternatives (such as survival vs. death [Janzen & Stern 1998]). Also, concerns have been raised about sensitivity of multivariate

Table 2.6 Examples of nonlinear correlational selection gradient (γ) estimates (±standard error when available) for pairs of phenotypic traits in several herbaceous species.

Species	*Trait pair*	*γ Estimates*	*Comments*	*Reference*
Cakile edentula	WUE × leaf size	0.12, **0.71****	Wet and dry conditions, respectively	Dudley (1996)
Chamaecrista fasciculata	Reproductive stage × leaf number	**0.356 ± 0.042*****, **−0.032 ± 0.018***, **−0.044 ± 0.028*****	Minnesota, Kansas, and Oklahoma sites, respectively	Etterson (2004a)
	Reproductive stage × SLA	**−0.091 ± 0.039*****, −0.073 ± 0.017, 0.030 ± 0.026		
	Leaf number × SLA	−0.010 ± 0.035, **−0.040 ± 0.017****, −0.063 ± 0.025		
Clarkia exilis	Photosynthetic rate × transpiration rate	−0.36 ± 0.27, 0.27 ± 0.30	Before and after flowering, respectively	Dudley et al. (2012)
	Photosynthetic rate × index of maturity	Not applicable, 0.03 ± 0.22		
	Transpiration rate × index of maturity	Not applicable, −0.04 ± 0.17		
	WUE × index of maturity	Not applicable, 0.09 ± 0.10		
Clarkia unguiculata	Photosynthetic rate × transpiration rate	−0.21 ± 0.25, 0.97 ± 1.08,	Before and after flowering respectively	Dudley et al. (2012)
	Photosynthetic rate × index of maturity	Not applicable, **−0.63 ± 0.42***		
	Transpiration rate × index of maturity	Not applicable, 0.59 ± 0.40		
	WUE × index of maturity	Not applicable, −0.05 ± 0.16		

(continued)

Table 2.6 [*continued*]

Species	*Trait pair*	*γ Estimates*	*Comments*	*Reference*
Helianthus anomalus	Leaf size × leaf succulence	0.06, 0.38	Mesic and dry sites, respectively (in July)	Donovan et al. (2007)
	Leaf size × N concentration	**0.27***, 0.32		
	Leaf size × WUE	−0.13, −0.26		
	Leaf succulence × N concentration	0.08, 0.21		
	Leaf succulence × WUE	**−0.24***, −0.09		
	N concentration × WUE	−0.03, −0.20		

For each *Clarkia* species, data were pooled across three to five populations (Dudley et al. 2012). Significant selection gradient estimates are in bold type. $^{*}p < 0.05$, $^{**}p < 0.01$, $^{***}p < 0.001$. N, nitrogen; SLA, specific leaf area; WUE, water use efficiency.

analyses to extensive intercorrelation among phenotypic traits (multicolinearity) and potential bias resulting from unmeasured environmental factors or phenotypic traits that affect fitness (Endler 1986; Mitchell-Olds & Shaw 1987; Rausher 1992; Fairbairn & Reeve 2001; Pigliucci & Kaplan 2006). These same problems can also apply to the additional multivariate technique to be addressed here: path analysis (Mitchell 2001). However, although standard multiple regression methods do not allow one to deduce the causal relationships that commonly exist among measured phenotypic traits and how they affect fitness directly or indirectly (Shipley 2000; Pigliucci & Kaplan 2006), path analysis shows promise for improving the understanding of selection by permitting a way to evaluate different hypothesized causal pathways of selection (Kingsolver & Schemske 1991; Scheiner et al. 2000; Byers 2005).

Like the methods already described, path analysis also uses multiple regression to account for the effects of phenotypic traits correlated with the trait of interest on fitness through the estimation of standardized partial regression coefficients, often simply called *path coefficients* (Wright 1934; Scheiner et al. 2000; Shipley 2000; Mitchell 2001). However, in path analysis, one specifies beforehand what particular set of interrelationships are likely to exist among a group of measured phenotypic traits and fitness. This path model is based on prior understanding of the biological system and species of interest. Alternative path models can be tested statistically, evaluated, and compared (Mitchell 2001; Byers 2005).

Path analysis was developed some time ago by Sewall Wright (1921, 1934; see Shipley [2000] and Matsueda [2012] for historical details). Agricultural researchers have long used it to distinguish the relative importance of different phenotypic components to seed yield (e.g., Dewey & Lu 1959; Gravois & Helms 1992; del Moral et al. 2003; and many more). However, it is only relatively recently that evolutionary ecologists have used path analysis to depict the relative importance of measured phenotypic traits to fitness in wild plants.

An early example is provided by Maddox and Antonovics (1983), who measured leaf area at multiple times during the development of two species of *Plantago* in a growth chamber, and related these measurements to the reproductive components of fitness using a series of linear structural equations. Note that path diagrams provide a visual representation of structural equation models (see Ho et al. [2012] for details). For both *Plantago aristata* and *Plantago patagonica*, the number of inflorescences and the number of capsules on an inflorescence had large, significant positive effects on fitness: the path coefficients were 0.85 and 0.63 for the two traits, respectively, in *P. aristata* (Fig. 2.9); and 0.71 and 0.66, respectively, in *P. patagoni*ca. The final measurement of leaf area (at 90 days) in *P. aristata* had a significant direct effect on reproductive fitness based on seed number and mass (at 90 days), but leaf areas at 50 days and 70 days did not (Fig. 2.9). However, analysis of the path model revealed that leaf areas at 50 days and 70 days *did* have significant *indirect* effects on fitness that were mediated via positive effects on the number of inflorescences and capsules in both species (Maddox & Antonovics 1983). Thus, selection may favor a greater leaf area during the middle and late stages of growth because this will, both directly and indirectly, improve fitness.

Jordan (1992) conducted a path analysis of growth and reproductive data collected on populations of the weedy annual *Diodia teres* transplanted into inland and coastal habitats in North Carolina. He constructed a path model showing the causal relations among the number of leaves on plants in June, July, and August, and seed

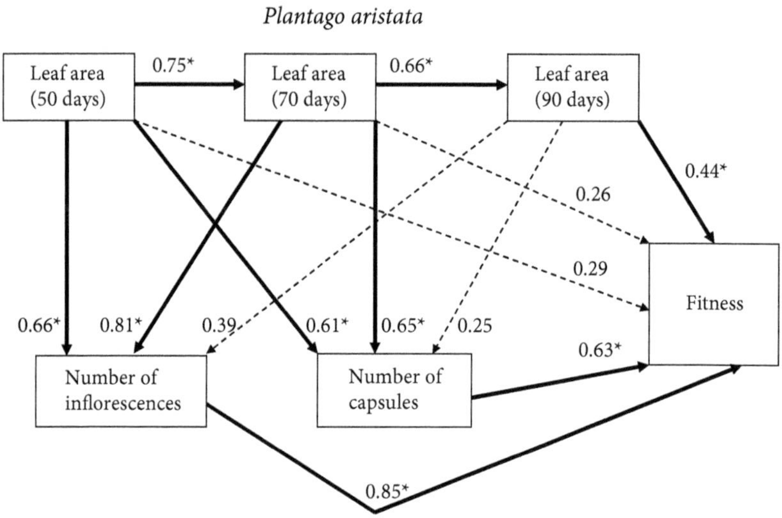

FIGURE 2.9
Simplified path diagram for *Plantago aristata* plants in a growth chamber experiment. Leaf areas at 10 days and 30 days, and seed number per capsule, were also included in the complete model but are omitted here for clarity. Significant ($p < 0.05$) path coefficients are indicated with an asterisk and solid line. Insignificant paths are denoted by a dashed line. Redrawn from Maddox and Antonovics (1983).

production during two growing seasons. Selection favoring larger plant size was evident, as leaf numbers in the different months tended to have substantial direct and indirect effects on seed production. Plant size was also important to fitness in the annual grass *Triplasis purpurea*, for which path analysis revealed that early size and final life span affected the number of tillers plants made significantly, which then affected seed production indirectly through a positive effect on vegetative mass (Cheplick & White 2002). Early growth and final dry mass were likewise shown to be important determinants of fecundity in a path analysis of *Avena barbata* populations in two habitats in California (Latta & McCain 2009).

As these examples show, and most plant ecologists recognize, size often shows a direct positive relationship to individual reproductive success, whether measured as per-capita seed or fruit production (fecundity), or the dry mass of reproductive structures. Most path analyses show that, typically, any phenotypic trait or external agent that affects plant size also affects reproductive fitness indirectly. Along an indirect causal path, the intervening variable is called a *mediator* (Shipley 2000, p. 125). Plant size may very well be one of the most important mediators in the path analysis of selection and the phenotypic traits that determine fitness indirectly.

Two additional examples that used path analyses to investigate phenotypic selection illustrate these points. The study of *Cakile edentula* (Dudley 1996) described earlier in this section, with regard to directional, stabilizing, and correlational selection, is one of the few that supplemented these analyses with a path model. Path analysis linked leaf size, WUE, and vegetative mass to relative fitness based on fruit mass (Fig. 2.10A). In both the wet and dry environments where plants grew, vegetative mass had a highly significant direct effect on fitness. In the dry environment, WUE had a significant indirect positive effect on fitness, mediated by its direct positive effect on vegetative mass (Fig. 2.10A). Shipley (2000) notes that, to quantify an indirect effect, one simply multiplies the path coefficients along the path. In the dry environment, the indirect effect of WUE on fitness is therefore $0.36 \times 0.72 = 0.27$. In the wet environment, this indirect effect was substantially smaller (from the path diagram in Dudley 1996): $0.16 \times 0.44 = 0.07$. Although leaf size did not affect fitness directly or indirectly, it correlated positively with WUE in the dry environment, indicated as a double-headed arrow in Fig. 2.10A (double-headed arrows imply an unknown causal relationship linking two variables [Shipley 2000]). Stronger selection on WUE in the dry environment (Table 2.4) was likely a result of the greater effects of WUE on vegetative mass, and a larger effect of vegetative mass on fitness, in this environment compared with the wet environment (Dudley 1996).

In the perennial herb *Solanum carolinense*, both plant density and insect herbivory were manipulated in a field experiment in Florida (Underwood & Halpern 2012). These two factors, which may function as important agents of selection, were incorporated into a path model, with stem size as a mediator (Fig. 2.10B). Stem size was measured as the length of the primary stem plus all its branches, and was highly

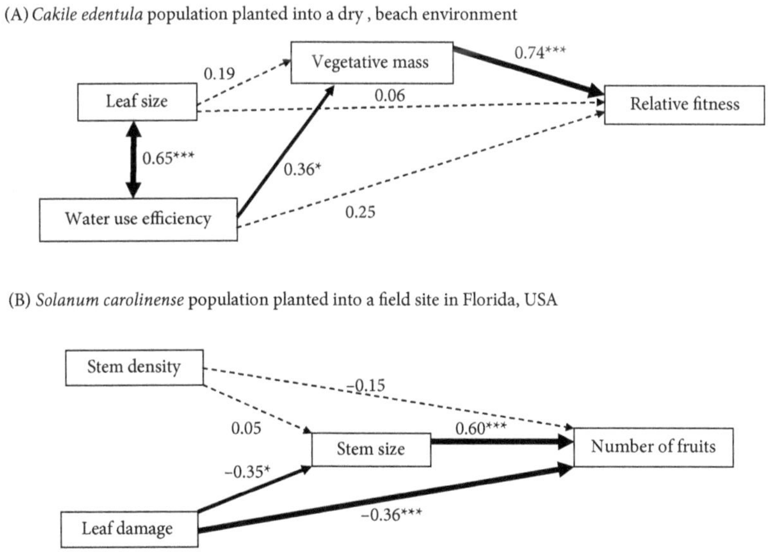

FIGURE 2.10
Examples of path diagrams with standardized path coefficients and their statistical significance (*$p < 0.05$, ***$p < 0.001$). Insignificant paths are denoted by a thin dashed line. (A) Model specifying the relationships of leaf size, water use efficiency, and vegetative mass to the relative fitness of *Cakile edentula* in an experimental population planted in a dry, beach environment at the Indiana Dunes National Lakeshore (Dudley 1996). (B) Model for the effects of herbivore damage to leaves, stem density, and stem size (= length) on fruit production (per stem) of the perennial herb *Solanum carolinense* in 2008 at a site in Vancouver, Canada (Underwood & Halpern 2012).

correlated with total biomass and fruit production. Data were taken in 2007 and 2008. In both years, leaf damage by insect herbivores affected stem size negatively and thus reduced fruit production indirectly (the 2008 path diagram is shown in Fig. 2.10B). There was also a significant negative *direct* effect of herbivore damage on fruit production as well. In 2008, stem density did not have direct or indirect effects on the number of fruits (Fig. 2.10B), although there was a negative effect of stem density on stem size in 2007 (path coefficient = −0.50, $p < 0.001$ [Underwood & Halpern 2012]). The effects of insect herbivory and intraspecific density on *S. carolinense* were primarily indirect and were mediated by stem size. The path analysis suggested that insect herbivores could be a powerful selection pressure in plant populations (see also Schemske and Horvitz [1988] and Section 9.2 in this book), reducing plant size and thereby reducing individual fitness.

2.2.4 Experimental Approaches to Natural Selection

Given the ease with which herbaceous plants can be manipulated and tracked, it is surprising how few studies have involved experimental investigations of natural selection explicitly, especially under field conditions (Kawecki et al. 2012). In actuality, it may be that it just *seems* that way because many plant population biologists have performed studies in which individuals (genotypes) were marked and subsequently monitored over one or more growing seasons. The big three of ecology—survival, growth, and reproduction—are often what is monitored for any of a variety of reasons. For example, the goal could be to compare population performance in distinct habitats or to decipher the effect of some manipulated environmental factor on population dynamics. However, often the results of past studies were never placed into the framework of natural selection and differential survival or reproduction among individuals, despite Harper's (1967) early emphasis on a "Darwinian approach" to plant ecology. In that article, he noted that "the theory of evolution by natural selection is an ecological theory" and that "Darwinian plant ecology has been largely neglected" (Harper 1967, p. 247). In all likelihood, if the raw data were available for many plant population studies, it would be possible to perform a thorough analysis of selection using some of the quantitative genetics techniques described earlier (e.g., Núñez-Farfán & Schlichting 2005). Although genetic identities were not always considered, phenotypic selection analyses (Section 2.2.3) could have supplied information on standardized selection differentials and, if multiple traits were recorded, selection gradients. In short, the apparent dearth of studies in experimental evolutionary ecology is, in part, simply the result of the disinclination of researchers to depict their research in evolutionary terms or to perform formal analyses of selection (the statistical techniques for which were not available for early studies). Much data collected by plant population biologists (i.e., the survival, growth, and reproductive output of individuals) are clearly very relevant to our understanding of natural selection and how it functions in nature (Harper 1977; Bradshaw 1984; Sarukhán et al. 1984; Silvertown & Charlesworth 2001).

What constitutes an experimental approach to natural selection? A "natural experiment" may involve the study of distinct, recognizable phenotypic variants that already exist in one or more populations. One can also categorize a continuously variable quantitative trait into groups (such as the bars in the histogram in Fig. 2.2). No putative selection agents are manipulated; rather, the fitness of discrete phenotypes (or groups with specific mean phenotypic trait values) is monitored under natural conditions. Thus, one quantifies the *results* of selection on a population of phenotypically different individuals, all of which have been presumably exposed to the same set of selection agents in the field.

Many plant species show substantial variation in floral color, size, and shape, some of which can be related clearly to pollinator-mediated selection pressures (see Section 9.3). However, other selection agents besides pollinators can and do affect the evolution of floral traits (Elle 2004; Strauss & Whittall 2006). Regardless of the agents responsible, an important first step to determining whether selection is favoring (or *has* favored) one particular floral variant over another is to estimate the fitness of each floral phenotype under natural conditions. Levin and Brack (1995) determined the relative fitness of red- and white-flowered plants of the annual *Phlox drummondii* growing in experimental populations in Texas. By performing crosses, true-breeding offspring were obtained from parents of each floral color type, and seeds of five populations were planted into different field sites. Demographic and life history data were collected on plants emerging from the seeds during a complete growing season. Relative fitness was based on net reproductive rate, estimated as the product of the proportion emerging, the proportion surviving to flowering, and mean fecundity. Plants with red flowers had the greatest relative fitness (set to 1.0) in all populations at all sites; fitness of plants with white flowers ranged from 0.41 to 0.88. Selection coefficients ($s = 1 -$ Relative fitness) estimate the strength of selection against a phenotype (Roughgarden 1979). For *P. drummondii* with white flowers, mean $s = 0.38$, averaged over all populations. This selective disadvantage to white-flowered plants was not simply a result of pollinator discrimination; other pleiotropic phenotypic effects of the petal pigment mutation were likely involved (Levin & Brack 1995).

The results of ongoing natural selection can sometimes be observed by monitoring changes in phenotypic traits in a variable population over multiple generations of interbreeding. Without conscious selection, the composite cross-populations of barley that were propagated for 18 generations, as described earlier (Section 2.2.2), showed significant temporal changes in a number of quantitative traits, including relative fitness based on seed production (Fig. 2.11 [Allard & Jain 1962]). Directional

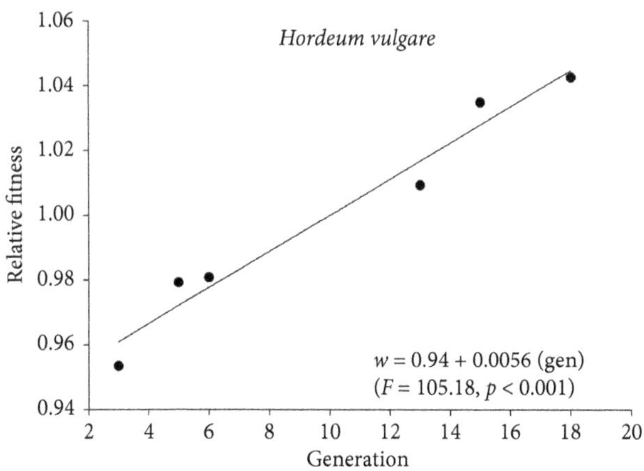

FIGURE 2.11
Relative fitness in a barley population propagated for 18 generations in a California field without conscious selection (drawn with data from Allard and Jain [1962]).

selection has apparently favored genotypes best able to survive, grow, and reproduce under the field conditions in California where these plants were grown each year. Thus, short-term microevolutionary responses to natural selection were documented as a general improvement in reproductive fitness in this annual species despite a high level of self-fertilization (Allard & Jain 1962).

Longer term evolutionary responses to selection in plant populations have been revealed by the Park Grass Experiment in the United Kingdom, which began in 1856 (Silvertown et al. 2006). Evidence of selective responses of populations of some species to edaphic conditions (see Chapter 6) is indicated by local adaptation and differentiation for phenotypic traits in relation to specific field plots receiving various mineral nutrient treatments (Snaydon 1970; Davies & Snaydon 1976; Snaydon & Davies 1976, 1982; Silvertown et al. 2005). Although common garden and reciprocal transplant experiments may not necessarily address natural selection *as it occurs*, they are powerful approaches for the investigation of the "ghost of selection past" (Bell 2008, p. 261)—namely, population differentiation and local adaptation, respectively. By comparing fitness measures of populations in their home site versus those that are not, selection coefficients against "alien" populations growing in a site away from their usual home site can be estimated from reciprocal transplant designs (Section 4.5). Experimental approaches to evolutionary ecology that involve common gardens or reciprocal transplants are explored in Chapters 3 and 4, respectively.

Other experimental approaches to investigating selection have involved the establishment of populations of known genotypic composition and subsequent resampling of the same populations to determine how genotypic frequencies have changed. One expects that, as selection agents exert their effects on a genetically variable population, a genotypic sorting process will ensue whereby the frequencies of genotypes will change over time. This process has sometimes been called *genotypic selection* (e.g., Pan & Price 2001; Stuefer et al. 2009), although I fail to see how this process differs from ordinary natural selection preserving some phenotypic variants while eliminating others (Darwin 1859). Beginning with experimental populations of *Potentilla reptans* containing 10 ramets of each of 10 distinct genotypes originally collected from diverse field sites in the Netherlands, Stuefer et al. (2009) documented pronounced differences in the frequencies of the genotypes after only five years. The genotypes were identified by unique DNA fingerprints. From an initial frequency of 10%, by the end of the experiment some genotypes became rare, one genotype went extinct, and another increased in frequency to more than 40%. The authors also measured a variety of phenotypic traits and found that genotypic-specific features important to competitive ability, such as large ramet size and high vegetative biomass, were important determinants of genotype success (Stuefer et al. 2009).

Experimental approaches to investigating microevolutionary processes such as natural selection that incorporate the demography of populations have long been advocated (Harper 1967, 1977; Solbrig 1980; Endler 1986; Travis & Reznick 1998; Metcalf & Pavard 2007). Demographic study includes quantification of survival, growth, and reproduction at multiple stages of the life cycle (Solbrig 1980; Metcalf & Pavard 2007), and plant ecologists have long used such techniques to investigate the dynamics of populations (Harper 1977; Clegg et al. 1978b; Sarukhán et al. 1984). Cohorts can be analyzed to determine whether specific patterns in demographic features in the field are associated with particular phenotypic trait values (Endler 1986). Of course, to integrate microevolution into such analyses, it is important to know something about the genetic composition of the cohorts being monitored.

A detailed study of *Plantago lanceolata* over several years in a field in Virginia by Shefferson and Roach (2012) provides a fine example of the interplay of genetics and the environment in determining population dynamics of a perennial species. The researchers replicated genetic composition by using a specific breeding design such that each cohort of seedlings planted began with the same gene pool. Cohorts were planted into different sections of the same field with minimal disturbance to the natural community in October of three years (2000–2002) and also in April of the final year (2002). Plants were categorized into eight size classes (ranging from seedling to large adult) for data collected annually for seven to nine years (depending on when the cohort was established). General linear modeling was used to analyze population growth rates (λ), which were influenced mostly by reproduction. Population dynamics were found to be variable genetically, but the demographic responses of genotypes varied among different cohorts (Shefferson & Roach 2012). The male parent (sire) used in the original breeding design had highly significant effects on λ, which depended on year. A stochastic simulation projection to 100 years revealed intriguing patterns in what the expected genetic composition would become. For the cohort established in October 2000, after 100 years about 87% of the individuals were derived from only two sires and one sire's lineage was no longer present. However, for the cohort established in October 2002, after 100 years the majority (~80%) of the individuals were derived from two different sires (compared with the 2000 cohort). This study, which couples demographic analyses to cohorts of known genetic composition, probably comes as close as possible to documenting natural selection *in the act* of discriminating among genotypes (and phenotypic variants). Unambiguous identification of the actual agents of selection responsible for these variable demographic patterns remains difficult. Carefully controlled experiments, in which an environmental factor thought to cause selection is manipulated by the investigator (Wade & Kalisz 1990; Conner 2003), can be instructive (for two examples using *Arabidopsis thaliana*, see Ward et al. [2000] and Mauricio & Rausher [1997]).

2.2.4.1 Artificial Selection Experiments

Artificial selection involves humans choosing which individuals in a population will mate and contribute genes to the next generation, with the goal of developing particular phenotypic characteristics. The power of such human-directed evolution to produce new types of plant varieties has long been recognized, and Darwin (1868) provided many examples of crop and horticultural varieties produced by artificial selection. The journal *Plant Breeding Reviews* provides a wealth of more recent examples of artificial selection changing phenotypic traits in domesticated crops successfully (e.g., Gepts 2004). Artificial selection experiments have also been useful to the investigation of phenotypic plasticity in several model plant species (reviewed in Callahan [2005]). The question here is: What can artificial selection experiments tell us about the evolutionary ecology of *wild* plant species under natural conditions?

One use of artificial selection experiments lies in their ability to reveal "what a given strength and form of selection can accomplish in terms of phenotypic change and how quickly this change can occur" (Conner 2003, p. 1650). In essence, by choosing which genotypes and phenotypes will provide the seeds of each generation, a researcher is mimicking directional selection as it might occur in nature (e.g., Miller 1995; Potvin & Tousignant 1996; Stanton et al. 2000; Burgess et al. 2007). Carey (1983) performed such an experiment with two species of *Plectritis* that differed in their breeding system (mostly outcrossed vs. highly selfed). Starting with two base populations, he selected for height (tall or short) and flowering time (early or late) at an intensity of about 10%, collecting 10 fruits from each of 20 plants out of a population of 200 to form each successive generation. For the outcrossed species *Plectritis congesta*, after five generations selected lines had diverged greatly for both height and flowering time (Table 2.7). The selfing species *Plectritis brachystemon* also responded to selection for flowering time, but not for height. In another artificial selection experiment using *Mimulus guttatus*, it was shown that the same selection regime (for larger flowers with quick development) could lead to divergent responses, depending on whether selected populations had been provided with a selfing, outcrossing, or mixed mating system (Holeski & Kelly 2006). These artificial selection experiments clearly show the potential for microevolutionary changes in life history traits in wild plant species, provided there is sufficient genetic variability. They also show how mating systems can influence the efficacy of selection.

Another important use of artificial selection is to develop specific phenotypes that can then be analyzed for their performance under field conditions while exposed to natural selection (Brakefield 2003). The production of distinctive groups with specific phenotypic trait values by artificial selection may be more realistic than directly "engineering" phenotypes by hormones or surgical procedures (Travis & Reznick 1998). Two examples of the use of artificial selection to generate particular phenotypes in wild plants are given here. Delph and Herlihy (2012) deliberately selected for small or large flowers via controlled crosses in the dioecious perennial *Silene latifolia*. Plants were subsequently grown in an experimental field at Indiana University, and sexual selection was examined by measuring a variety of phenotypic traits important to reproductive fitness. Larger flowers in general tended to be favored in both sexes. In the second example, three generations of artificial selection were used to generate groups of early- or late-flowering plants in the short-lived, outcrossing herb *Campanulastrum americanum* (Galloway & Burgess 2012). Plants in the early-flowering group flowered, on average, 25 days earlier than those in the late-flowering group. To observe the ecological consequences of this variable phenology in relation to environment, plants were placed into understory and light-gap areas at a site in southwestern Virginia. Reproductive phenology was integrated closely in the

Table 2.7 Results of an artificial selection experiment after five cycles of selection for tall versus short height and early versus late flowering in the outcrossing plant *Plectritis congesta*.

Variable	*Start*	*Selection for*	
		High	*Low*
Height (mm)	223.9	315.9 (+41%)	167.9 (–25%)
Days to anthesis	95.0	114.0 (+20%)	82.2 (–13.4%)

Mean values are shown with percentage change relative to the start of the experiment in parentheses. Data from Carey (1983).

artificially selected groups, and selection on flowering time altered the timing of all subsequent reproductive stages (such as the time when seeds dispersed), having the potential to affect multiple stages of the species' life history.

2.3 NATURAL SELECTION IN PLANTS: WHAT HAVE WE LEARNED?

That natural selection invariably occurs in plant populations is now a well-established fact. Because of their sedentary nature, the local environment of a plant includes multiple factors that might act as agents of selection. These abiotic and biotic factors (Table 2.1) obviously vary in space and time, and so do the quantitative estimates of selection (Siepielski et al. 2009, 2013) . The agents of selection cause microevolution by changing a population's gene pool (i.e., changing the frequencies of alleles and genotypes) and its phenotype pool (i.e., changing the frequencies of phenotypes with different trait values). Traditionally, the agents of selection are thought of as external to the organism, but internal factors that influence the development of the phenotype (e.g., biochemicals, gene regulatory proteins, endosymbiotic microbes) also influence selective processes.

Based on 656 estimates, the median linear selection gradient for plants had an absolute value of 0.14, whereas that of quadratic (nonlinear) selection gradients based on 326 estimates was only 0.05 (Kingsolver & Diamond 2011). Based on 653 estimates in 28 plant species, the median linear selection gradient was given as 0.12 (mean ± SE, 0.20 ± 0.01) by Geber and Griffen (2003), whereas the median quadratic selection gradient based on 214 estimates was 0.03 (mean, 0.08 ± 0.01). These values indicate that low to moderate directional selection is likely to be a common feature for many phenotypic traits in plant populations. For example, a meta-analysis of phenotypic selection on flowering phenology in 87 plant species produced a linear selection gradient of −0.14, suggesting that selection tends to favor earlier flowering plants, especially in temperate plants at higher latitudes (Munguia-Rosas et al. 2011). Note the remarkable agreement of this estimate with that of Kingsolver and Diamond (2011) noted earlier. For living things in general, directional selection estimates for size-related variables tend to be positive, whereas those for seasonal timing of life cycle events tend to be negative (Kingsolver et al. 2012). This is likely to be true for plants as well (Geber & Griffen 2003). Lower estimates of stabilizing and disruptive selection (compared with directional selection) may have biological and/or methodological explanations and "our understanding of nonlinear selection remains rudimentary at best" (Kingsolver et al. 2012, p. 1114).

Many possible selection agents responsible for the measured selection gradients in plants have been identified, and much of this book is concerned with exploring them in more detail. Although a few examples were provided in this chapter, more experimental approaches, such as the manipulation of putative selection agents, are needed to understand the ecological causes of microevolution more completely (Wade & Kalisz 1990; MacColl 2011). We might conclude using the prescient words of Stebbins (1950, p. 107): "while the demonstration that selection has occurred is not excessively difficult, the nature of action and the causes of the selective processes are much harder to discover or prove." For now, natural selection can be considered to be an important part of a basic law of evolutionary ecology: Whenever organisms with heritable variation reproduce in nature, natural selection *will* take place (Reed 1981)!

3 The Common Garden Approach

3.1 INTRODUCTION

In a discussion of methods used to study genetic differentiation among plant populations, Mazer and LeBuhn (1999) addressed three experimental approaches:

1. Greenhouse or growth chambers
2. Common gardens
3. Reciprocal transplants

Because the first two approaches have the minimization of environmental variation as a common goal to expose all putative genetic groups to conditions that are as homogeneous as possible, they are considered together here. Indeed, Gibson (2015) considers common garden experiments as those conducted either in the field or greenhouse (e.g., Gibson & Risser 1982). The shared element in the two approaches is the testing for differentiation among any set of genetically distinct groups in a relatively homogeneous environment. These genetically distinguishable groups can be (1) clonally replicated genotypes, (2) sibships (families) derived from separate maternal parents, (3) populations, (4) ecotypes, (5) varieties (or "races" in the older literature), (6) cultivars or agricultural accessions, (7) subspecies, or (8) hybrids.

Many types of common garden experiments have been conducted by plant ecologists, and no attempt is made here to review all of them. Instead, examples are used to illustrate several methodological variations of common garden designs and the types of issues they are used to address. Some of the important practical applications of these techniques are also highlighted.

3.2 SINGLE COMMON GARDEN, NO ENVIRONMENTAL FACTORS VARIED

In this simple design, seeds, seedlings, or adult plants of putative genetically distinct groups are used as source material for growth in one homogeneous set of conditions. None of the environmental factors that might be selectively responsible for genetic differentiation are manipulated. Thus, from a statistical standpoint, the only factor that varies is the source group, whether a population or genotype or other genetically cohesive unit. If there is statistically significant variation among groups for the measured quantitative traits, then groups show genetic differentiation.

This type of experiment can be useful in an initial screening process to determine the relative levels of quantitative trait variation within and/or among populations. Data on genetic variation in life history traits can give some perspective on recent evolutionary history of the populations and the potential for further evolutionary change (Mazer & LeBuhn 1999; Conner & Hartl 2004). Heritabilities can be estimated using standard techniques of quantitative genetics, but because environmental variance is likely to be reduced in the common garden, these may be overestimates (Mitchell-Olds & Rutledge 1986). Nonetheless, the correlation between lab and field estimates of heritability is often fairly good (Weigensberg & Roff 1996), and thus some useful information about the operation of natural selection in wild populations can be obtained (Primack & Kang 1989).

A study of the common evening primrose (*Oenothera biennis*) provides an example of the use of a common garden environment to estimate quantitative genetic parameters (Johnson et al. 2009). This herb has a genetic system that maintains a permanent translocation and is, therefore, functionally asexual. Thus, seed groups, each derived from a different maternal parent, provided replicates of genotypes to be distinguished statistically. Plants were grown outdoors in an abandoned agricultural field. Measurements were made of 24 physiological, biochemical, and life history traits, several of which are important to fitness in this species. By using the fraction of the total phenotypic variation that was the result of differences between genotypes, heritability in the broad sense was estimated from mixed-model analyses of variance for each trait.

In the common garden with 39 genotypes of *Oenothera biennis* (each with 5–11 replicate plants), Johnson et al. (2009) found statistically significant genetic variation in all 24 phenotypic traits. Mean heritability of four life history traits was 0.36, whereas the concentration of various secondary compounds (several of which function in herbivore deterrence) was 0.68. Lifetime fruit production, a metric of reproductive fitness, also showed significant heritability (= 0.38). Thus, the common garden approach has revealed the genetic basis to a diversity of phenotypic traits and indicates much potential for natural selection to cause evolutionary change (Mazer & LeBuhn 1999).

A single common garden experiment with no manipulated environmental factors was used recently to demonstrate how communities of soil microorganisms could be affected differentially by the genotypes of *Populus angustifolia* trees they are beneath (Schweitzer et al. 2008). In this innovative study, tree genotypes were replicated as stem cuttings and planted into a common garden in Ogden, Utah. After the trees had been growing for 13 years, soil samples were collected within 0.25 m of the trunks, and microbial biomass and community composition were assessed using phospholipid fatty acid biomarkers and microbial biomass nitrogen (methodological details in Schweitzer et al. [2008]). Broad-sense community heritability was estimated by treating microbial biomass and community composition as quantitative traits. Both of these quantitative measures of the microbial community showed significant variation among tree genotypes; for example, up to 70% of the variation in the composition of the microbial community was explained by the genotypes of *P. angustifolia*. This study is a good example of the use of the common garden to minimize environmental variation so that genotypic differences within the plant population can be readily detected. However, in this instance, the measured quantitative "traits" were actually properties of the biotic community in the soil surrounding, and thus being affected by, plant genotypes.

Common garden experiments provide a simple way to reveal genetic differentiation among widely distributed, divergent populations. Often, the differences among groups that represent population samples from diverse geographic regions can be correlated with environmental differences among these regions. Endler (1986, p. 56) called this "the commonest and oldest method" for the detection of natural selection in the wild. However, he also notes that such correlative methods can only provide indirect evidence of past selection. Nevertheless, calibration of the relative performance of populations to the degree of environmental difference between the sites of origin and the site of the common garden has been used to detect local adaptation (Rutter & Fenster 2007; Climent et al. 2008). For example, using an environmental distance metric that summarizes and scales the climatic differences between sites, Rutter and Fenster (2007) showed how this metric was related significantly to the fitness of *Arabidopsis thaliana* populations in a single common garden in Maryland. As might be expected, fitness was greatest when historical conditions normally experienced by the population at its site of origin were similar to the environmental conditions at the site of the common garden.

Any tightly controlled homogeneous environment can constitute a common "garden" in which genetic differentiation among individuals or populations can be documented. Kauth and Kane (2009) used a technique they termed "in vitro ecology" to assess "ecotypic differentiation" in an orchid native to eastern North America. The species forms swollen basal stems, known as *corms*, throughout its range. Because they store carbohydrates, corms made by various plant species are thought to represent an adaptation important to survival, growth, and reproduction after unfavorable growth periods (Zimmerman & Whigham 1992; Cheplick 2003). Kauth and Kane (2009) obtained seeds from four widely separated populations of *Calapogon tuberosus* var. *tuberosus* and grew seedlings from them in culture boxes under constant growth-chamber conditions. Corm diameter was one of a number of traits recorded over 20 weeks of in vitro culture.

Plants of *Calapogon tuberosus* var. *tuberosus* from the northernmost population (Michigan) were the first to produce corms, whereas plants from two southern populations (Florida) initiated corms several weeks later (Fig. 3.1). A much shorter growing season in the north probably selected for faster allocation of resources to corms (Kauth & Kane 2009). The final size attained by the corms was greatest for populations from north-central Florida and South Carolina (Fig. 3.1). Because no environmental factors were explicitly varied and a single common environment was used, precise explanations for trait divergence patterns among the four populations are not possible. Nonetheless, the controlled environment provided by such in vitro ecology studies can be useful in identifying and evaluating specific genotypes, populations, or ecotypes of a particular plant species in terms of their life history features.

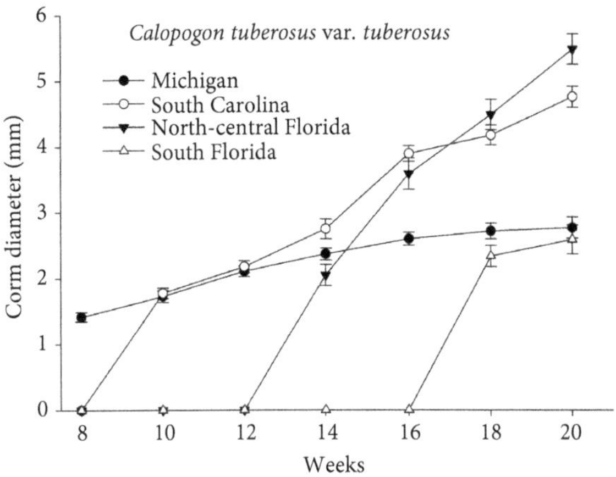

FIGURE 3.1 Mean diameter ± standard error of corms produced by four populations of *Calopogon tuberosus* (Orchidaceae) over 20 weeks of in vitro culture. Data from Kauth and Kane (2009).

There is a long, 200-year history of foresters using the common garden technique as a practical tool to determine the extent to which populations are differentiated and locally adapted in many tree species (Langlet 1971; Mátyás 1996; Boshier & Billingham 2000; Savolainen et al. 2007). Genetically based geographic variation among latitudinal and climatic gradients for populations established in a common garden is often demonstrated in such "provenance" tests. For a few examples, see the 15-year provenance trial conducted by O'Brien et al. (2007) in southwestern Australia using *Eucalyptus marginata*, and the common garden experiment conducted by Bower and Aitken (2008) in northwestern North America using *Pinus albicaulis* seedlings collected from 48 populations throughout its range.

In the study by Bower and Aitken (2008), quantitative traits important to cold adaptation (date of leaf emergence and an index of cold injury) were strongly correlated with the mean temperature of the coldest month found at the population collection site. Growth-related traits such as height and biomass were correlated with growing season length at the original collection sites. The multivariate analyses the authors presented typify these types of investigations in which many variables are tabulated. Because eight climatic (e.g., MAT, length of frost-free period) and three geographic (latitude, longitude and elevation) variables were recorded, canonical correlation analysis was used to examine the interrelationships among these variables and the measured quantitative traits. This permitted the researchers to determine the proportion of variation in the quantitative traits that was accounted for by canonical correlations with climatic and geographic variables. The significant relationships that emerged between some quantitative traits and canonical climatic variables provided solid evidence that climatic factors have acted collectively as agents of natural selection and driven local adaptation of *Pinus albicaulis* populations (Bower & Aitken 2008). Other studies that provide evidence of local adaptation from provenance trial experiments with tree species can be found in the review by Savolainen et al. (2007).

Single common garden experiments with herbaceous plants also have a long history in plant ecology, beginning with the well-known work of Turesson in the 1920s (and beyond) in which he developed the ecotype concept (see historical overviews in Heslop-Harrison [1964], Langlet [1971], Briggs and Walters [1997] and Lowry [2012]). Turesson's remarkable transplant garden at the Institute of Genetics in Åkarp, Sweden, occupied more than 10 ha, with plants from many populations and species planted at 40-cm intervals in rows 50 cm apart (Turesson 1922). By 1925 (Turesson 1925, p. 148), he noted that "the number of transplants . . . now in culture in the experimental fields of the Institute exceeds 10,000!" His work focused on determining to what extent different morphological types observed in plants from specific habitats were genetically based. These "forms" he described as "shade forms, dwarfs, succulent shore forms of inland species . . . inland and coastal forms" (Turesson 1922, p. 215). His article, "The Genotypical Response of the Plant Species to the Habitat," is long (139 pages!) and full of photographs showing the general morphology and leaves of ecotypes of multiple species. He summarized his observations: "the differentiation of the species-population into different hereditary variations in the various habitats was found to be the rule in the majority of cases" (Turesson 1922, p. 331).

In another lengthy paper (89 pages [Turesson 1925]), he distinguished ecotypes of 15 species, presenting 29 data tables of morphological measurements and 50 figures (mostly plant photographs and several drawings of leaf internal anatomy). Again, the conclusion is that the ecotype represents a genetic response of a species to a specific set of habitat conditions. Thus, Turesson was one of the earliest practitioners of the common garden approach, clearly demonstrating the existence of habitat-correlated genetic variation in many herbaceous species.

Using common gardens to grow plants collected from a diversity of habitats, early researchers continued to document genetically based differentiation in morphological traits among populations (commonly called *ecotypes*) in several perennial grasses and herbs (Gregor & Samsone 1927; Kemp 1937; Gregor 1946; Böcher 1949). Variation among populations in a common environment was often depicted in tables of morphological measurements and plant photographs. Phenotypic differences among populations were usually attributed to evolutionary responses to different natural selection pressures in the original source habitats. For example, the prostrate growth forms of some grass species' populations maintained in the common garden were attributed to heavy grazing pressure in the original habitat (Gregor & Samsone 1927). The detailed studies of *Prunella vulgaris* in Denmark by Böcher (1949) are notable in that several experimental gardens were used and experimental growing conditions were sometimes purposefully manipulated (e.g., shady vs. sunny conditions). The clinal nature of much ecotypic differentiation was also acknowledged by Böcher (1949), expanded on by Stebbins (1950), and emphasized later by Gregor and Watson (1961).

The numerous transplant experiments of McMillan during the 1950s and 1960s using perennial grasses from the central grassland of North America provide additional examples of the use of a single common garden to document clinal differentiation of plant species (McMillan 1956, 1959, 1964, 1965, 1967). Typically, McMillan would collect clonal plant material from multiple populations distributed along north–south and east–west axes within the central part of the United States (and sometimes south-central Canada). These plants were then grown as spaced plantings in a common garden near to wherever he worked at the time: Lincoln, Nebraska (McMillan 1956, 1959), or Austin, Texas (McMillan 1964, 1965, 1967). Although he collected data on a number of quantitative traits such as height and leaf pubescence, the timing (and duration) of flowering were emphasized, perhaps because reproductive phenology showed distinct differences among the source populations. These differences could often be related to characteristics of each population's habitat of origin.

An example of a geographic cline in a quantitative trait important to Darwinian fitness is flowering date. In nature, individuals in northern regions tend to flower

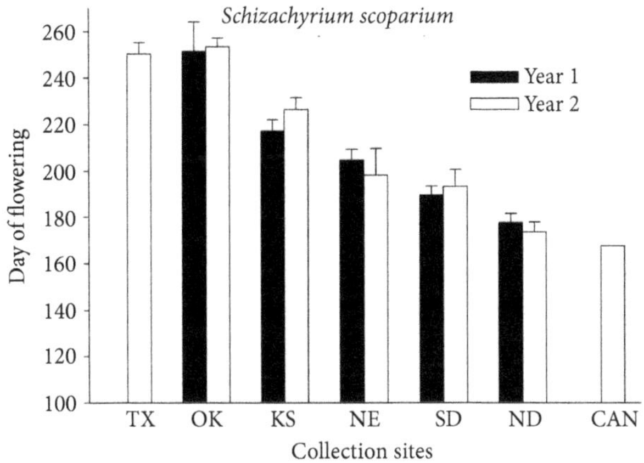

FIGURE 3.2
Number of days to flowering for clones of *Schizachyrium scoparium* collected from populations from northern Texas to southern Canada and grown for two years in a common garden in Lincoln, Nebraska. Bars show mean ± standard error for three to five populations, except for Canada, where two populations were sampled. CAN, Canada; KS, Kansas; ND, North Dakota; NE, Nebraska; OK, Oklahoma; SD, South Dakota; TX, Texas. Data from McMillan (1959).

earlier than those in southern regions. Earlier flowering in the north allows completion of reproduction, including seed maturation, within a shorter growing season and, presumably, this phenological pattern should be favored by natural selection. Using a common garden in Lincoln, Nebraska, McMillan (1959) showed how clones of a number of species originally collected in a south-to-north gradient from northern Texas to southern Canada maintained the field pattern of flowering behavior. The day of first flowering in the common garden is plotted for *Schizachyrium scoparium* in relation to the sites from which clones were collected (Fig. 3.2). For the two years shown, plants from northern sites such as Canada and North Dakota flowered earliest. McMillan (1959, p. 302) clearly recognized that this result meant that there was a "genetic basis for differences in flowering time of several species of grasses." He also repeatedly invoked selection as being responsible for the genetic differentiation, especially the "selective influence of climate" (McMillan 1959, p. 305). Unfortunately, these interpretations are not quite as powerful or direct as those provided by the transplant experiments of Clausen and associates (1948) because it cannot be inferred that a local population is more closely adapted to its immediate habitat (site) than a more distant population. Single common garden experiments also do not allow the detection of population-by-environment interactions. However, additional insight can be gained into the putative selective agents responsible for genetic differentiation when multiple gardens are used or when one or more environmental factors are varied experimentally within a single common garden.

3.3 MULTIPLE COMMON GARDENS, NO ENVIRONMENTAL FACTORS VARIED

The primary rationale for placing genetically distinct groups into multiple gardens is to detect population-by-environment or G × E interactions (Nuismer & Gandon 2008; Williams et al. 2008). The presence of G × E interactions is an important prerequisite for the evolution of local adaptation (Kawecki & Ebert 2004). In these studies, the gardens normally differ in one or more environmental factors that can be measured by the investigators. However, in this class of experiments, environmental factors are not manipulated explicitly. In general, populations or genotypes are most likely to perform best (i.e., show the greatest growth, survival, and reproduction) in gardens where the environmental factors closely match the habitat conditions to which they are adapted. In this regard, experiments that involve multiple common gardens are very similar to, and intergrade with, reciprocal transplant experiments (for example, see the model presented by Nuismer and Gandon [2008]).

Early studies by Frederic E. Clements and associates typically involved multiple gardens to investigate complex edaphic and climatic gradients in the western United States (Clements [1929] and references therein). The familiar studies by Clausen and coworkers (Clausen et al. 1940, 1948; Clausen & Hiesey 1958; reviewed by Núñez-Farfán and Schlichting [2001]) using plants collected along an elevational transect in central California provide another early example of the use of multiple common gardens to study the "genetic structure of ecological races." It should be noted that the concept of an ecological race is fundamentally similar to the ecotype (i.e., locally adapted populations). Clausen (1951, p. 30) refers to the race as "composed of a considerable number of variable local populations existing within a given ecological zone." He considered it to represent the next evolutionary stage in differentiation, above that of the local population. It is no surprise that the climate along this 322-km transect varied greatly, ranging from mild, with a lengthy growing season (coast) to alpine regions with long winters and short summers (Clausen et al. 1940, 1948). Thus, Clausen (1951) referred to the populations of a particular species collected from different elevations that maintained their distinctive phenotypic characteristics in the common garden as "climatic races." As with McMillan's (1959) work described earlier, climatic factors were usually proposed to be the primary selection agents responsible for the evolution of genetically distinct populations (Clausen et al. 1948). However, although the gradients of McMillan (1959) were latitudinal, those of Clausen et al. (1948) were elevational.

One of the species analyzed thoroughly along their elevational transect was the perennial herb *Potentilla glandulosa*, collected from more than 20 locations and transplanted into three common gardens at contrasting altitudes. The Stanford garden was near sea level (30 m), whereas Mather was at 1,400 m and Timberline was at 3,050 m. Data from Table 2 in Clausen and Heisey (1958) on leaf length and width were multiplied to obtain an estimate of leaf area, and were plotted against the elevation of origin for populations in the Stanford common garden (Fig. 3.3A). Stem height is similarly plotted in Figure 3.3B. Note that each point in Figure 3.3 represents the mean of a group of plants originally collected at a particular elevation along the transect. These types of quantitative data within multiple common gardens allowed Clausen and Heisey (1958, p. 2) to recognize "four morphologically distinct subspecies having ecologically distinct ranges." It is not clear whether the term "subspecies" is appropriate here, and in their earlier work, Clausen et al. (1940) referred to these as "ecological races." Indeed, both "race" and "ecotype" were still used freely in Clausen and Heisey (1958)! However, Clausen (1951, p. 48) makes it clear that each subspecies could be subdivided into "two or more ecological races." Therefore, "subspecies" encompasses a broader array of populations and represents a greater degree of intraspecific differentiation than "race" or "ecotype."

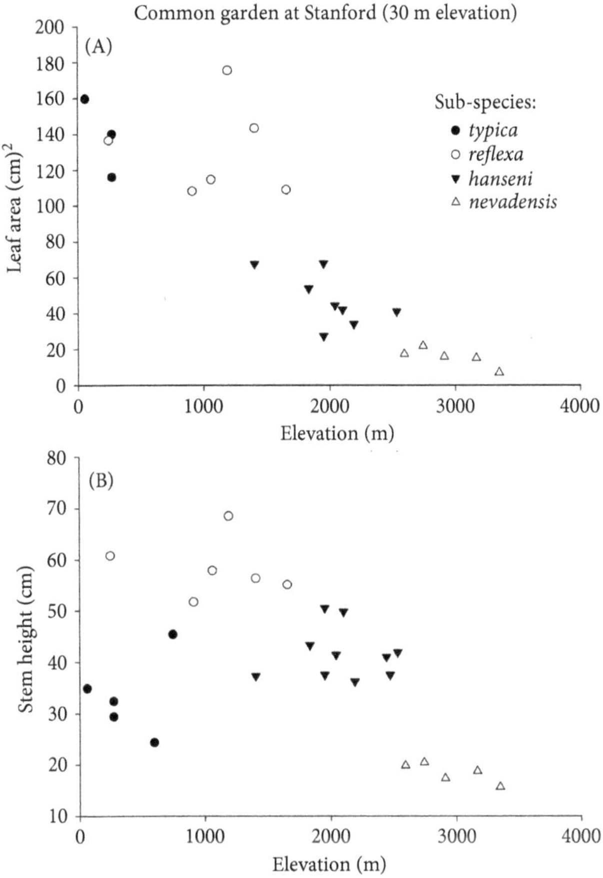

FIGURE 3.3
(A, B) Leaf area (A) and stem height (B) of subspecies of *Potentilla glandulosa* collected from populations along an elevational transect in central California and grown in a common garden in Stanford, California. Each point is the mean for a group of plants collected at a particular elevation. Data from Clausen and Hiesey (1958).

A substantial amount of genetic differentiation in quantitative traits was detected among the many populations of *Potentilla glandulosa* planted into the Stanford garden. Plants from the highest elevations (ssp. *nevadensis*) were much shorter and had smaller leaves that those from lower elevations (Fig. 3.3). Even among the five populations of subspecies *nevadensis*, there were distinguishable "climatic races" (Clausen & Heisey 1958, p. 23). As was true of the other subspecies, transplants of *nevadensis* tended to show greater growth and vigor at the garden (Timberline) with an elevation that matched most closely the elevation from which the populations had been collected. The multiple common garden studies of Clausen and colleagues provided solid evidence of local adaptation of *P. glandulosa* to the prevailing climatic conditions resulting from natural selection (Núñez-Farfán & Schlichting 2005). Common garden studies continue to be used to characterize adaptive trait differences of plants collected along complex climatic gradients (Etterson 2004a; Maron et al. 2004; Rutter & Fenster 2007; Oyarzabal et al. 2008; Scheepens et al. 2010; Scheepens & Stöcklin 2013).

Potential divergence in life history traits were investigated for northern and southern populations of the introduced species *Senecio squalidus* in Britain (Allan & Pannell 2009). In one experiment, common gardens were established at two sites several kilometers apart (Edinburgh and Oxford) with measurable climatic differences. Seeds were collected in bulk from wild populations in Leeds, Newcastle, and Edinburgh. Seedlings that germinated from these seeds were planted into the two gardens. The seed source-by-garden interaction shown for an important component of reproductive fitness—the number of seed heads produced (Fig. 3.4)—indicates that population performance in the field depended on the local environment. The authors also found that populations tended to perform best in the gardens that were most similar to their source regions in terms of climate (Allan & Pannell 2009). Plants from Edinburgh, for example, showed the typical "home-site advantage" (i.e., greatest fitness in the Edinburgh garden; Fig. 3.4) expected from reciprocal transplant experiments (see Chapter 4).

The study of Maron et al. (2004) is of particular note because it addresses the question of rapid, adaptive evolution in a widespread, invasive species. This is a potentially important and useful aspect of the multiple garden approach (Bossdorf et al. 2005). The common gardens can be situated within both the native and introduced portions of an invasive species' range (Williams et al. 2008; Moloney et al. 2009). However, these studies can be very labor intensive, especially when many populations are used (witness the 53 people in the acknowledgments of Maron et al. 2004). Maron et al. (2004) investigated the performance of 50 populations of the exotic perennial *Hypericum perforatum* (St. John's wort) within four common gardens, although not every population was planted into each garden. Seeds were collected from 18 native European and 32 introduced North American populations. Two gardens were in Europe (Sweden and Spain) and two were in the western United States (Washington and California). Quantitative traits such as plant size and fecundity were recorded. Genetic relationships were explored using polymorphic DNA markers. Latitudinally based clines in quantitative traits were found in most gardens. Both native and introduced populations from northern latitudes showed greater growth and reproduction in northern-latitude gardens (relative to populations from southern latitudes)—that is, they showed a home-site advantage. Most of the molecular variation (65.6%) was partitioned among populations, and introduced populations

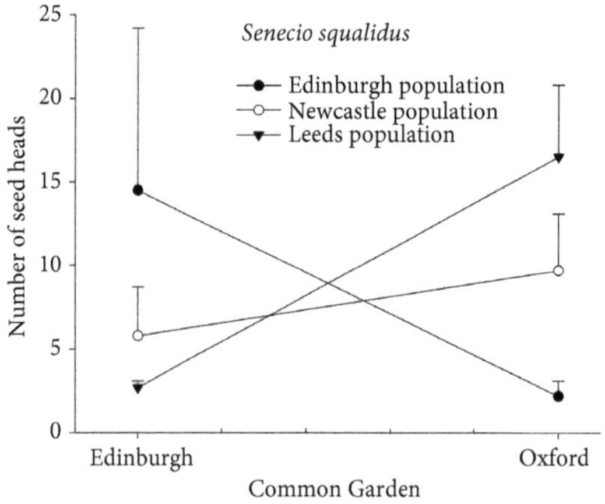

FIGURE 3.4
Mean ± standard error of the number of seed heads produced by three populations of *Senecio squalidus* grown in common gardens at Edinburgh or Oxford, England. The use of two common gardens reveals substantial population-by-environment interaction. Data from Allan and Pannell (2009).

showed genetic differentiation for quantitative traits. Although not all of the genetic differences among populations could be related to climatic conditions prevailing at their site of origin, the data revealed that the introduced populations of this invasive species were showing rapid adaptive evolution in relation to environmental conditions in their adopted home (Maron et al. 2004). Clearly, the use of multiple common gardens can be very informative in studies of the evolutionary changes thought to be important to plant invasions (Bossdorf et al. 2005; Williams et al. 2008; Moloney et al. 2009; Flory et al. 2011).

Multiple common gardens can provide much more powerful evidence than a single garden that populations from different geographic regions are genetically distinct (Scheepens et al. 2010). This is because the different gardens used are likely to vary in environmental features and consistency in the overall performance of populations across diverse conditions can be assessed. In this way, six common gardens (five in the northeastern United States and one in Korea) were used to verify the strong genetic component to the distinctive growth form and life history attributes of *Pinus rigida* (pitch pine) populations from highly fire-prone Pine Plains, New Jersey, relative to those from surrounding areas (Ledig et al. 2013). Flory et al. (2011) explored potential genetic differences between native and introduced populations of *Microstegium vimineum* in 22 common gardens distributed among sites in southern Indiana. These sites varied widely in environmental factors such as light availability, soil moisture, and inorganic nitrogen. This annual grass, which is highly invasive in the eastern United States, is native to eastern Asia. Flory et al. (2011) found that, on average, introduced populations (collected from 10 states in the United States) outperformed native populations (collected from 10 sites in China) in terms of survival and biomass accumulation in every common garden. They suggested that postintroduction evolutionary changes resulting in greater plant vigor caused genetic differentiation of the invasive populations in the United States from the original, source (native) populations of eastern Asia.

A complex variation on the "replicated" common garden approach was used in a detailed investigation of population differentiation and possible local adaptation in the annual legume *Chamaecrista fasciculata* (Galloway & Fenster 2000). Gardens were located at sites in Kansas, Illinois, and Maryland. Seeds from glasshouse-grown plants (to reduce maternal environmental effects) of the "target" populations at these sites were transplanted reciprocally among the sites. Because the researchers were also interested in the spatial scale of local adaptation, four additional populations collected 0.1, 1, 10, and 100 km from the target populations were also transplanted. Note that this design actually combines the traditional reciprocal transplant approach for the three target populations with multiple common gardens for the four additional populations. Germination, survival, vegetative mass, and fruit production were recorded for two years. The researchers found little evidence of local adaptation except at the farthest spatial scales (>1,000 km [Galloway & Fenster 2000]).

3.4 SINGLE OR MULTIPLE COMMON GARDENS, ONE OR MORE ENVIRONMENTAL FACTORS VARIED

In these experiments, one or more environmental factors are identified, based on past studies or observations of current field conditions, as the selection agents most likely responsible for the adaptive differentiation of populations. For populations distributed along an aridity gradient, for example, soil moisture levels would be an obvious candidate for manipulation in a common garden study. One might designate three soil moisture levels to be maintained within separate plots at a field site (or among pots in a greenhouse), and then one would plant individuals from each population into each plot. Note that each plot within the same field site could be considered a separate common garden, with the recognition that the primary difference between the three gardens is available soil moisture. This is the reason this class of experiment incorporates single or multiple gardens; the distinction between them is decidedly artificial. The simple factorial design includes population, moisture level, and their interaction as the sources of variation. Many types of factorial experiments in both the field and glasshouse fall into this category; however, the primary distinguishing feature of the common garden design is that one of the factors consists of genetically distinct groups such as populations, families, or replicated genotypes.

As with the previous category of multiple gardens but with no environmental factors manipulated explicitly, G × E interactions can be detected readily because the genetically distinct groups are exposed to different environmental conditions (or, if one prefers, different gardens). The difference is that the environmental conditions in the current category are being manipulated by the investigator and are well defined. If adaptive evolution has occurred, the expectation is that the plants that perform best under one set of experimental conditions (e.g., low soil moisture) will be those from habitats that most closely match those conditions (e.g., sites with a history of regular drought episodes).

An extensive single common garden study was conducted using seeds of *Elymus glaucus* collected from 153 sites throughout northeastern Oregon and southeastern Washington (Erickson et al. 2004). The goal of the study was to describe "geographic patterns of potentially adaptive differentiation" and to use this information to guide decisions on the use of this native forage grass for ecological restoration in the region. Plants were grown for three years in a garden in Pullman, Washington. The manipulated environmental factor was precipitation; some plants received ambient rainfall, whereas others received ambient rainfall plus supplemental irrigation with a sprinkler three to four times during the growing season (to maintain soil field capacity). A large variety of quantitative growth and reproductive traits were recorded. As with other studies of multiple quantitative traits in many populations, principal component and cluster analyses were used to depict population variation and to classify seed sources by the phenotypic similarity of plants grown from them (details in Erickson et al. [2004]). In all, 81% of the phenotypic variation was related to differences among sites. The effects of irrigation on phenotypic variation among source populations were minor and often insignificant. Longitude and ecological region best described three distinct geographic groups of sites. These two features were related to multiple climatic and edaphic factors. Some differences among groups could be related to climatic and edaphic conditions prevailing at the sites from which seeds had been collected, provoking speculation about adaptive variation within this species. The utility of the common garden approach for categorizing phenotypic variation across a broad geographic area in this ecologically important forage species was nicely illustrated by a restoration framework and seed management guidelines provided by the authors (Erickson et al. 2004).

Some common garden designs permit a finer level of genetic resolution than populations. Investigators may wish to examine the performance of subpopulations, genetic relatives (i.e., families), or replicated genotypes. These studies are normally pursued in relation to one or more environmental factors predicted to be significant selection agents in the usual habitat of the species. For example, Cheplick and White (2002) investigated the possibility of subpopulation differentiation within a population of the coastal annual grass *Triplasis purpurea* on Staten Island, New York. The environmental factor chosen as a putative selection agent was airborne salt spray because prior work had demonstrated that salt deposition onto leaf surfaces of *T. purpurea* was related directly to distance from shore in the same population (Cheplick & Demetri 1999). Seed families were collected from individual *T. purpurea* plants "near" to shore (15 m) where salt spray exposure was most intense and "far" from shore (80 m) where salt spray was negligible. There were 13 families from the near subpopulation and 11 families from the far subpopulation, with 8 to 12 siblings per family planted in the glasshouse that served as the common "garden." Half the siblings in each family were subjected to weekly sprays of saltwater, whereas the other half received sprays of distilled water. If there was prior adaptive differentiation between these two subpopulations, it was predicted that the near subpopulation would not be affected as negatively by salt spray as the far subpopulation because only the former regularly experiences significant levels of natural salt spray (Cheplick & Demetri 1999). Improved salt tolerance might evolve rapidly in an annual species in which generation time is short and reproduction by seed is essential to fitness.

However, for the quantitative traits measured, there was no evidence of subpopulation differentiation in relation to growth ability after salt sprays (Cheplick & White 2002). In common with most traits recorded, there was highly significant family variation within subpopulations in tiller production with and without salt spray exposure. Salt spray caused a significant reduction in tiller number; however, analysis of variance did not reveal any significant differences between subpopulation means or any significant subpopulation-by-treatment interaction (Cheplick & White 2002). Thus, in this experiment with a single abiotic factor manipulated, local adaptation to salt spray was not demonstrated. The authors reasoned that either salt spray levels along the Staten Island shore were not great enough to act as a significant selection agent or that phenotypic plasticity of growth and reproduction was sufficient to buffer potentially negative effects of salt spray that vary greatly in time and space.

Single common gardens have also been useful in describing differences among genotypes or populations in relation to a manipulated biotic factor. Genotypes of perennial ryegrass (*Lolium perenne*) were replicated by separating tillers manually such that new individuals could be grown for each genotype (Cheplick 2008). Leaves of this important forage grass are commonly infected by a clandestine endosymbiotic fungus, which was the biotic factor of interest and a potential internal selection agent. This type of host–endophyte symbiosis is widespread among cool-season grasses, and the interaction may range from antagonistic to mutualistic (Cheplick & Faeth 2009). The infected genotypes of *L. perenne* used in the common garden were also available free of the endophyte because some cloned tillers used for replanting had been exposed to a systemic fungicide years earlier. Thus, infected and uninfected plants of nine host genotypes were planted in an outdoor garden in central New Jersey and monitored for almost three years (Cheplick 2008). The primary goal was to determine the relative importance of host genotype and endophyte infection to survival, tiller production, and flowering over time. Might the fungal endophyte function as an agent of natural selection on a perennial ryegrass population by affecting the survival, growth, or reproduction of host genotypes differentially?

The general result of this common garden study (Cheplick 2008) was that, although endophyte infection did improve tiller and biomass production of some host genotypes, a far greater proportion of the variance in phenotypic traits was explained by host genotype. Also, host survival, percentage of plants that flowered, flowering time, number of flowering tillers, and mean tiller mass were not affected by endophytes. Thus, in the benign environmental conditions of the experimental garden (no competition, moist soil), selection could differentiate readily among host genotypes, but the lack of genotype-by-infection interactions implies that endophyte infection was not influential in this process.

A more complex design that involved manipulation of one biotic factor (competition) within two common gardens was used by Leger and Rice (2007) to assess local adaptation in the invasive California poppy (*Eschscholzia californica*). The first garden involved plants of 20 populations: 10 from California where it is native and 10 from Chile where it is invasive. Plants were grown in containers without competition or with four individuals of *E. californica* of horticultural origin. The containers were kept outdoors in Davis, California, and a variety of size and fecundity traits were measured. In the second common garden experiment, seeds were sown directly into the soil in Montara, California, and plants experienced ambient field conditions. The background vegetation provided a competition treatment; in a second treatment, plants grew without competition. Many measured plant traits were intercorrelated, and principal component analysis was used to explore the relationships between traits and environmental variables found at the site of origin.

Populations of *Eschscholzia californica* from both California and Chile appeared to be adapted in a similar way to the local environments associated with their original sites (Leger & Rice 2007). Size and fecundity traits of populations from both invasive and native parts of its range in the common garden correlated with average precipitation—larger, more fecund plants originated from regions with low levels of summer and winter precipitation. Although the competition treatments reduced size and fecundity significantly, the competitive environment did not affect the correlations between traits and environmental principal components. Thus, the researchers had found solid evidence for adaptive differentiation to prevailing climatic conditions in populations of the California poppy that had been introduced into Chile about 150 years earlier.

3.5 NATURAL SELECTION IN THE COMMON GARDEN

Although one uses a common garden to provide a relatively homogeneous set of environmental conditions, some (often-unmeasured) level of spatial and temporal heterogeneity should be expected, especially if the garden occupies a large area or is maintained for a long time. Although climatic conditions (e.g., precipitation) can be presumed to be the same for all plants, there can be subtle variations in soil moisture and mineral availability as well as the amount of diurnal solar radiation that affects different areas within a single common garden. In addition to spatial variation in potentially important abiotic factors, variation in biotic factors (e.g., competition) may also occur, especially if the garden is not maintained to be free of weeds. Many common garden experiments do use a weed-free environment, but if the objective of a study is to examine the performance of diverse populations or genotypes in a quasi-natural competitive environment similar to that typically experienced by the species in its usual habitat, then the garden should be established with minimal disturbance to the background vegetation. If this is the case, then the intensity of competitive interactions may show spatial heterogeneity within a single garden. Other biotic factors such as pathogens or herbivores can show similar variation in space and time.

If the spatial heterogeneity in abiotic and biotic factors affects the survival, growth, and/or reproduction of individuals, then these factors are functioning as agents of natural selection within the common garden. Of course, variation in the strength of natural selection and local adaptation may be especially evident when quantitative trait values for different populations or genotypes are compared among multiple common gardens in which abiotic conditions and interacting species will differ (Nuismer & Gandon 2008). This variation would be detected statistically as a significant garden-by-population (or -genotype) interaction (Maron et al. 2004; Williams et al. 2008).

To control for possible unidentified variation in environmental conditions within a common garden, researchers have typically used block-type experimental designs (Underwood 1997; Gotelli & Ellison 2004). This can be a good way to separate statistically the potential effect of spatial heterogeneity on measured quantitative traits. When block effects (resulting from location within a garden) are statistically insignificant, researchers have better confidence in the assumption that the garden environment is relatively homogeneous.

Completely random allocation of plants to different locations (blocks) within a garden can be one way to deal with spatial heterogeneity. However, this can result in different numbers of individuals of a population or genotype being placed into different locations, and so the explicit incorporation of blocks into the experimental design is probably most desirable. An effort should be made to ensure that the same numbers of individuals of a particular population or genotype are allocated to each block within the common garden. The blocks can be situated randomly or delimited systematically (Gotelli & Ellison 2004), as in the example that follows.

As an example of how spatial heterogeneity within a common garden can affect reproductive fitness and result in natural selection, the study described earlier with *Lolium perenne* genotypes in a garden in central New Jersey (Cheplick 2008) is reconsidered. In that study, flowering tillers, each producing an elongate spike typical of this species, were counted during 3 growing seasons for 10 replicated genotypes. The cumulative number of flowering tillers will serve as an estimate of reproductive fitness. Because flowering tiller production was not affected significantly by fungal endophyte infection (Cheplick 2008), 20 plants per each of 10 genotypes were available for analysis within this small common garden (3 × 4 m). The rectangular garden was situated with the long axis of one side perpendicular to, and 1 m from, an adjacent woodland. The other side was adjacent to a mown lawn (Fig. 3.5). Individual *L. perenne* plants (1–3 tillers) were planted into a grid of 15 rows and 15 columns. Plants of each genotype were distributed haphazardly throughout the garden.

Continuous observations made each spring when plants flowered suggested that flowering tiller production was reduced in the back of the garden near the woodland relative to the front near the lawn. It was suspected that this could be the result of heterogeneity in the duration of direct solar radiation across the garden from front to back. Sunlight was available during the flowering period for about seven to eight

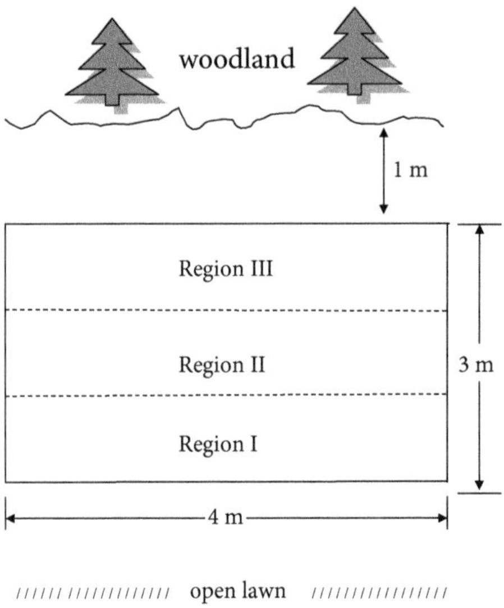

FIGURE 3.5
Small common garden used to investigate long-term growth and reproduction of *Lolium perenne* at a field site in central New Jersey (Cheplick 2008). To investigate heterogeneity of flowering, the garden was subdivided (dotted lines) into three regions for further analysis.

hours per day for plants nearer to the lawn and for about four to five hours per day for plants nearer to the woodland. To quantify the suspected heterogeneity in flowering, the garden was subdivided into three 1 × 4-m regions, as depicted in Figure 3.5. The cumulative number of flowering tillers (summed over the three years) was determined for each mapped plant ($n = 203$) of all genotypes. Prior analyses (Cheplick 2008) had shown that flowering tiller production within the garden was strongly affected by genotype.

A new two-way analysis of variance was performed on the cumulative number of flowering tillers using region ($df = 2$), genotype ($df = 9$), and their interaction ($df = 18$) as the sources of variation. Region within the garden had a highly significant effect on flowering tiller number ($F = 51.2$, $p < 0.0001$), and the relative magnitude of its effect (Underwood 1997) was 57.8%. The effect of genotype was also significant ($F = 5.9$, $p < 0.0001$), but its magnitude was less (25.4%). Genotype interacted significantly with region ($F = 2.6$, $p = 0.0007$). The strong influence of region within the garden on flowering tiller production by most genotypes is visually apparent in Figure 3.6. As suspected, most genotypes showed the greatest flowering tiller output in region I near the lawn (Fig. 3.5), where sunlight was readily available. Near the woodland at the back of the garden, where light availability was less, flowering was much reduced (Fig. 3.6).

Using the cumulative mean number of flowering tillers made by a genotype over three years as an estimate of absolute fitness, relative fitness may be defined as the number of flowering tillers made by a genotype divided by the mean number of flowering tillers made by all genotypes. Relative fitness clearly varied greatly with garden region, and only one genotype (D in Fig. 3.7) showed a consistently high fitness in all regions compared with other genotypes. In general, the rank order of genotypic fitness changed from one region to another (Fig. 3.7).

The lesson to be learned from this simple analysis is that selection can be expected within a common garden when heterogeneity in one or more environmental factors prevails. The performance of many species is affected profoundly by small-scale heterogeneity, as exemplified by numerous studies of clonal plants (e.g., Fitter et al 2000; Hutchings & Wijesinghe 2008). Furthermore, not all genotypes or

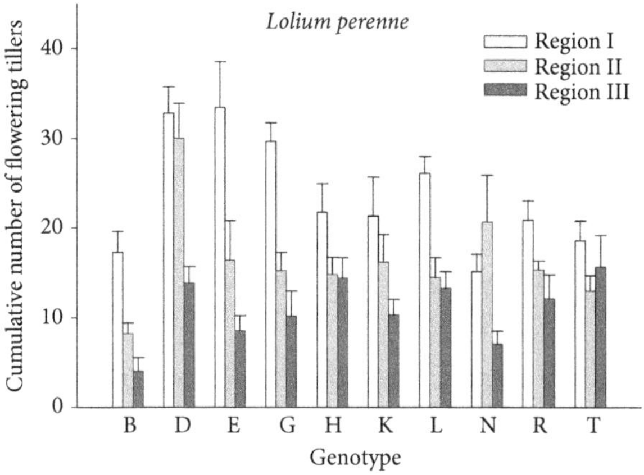

FIGURE 3.6
Cumulative number of flowering tillers produced over three consecutive years for 10 genotypes of *Lolium perenne* in a common garden in central New Jersey (Cheplick 2008). Mean ± standard error for 3 to 10 replicate plants of each genotype per region of the common garden (see Fig. 3.5) are shown.

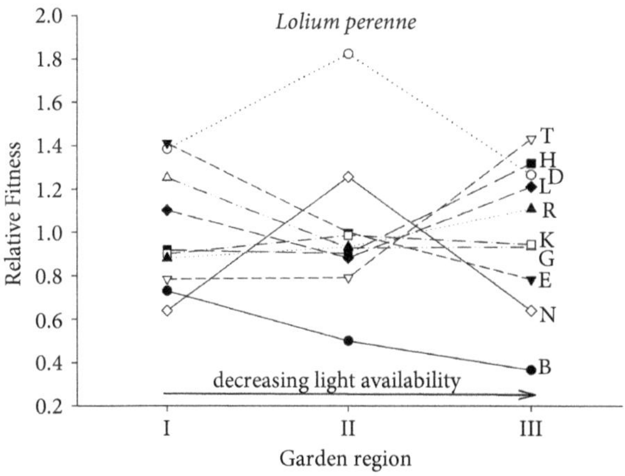

FIGURE 3.7
Relative fitness of 10 genotypes of *Lolium perenne* in the three regions (see Fig. 3.5) of the common garden used by Cheplick (2008). Relative fitness was estimated for each genotype in each region based on the cumulative number of flowering tillers produced over three years (see Fig. 3.6).

populations will be affected by heterogeneity in the same way. The assumption of a common environment within this small garden was clearly violated. Practitioners of the common garden experiment should probably give more attention to this assumption in the future.

3.6 QUESTIONS AND CONSIDERATIONS IN USING COMMON GARDEN EXPERIMENTS

Our exploration of the many types of common garden experiments and the kind of questions they address leaves no doubt that this approach offers much to evolutionary ecology research. Common gardens can reveal genetically based differences among any genetically cohesive group such as different clones, sibling families, or populations. They can also provide clues as to the types of selection agents (Table 2.1) in the original source habitats that may be responsible for putative adaptive differentiation.

Given these useful aspects, there are several questions and considerations to address when planning and using the common garden technique (many of these apply to the reciprocal transplant design as well, discussed in Chapter 4).

3.6.1 What to Use: Seeds, Seedlings or Ramets?

The choice of what to plant in the common garden will delimit the types of phenotypic traits that can be recorded. If early stages of the life cycle are of interest (e.g., germination probability, seedling growth), then seeds should be used. However, one should always be aware of potential maternal effects (see the following point) resulting from environmental conditions at the source sites of seed collection. Seedlings of a similar size or age may be planted to ensure greater establishment success in the experimental garden(s). This is useful if germination and recruitment variation among populations or genotypes are not part of the study's objectives. Ramets are most appropriate when replicated copies of specific genotypes are desired within and between gardens to investigate genetic and environmental variance components. However, care must be taken to ensure ramets from different source plants have not experienced different preconditions before planting. Growth of source plants for one generation in a common environment (e.g., greenhouse) can minimize this potential problem.

3.6.2 Can and Will Potential Maternal Effects be Controlled?

Plant ecologists are generally aware that the microenvironment of a plant can influence profoundly the quantity and quality of seeds it manages to produce. The phenotypic features of seeds collected directly from a field habitat, for example, their mass and germinability (Baskin & Baskin 2014), reflect not only the maternal and paternal genotypes that combined to create the embryo and endosperm, but also the conditions of the microhabitat (e.g., light and mineral nutrient levels) in which the maternal parent grew as it matured and provisioned its seeds (Wulff 1995). Environmental maternal effects (Roach & Wulff 1987; Weiner et al. 1997) carry over from the mother to the offspring, which of course are carried as embryos within seeds made by the mother plant. Seed features, such as mass and nutrient content, strongly influenced by the environment of the mother plant, can affect the size and growth of seedlings that emerge (e.g., Cheplick & Sung 1998). For the researcher using seeds collected directly from maternal plants in the source populations for a common garden, differences among populations in phenotypic traits, especially those recorded during early stages of the life cycle, may not be simply the result of adaptive genetic differentiation. Rather, the differences could reflect environmental maternal effects, specific to each source population, on offspring establishment and growth.

There exist three methods for controlling possible maternal effects in common garden studies (Bischoff & Müller-Schärer 2010). First, one can account for seed

Table 3.1 Examples of sample sizes used in a variety of common garden experiments.

Species	*Populations per species*	*No. of gardens*	*No. planted*[a]	*Reference*
Annuals				
Ambrosia artemisiifolia	34	5[b]	7–10 seedlings	Hodgins and Rieseberg (2011)
Arabidopsis thaliana	21	1	18 seedlings	Rutter and Fenster (2007)
Geranium carolinianum	6	2	30 seedlings	Bell and Galloway (2008)
Microstegium vimineum	20	22	2 seedlings	Flory et al. (2011)
Perennial (and biennial) herbs				
Campanula thyrsoides	18	3	4–19 seedlings	Scheepens et al. (2010)
Cynoglossum officinale	20	2	10 seedlings	Williams et al. (2008)
Eschscholzia californica	20	2	20 seeds	Leger and Rice (2007)
Hypericum perforatum	50	4	9–14 seedlings (?)	Maron et al. (2004)
Potentilla glandulosa	26	3	? plants	Clausen and Hiesey (1958)
Senecio inaequidens	10	1	10 seedlings	Monty et al. (2009)
Senecio squalidus	3	2	10 seedlings	Allan and Pannell (2009)
Four species	4–5	1	25 juveniles (60 for one species)	Bischoff and Müller-Schärer (2010)
Four species	5	2	144 seeds (240 for one species)	Bischoff et al. (2010)
Perennial grasses				
Elymus glaucus	153	1	8 juveniles	Erickson et al. (2004)
Lolium perenne	6	2	10–14 plants	Cheplick (2011)
Four species	7	1	10 plants	McMillan (1967)
Twelve species	40	1	3 plants	McMillan (1959)
Trees				
Picea sitchensis	17	1	~25 seedlings	Mimura and Aitken (2010)
Pinus albicaulis	48	1	1–93 seedlings	Bower and Aitken (2008)

[a]Numbers planted represent seeds, seedlings, juveniles, or adults from each source population placed into each garden. [b]Five experimentally imposed conditions in a greenhouse.

mass by including it as a covariate in the statistical analysis. However, seed mass may not be the sole mediator of environmental maternal effects. Second, one can correct statistically for initial plant size in a similar manner or make sure that all seedlings used are of similar size and age at the beginning of the experiment. Third, one can grow plants for one or more generations under standardized, homogeneous conditions before collecting the seeds to be used in the common garden. Because correcting for seed mass may be insufficient to eliminate potential maternal effects completely, Bischoff and Müller-Schärer (2010, p. 452) reasonably recommend that "the most straightforward method to account for maternal effects is the growth of at least one generation in a standard environment." Some studies of genetic differentiation among populations in common gardens that have examined maternal effects have found no evidence for such effects (Monty et al. 2009; Hodgins & Rieseberg 2011); however, in several herbaceous species, environmental maternal effects were significant for germination and seed mass, but not growth or reproduction (Bischoff & Müller-Schärer 2010).

3.6.3 How Many Gardens Will be Used and Where Should They be Placed?

As we saw in the overview of the different types of common garden experiments, although a single garden can be of some use in detecting significant genetic variation for phenotypic traits, multiple gardens in different environments provide the opportunity to examine genotype (or population)-by-environment interactions. There have been calls to use multiple gardens in studies of invasive species (Williams et al. 2008; Moloney et al. 2009); yet, 85% of 54 studies reviewed by Colautti et al. (2008) used a single common garden.

The question of where to place several gardens may be a matter of how environmentally diverse one wishes the gardens to be, which can be important when assessing consistency of performance of multiple populations (e.g., Flory et al. 2011). Obviously, gardens separated by only a few dozen meters are likely to be much more similar than those separated by tens of kilometers. Nonetheless, the choice of garden location is also, in part, a matter of convenience, and multiple gardens within a geographically limited area can be selected intentionally to encompass a wide range of habitat conditions, as in the 22 garden sites used by Flory et al. (2011). The two gardens used in the study of population differentiation in *Lolium perenne* were only 30 m apart, yet differed considerably in exposure to direct sunlight, soil moisture, and co-occurring species (Cheplick 2011).

The number of gardens that can be used realistically depends on how many seeds, plants, or ramets from the source populations are available and how many replicate plants per population (or replicate ramets per genotype) are desired in each garden. The location of the gardens necessarily depends on convenience and ease of access, but also on whether specific habitat conditions are being subjected to study (e.g., effects of light limitation or drought on population or genotype performance). Last, if one is hoping to demonstrate local adaptation of populations to their original source habitats, then a garden should be established at each collection site and populations should be grown in all these sites—that is, one should use the reciprocal transplant technique (Chapter 4).

3.6.4 Given Space and Time Limitations, What Sample Sizes (Number of Individuals, Populations, and so on) Can be Used?

This is the perennial question asked by biological researchers everywhere. Clearly, plants are variable phenotypically, and the more individuals sampled from the source populations and placed into the common garden(s), the more precise the estimates of the true population mean and variance (Underwood 1997). Thus, the number of seeds, seedlings, or adults planted into the garden is an important consideration. Table 3.1 tabulates sample sizes for some common garden studies, including many of those discussed in this chapter. As might be expected, often there is a trade-off between the number of populations sampled and the number of individuals (per population) planted into the garden(s).

For those trying to estimate quantitative genetic parameters, Conner and Hartl (2004, p. 133) suggested 20 families (sibships) as a "minimum" number to obtain reasonable statistical power. Similarly, a well-known text on quantitative genetics proposes using 20 to 30 families for half-sib analyses (Falconer & Mackay 1996). The number of siblings to use per family depends on the expected level of phenotypic variation within families, although this is typically not known. Ten to 20 siblings per family appears to be widely used (see the dozens of studies tabulated in Appendix 4.1 of Mazer and LeBuhn [1999]). The splitting of genotypes into ramets to create replicates of each genotype is possible with some species (e.g., bunchgrasses), but may be difficult with others (e.g., trees). The ability to replicate genotypes in this way obviously limits the number of individual ramets of each genotype that can be placed in the garden(s). In a common garden study with perennial ryegrass (Cheplick 2008) described in the previous section, 20 replicate ramets of each of 10 genotypes were obtained (10 ramets had endophyte infection, 10 did not). Although it is not difficult to take many ramets from a large individual of a caespitose grass species, it can take considerable time for a seedling to reach the size necessary to provide sufficient ramets for experimental purposes. For plant species with asexually produced apomictic seeds such as *Erigeron annuus* (Stratton 1992 a,b), or functionally asexual species such as *Oenothera biennis* (Johnson et al. 2009), seeds from one maternal parent can be used as clonal replicates of a single genotype. If each maternal plant produces a lot of seeds, there should be adequate clonal replicates to distribute across several gardens.

3.6.5 How Can Blocks be Used to Control Statistically for Environmental Heterogeneity Within the Garden?

As described in Section 3.5, a single field garden can be quite heterogeneous in time and space, and variation in environmental factors becomes manifested as variation in plastic phenotypic traits. The best way to correct for possible variation within a garden is to partition it into blocks (subplots), and many common garden studies use this technique. Then, equivalent numbers of individuals (replicates) from each source population (or genotype) should be allocated to each block. The number of blocks and how they are to be arranged must be determined before planting the garden. In studies in which the common environment is a greenhouse, different benches (or locations on them) could constitute the blocks. If different growth chambers set to the same conditions are used to grow plants from multiple source populations, equivalent individuals of each population should be allocated to each chamber if possible, and the chambers should be considered as blocks in the statistical analysis.

3.6.6 Will Naturally Occurring Vegetation in a Field Garden be Left Intact or Will the Garden be Weeded?

Weed problems can sometimes be circumvented through the application of a short-lived herbicide before establishing the garden. Preparing a garden by plowing up the area disturbs the soil and encourages weed establishment, which can create an unnatural situation for the study species if it normally does not co-occur with dense populations of weeds in its source habitat. However, it is also not a natural situation either for plants of one species to be growing in a habitat completely free of other vegetation! Hence, the question of whether to remove co-occurring vegetation from the common garden arises (Clements 1929).

The resolution of the issue depends on the experimental objectives. If one hopes to characterize genetically cohesive groups such as replicated genotypes or populations to detect differentiation, then a standard, homogeneous environment in the garden is desirable. This can only be accomplished by removing other vegetation both before and during the experiment. Volunteer plants such as weeds are likely to be aggregated spatially and distributed heterogeneously (Cousens & Mortimer 1995), adding environmental variation within the garden that likely causes greater phenotypic variation among plants than would otherwise occur.

Alternatively, if one hopes to characterize and differentiate plant groups in deference to their usual habitat conditions, which often includes other vegetation, one may wish to leave the other vegetation intact within the garden (e.g., Allan & Pannell 2009). Unfortunately, this results, at best, in an imperfect mimic of the conditions prevailing at the source collection sites, which are likely to vary among locations as well. In any event, it might be prudent also to include a second garden with vegetation removed, so that comparisons can be made of plant performance both with and without competitive pressure (see Leger & Rice 2007; Bischoff et al. 2010).

3.6.7 Will Environmental/Climate Data be Obtained for the Sites of the Source Populations?

Environmental/climate data can be critical to the interpretation of common garden data. Climatic and edaphic factors have long been recognized as abiotic selection agents (Table 2.1) especially likely to guide the evolution of plant populations and adaptive characteristics (Turesson 1925, 1930; Clausen et al. 1948). Climatic information can (and has) been obtained from weather stations near the source populations (e.g., Oyarzabal et al. 2008; Monty et al. 2009) or downloaded from databases available on the internet such as WorldClimate (www.worldclimate.com) or the National Climate Data Center (www.ncdc.noaa.gov) (e.g., Erickson et al. 2004; Franks et al. 2007; Livshultz et al. 2011). Methods are available to calibrate the performance of populations in a common garden with the extent of environmental differences between the source and garden locations (Rutter & Fenster 2007; Climent et al. 2008).

3.6.8 Will any Environmental Variables be Purposely Manipulated?

As noted in Section 3.4, if there are suspected environmental variables acting as selection agents in the habitats of the source populations, then experimental manipulation of one or more of these variables in otherwise homogeneous gardens may be the way to proceed. Of course, the more variables manipulated and the more populations studied, the more complex and labor intensive the experiment becomes.

Hodgins and Rieseberg (2011) used five common "garden" environments established in a greenhouse, applying environmental stresses to each in a detailed study of 22 native and 12 introduced populations of common ragweed (*Ambrosia artemisiifolia*). The factors examined were light stress, herbivory (simulated), mineral nutrient stress, and drought. In general, introduced ragweed populations tended to show greater growth and reproduction, especially under nonstressful conditions. Except for nutrient stress, all manipulated variables significantly affected measured traits such as survivorship and reproductive biomass, which were typically greater in introduced populations (Hodgins & Rieseberg 2011). Although undoubtedly very laborious to perform, this type of study illustrates well the power of the common garden approach to distinguish among the selection agents that probably affected the microevolution of source populations in their original habitats.

3.7 UTILITY AND APPLICATIONS OF THE COMMON GARDEN APPROACH

The common garden experiment, like the reciprocal transplant design modeled by Nuismer and Gandon (2008), has been extremely useful to studies of adaptive differentiation in plant evolutionary ecology and is likely to continue to be used in future research. As long as careful planning with regard to the experimental design of the common garden study is made, including consideration of the environmental factors that may impinge on the results, many fruitful avenues of basic and applied research can be realized. Examples of the utility and application of common garden approaches are widespread in the ecological literature.

Reference has already been made to the utility of the common garden in investigations of invasive plants (Maron et al. 2004; Bossdorf et al. 2005; Williams et al. 2008; Flory et al. 2011). The common garden can reveal the quantitative traits crucial to the success of invasive species that show genetic differentiation between invasive and noninvasive populations (or genotypes). Replicated common gardens in different habitats across both the invaded and native ranges are important for demonstrating local adaptation of a species to parasites, competitors, or other biotic agents (Nuismer & Gandon, 2008).

Species adaptation to predicted environmental factors likely to be altered in the face of global climate change (Jump & Peñuelas 2005) may be studied using common garden approaches (Etterson 2004a; Savolainen et al. 2007). In these experiments, climatic variables at the site of population origin can be calibrated to those found at the test site (Rutter & Fenster 2007; Climent et al. 2008). This information can be used to detect local adaptation of populations to their sites of origin.

The information gleaned from common garden experiments on the adaptedness of populations to their local (home) habitats can inform practical decisions regarding the use of plant materials for restoration efforts (McKay et al. 2005; O'Brien et al. 2007; Bower & Aitken 2008). The geographic and/or environmental scale at which plant populations show local adaptation is an important issue in restoration ecology (McKay et al. 2005; Bischoff et al. 2010). Hufford and Mazer (2003) described the relevance of the common garden and reciprocal transplant approaches to planning the translocation of organisms and to delineating seed transfer zones during ecosystem restoration.

Common gardens can also be used to assess the fitness of hybrids formed when related plant species interbreed. Snow et al. (2001), for example, used multiple common gardens to determine the relative fitness of hybrids between cultivated radish (*Raphanus sativus*) and the weedy congener, wild radish (*R. raphanistrum*). Hybridization between other crop species and their weedy relatives presents the opportunity for transgene escape from genetically modified crops (Ellstrand 2014). Common gardens can be used to determine the fitness effects of crop genes established within wild relatives and vice versa (e.g., Snow et al. 1998; Campbell & Snow 2009).

Last, theoretical questions regarding the relative importance of natural selection and genetic drift to population differentiation in quantitative versus neutral molecular markers (Merilä & Crnokrak 2001; Steinger et al. 2002; Leinonen et al. 2008; Section 5.2.4) may be best explored when populations and genotypes are grown in a homogeneous, common garden environment (Jaramillo-Correa et al. 2001; Chun et al. 2009, 2011). As with so many of the examples presented in this chapter, exploration of the genetic basis for many of the phenotypic differences revealed by the common garden approach can be best conducted under carefully controlled environmental conditions.

4 Reciprocal Transplant Experiments

In spite of inherent significance and interest, the experimental study of adaptation has developed more slowly than the field of ecology in general (Clements 1929, p. 357).

4.1 INTRODUCTION

Exposure of a population to a diverse set of natural selection pressures results in the differential survival and reproduction of individuals. As one or a few predominant agents of selection continually affect the population, some genotypes and their corresponding phenotypes show greater survival and reproductive success than others. Thus, the gene and phenotype pool changes over time during a process of adaptive evolution. The ultimate result of this process is a population of individuals better able to survive, grow, and reproduce under a particular selective environment. In other words, present-day individuals in a population are better adapted to local environmental conditions that have regularly acted as selection agents in the past compared with individuals of previous generations. Following adaptive evolution, a population is expected to show adaptation to the prevailing agents of selection to which it has been exposed repeatedly in the past.

In the previous chapter we saw how the common garden approach has long been used to reveal differentiation among genetically distinct groups, including populations and ecotypes. We also saw how the purposeful manipulation of one or more environmental factors in different gardens could provide evidence with regard to which selection agents might have been responsible for genetic differentiation. The working hypothesis was that variation in selection pressures among habitats led to the evolution of populations that were adapted to the local environmental conditions that normally prevailed in their source (home) habitats.

A similar line of reasoning prevails in the practitioners of the classic reciprocal transplant experiment. If a population is adapted to its local habitat conditions, then individuals from that habitat that are transplanted to a different, "alien" habitat to which they are not adapted show reduced survival, growth, and/or reproduction relative to their performance in their native, "home" habitat (Mazer & LeBuhn 1999; Joshi et al. 2001; Kawecki & Ebert 2004; Leimu & Fischer 2008; Hereford 2009).

This idea of transferring species between habitats to study evolutionary adaptation has a lengthy history. Consider this quote from Frederic E. Clements published in the first edition of his text on *Research Methods in Ecology* in 1905:

All field experiments in evolution are based upon a change of habitat. The latter is accomplished by the modification of the habitat itself or by the transfer of the species to one or more different habitats, or to different areas of the same habitat. (Clements 1905, p. 153)

He also explicitly described the standard reciprocal transplant experiment for plant ecology, referring to such operations as "reciprocal transfers" of seeds or entire plants (Clements 1905, p. 155). As a result of the ease of moving them around, very many transplant experiments have been conducted using plants throughout the history of ecological research. This chapter provides examples of how the reciprocal transplant approach can reveal local adaptation in plant populations.

4.2 A BRIEF ASIDE ON ADAPTATION

Traditionally, "adaptation" has been used by evolutionary biologists in two ways. First, it is used to denote a process that is functionally equivalent to the evolution of a population by natural selection (Stern 1970; Larson 2009). During adaptation in this sense, the characteristics of a population change over generations as the population undergoes adaptive evolution in response to natural selection (Latta 2010). The result is a better "fit" or match between organisms and their immediate environment. It is actually the population-specific *result* of the process of adaptation (i.e., adaptive evolution) that is being examined in a reciprocal transplant experiment. However, finding evidence of local adaptation tells us very little about the underlying evolutionary processes and environmental factors that were responsible for it (Kawecki & Ebert 2004; Salmela 2014). Nonetheless, reciprocal transplants offer a practical experimental tool to detect the results of a process important to evolutionary ecology: adaptive evolution by natural selection.

In a second, common use of the term, an adaptation refers to a trait that improves the fitness of individuals that have it relative to others that do not (Stern 1970; Latta 2010). Larson (2009, p. 93) refers to these traits as "character adaptations" to distinguish this use of the term from its use to describe an adaptive evolutionary process. Character adaptations arise as a consequence of the differential ecological success of phenotypic variants in a population (Reznick & Travis 2001). Thus, character adaptations are expected as a result of long-term evolution by natural selection.

A key problem with the study of character adaptations is that most phenotypic traits show continuous, quantitative variation in nature, and thus individuals do not differ in a discrete way with regard to whether they have a particular adaptation. Furthermore, it is an integrated phenotype reflecting a combination of traits that responds to selection pressures and expresses its reproductive fitness.

What can the reciprocal transplant experiment tell us about character adaptations? We might be able to surmise that some combination of phenotypic trait values is better suited to a specific environment than other combinations, provided that the relevant traits are measured. For example, genotypes with more pubescent, succulent leaves that show greater WUE may have greater fitness in an arid environment relative to genotypes with lower values for these traits. A transplant experiment that involves moving populations between an arid habitat and a mesic habitat should reveal that the former genotypes perform better than the latter in the arid (home) habitat, indicating population adaptation to local conditions. The phenotypic trait values of the measured characters are adaptive in the sense that they improve fitness in the arid habitat more than the phenotypic trait values found in genotypes transplanted into the arid habitat from elsewhere. In a general way, the reciprocal transplant technique provides evidence of prior adaptive evolution in a population over some interval of time (often unknown).

4.3 TESTING HYPOTHESES WITH THE STANDARD DESIGN

The standard reciprocal transplant design begins with locating two or more populations inhabiting different environments (i.e., habitats). These environments are often chosen a priori because they are distinct or of specific interest, and typically differ in one or more measureable characteristics. Depending on the objectives (and often the convenience!), seeds, seedlings, or ramets of each population from each environment are (1) transplanted into a different environment and (2) replanted back into their original environment of origin (Fig. 4.1). Plants in the habitat where their population was originally found are said to be in their "home" site. Adaptation to this site is expected to be revealed as a "home-site advantage" (e.g., Bennington et al. 2012), whereby plants at home show greater survival and reproductive fitness compared with plants that have been transplanted into the site from elsewhere. These latter plants transplanted to a site that is not their original habitat are said to be in an "away" site (Fig. 4.1). Other terms that have been used in reference to plant sources

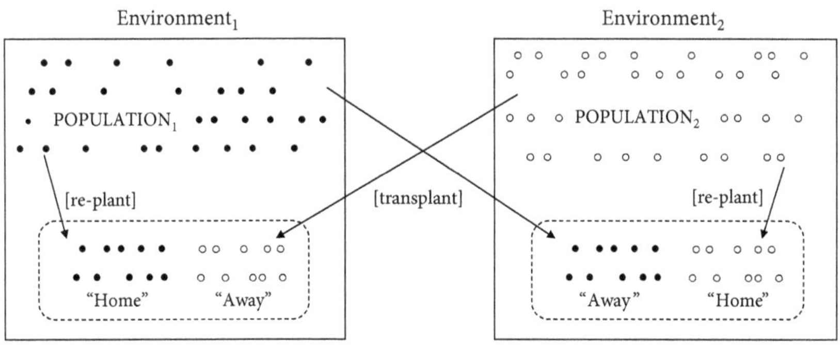

FIGURE 4.1
Diagram illustrating the classic reciprocal transplant design. For ease of presentation, only two environments (sites/habitats) are shown, each originally supporting a natural population of the study species. If local adaptation to the environment has occurred, it is expected that each population will show the greatest survival, growth, and/or reproduction in the garden of its "home" environment.

and planting sites include *local* or *native* for a population in its home site and *alien* or *foreign* for a population not in its home site.

Data recorded in a reciprocal transplant experiment are usually measures of phenotypic traits likely to be important to Darwinian fitness: survival, growth, and reproduction. This is sensible because individuals in a population adapted to the environmental conditions that regularly occur in its habitat should show greater fitness than those not as well adapted to the same conditions (Hereford 2009). From the perspective of showing local adaptation, the key comparison is between the fitness of local and foreign populations in each habitat, not between that of home and away populations (Kawecki & Ebert 2004, but see Blanquart et al. [2013] for a different view). This is because a population in an away site may outperform the same population in its home site if the away site is more conducive to growth and reproduction (e.g., it has greater mineral, water, or light availability). The greater fitness of the population in its away site in this case would have nothing to do with local adaptation.

Two examples that help in visualizing the results from standard reciprocal transplant experiments are presented to illustrate what is expected if local adaptation has occurred. The first involved reciprocal transplanting of seeds of the cleistogamous annual *Impatiens pallida* between forest interior and edge habitats in central Illinois (Schemske 1984). Seeds used were from greenhouse-grown families, so maternal environmental effects were unlikely. More than 1,000 seeds were planted per site, and germination, survival, and reproduction were censused from early spring, when seeds germinated, until later in the year, when all plants had died after reproduction. The percentage of the seeds planted into each site that eventually gave rise to reproductively mature adults showed a significant source population-by-planting site interaction ($\chi^2 = 6.6$, $p < 0.01$). The differences between populations within each site were in the direction expected if local adaptation had occurred (Fig. 4.2A).

In the second example (Lovett Doust 1981b), the dynamics of the clonal perennial *Ranunculus repens* (buttercup) were followed in ramets reciprocally transplanted between a mixed woodland and an adjacent park "grassland" (established by sowing commercial grass seeds about 10 years before the study began). *R. repens* was originally common in both sites. Sixty ramets were excavated from each site; 30 were placed into the away site and the other 30 were replanted into their home site. As in the study by Schemske (1984), statistical analyses revealed a significant interaction between source population and planted site for several traits (Lovett Doust 1981b). Figure 4.2B shows this interaction for the number of leaves (per ramet). At both the woodland and grassland sites, the number of leaves for the local population was significantly greater than that of the foreign population, suggesting local adaptation. Note that a comparison of leaf number in home versus away groups of the grassland population was not significant. Plants from the grassland grew as well in the woodland as they did in their home habitat (Fig. 4.2B). Nonetheless, local adaptation is indicated by the valid comparison of local and foreign populations *within* each habitat (Kawecki & Ebert 2004).

Using a data set of reciprocal transplant experiments on 32 plant species, meta-analysis showed that local plants in their original site performed significantly better than foreign plants in 71% of the sites studied (Leimu & Fischer 2008). However, pairwise comparisons of local versus foreign population performance at *both* of two sites involved in reciprocal transplantation showed local plants performed better at both compared sites in only 45.3% of the cases (Leimu & Fischer 2008). This latter

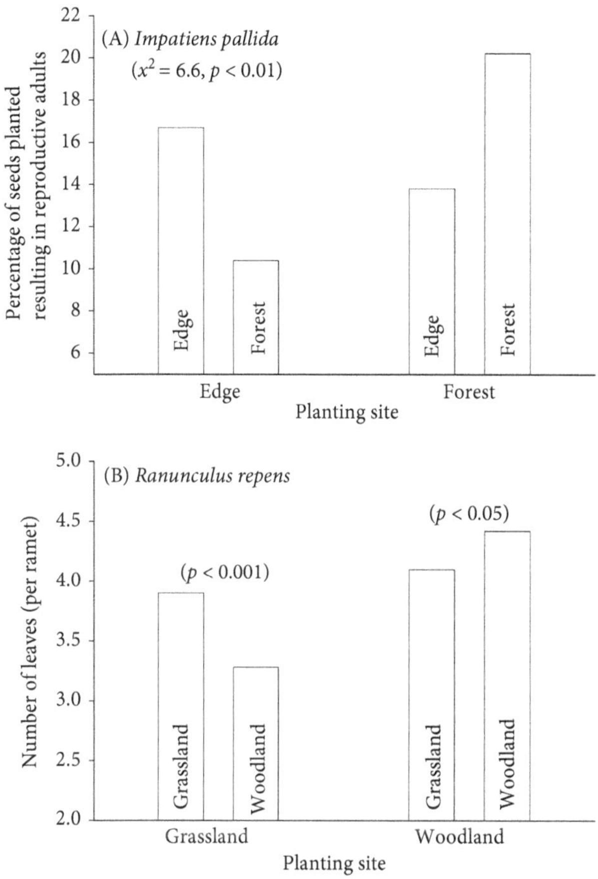

FIGURE 4.2
(A, B) Illustrations of local adaptation for plants from reciprocally planted seeds of the annual *Impatiens pallida* between an edge and forest site (A) (Schemske 1984) and ramets of the perennial *Ranunculus repens* reciprocally planted between a grassland and woodland site (B) (Lovett Doust 1981b).

comparison reflects a somewhat stricter view of adaptation (Kawecki & Ebert 2004) in which home-site advantage is shown to be true for each pair of populations (and sites) investigated.

4.4 DIVERSITY OF RECIPROCAL TRANSPLANT APPROACHES

A thorough survey of the many reciprocal transplant experiments that have been performed by plant ecologists would be an arduous undertaking and could easily fill another book of this size. Sometimes the objective is simply to demonstrate local adaptation to the habitats chosen without regard to which selection agents might have been responsible. Speculations regarding selection pressures are often offered, however, based loosely on observations or, in some cases measurements, of environmental differences between the habitats. Even when the selection agents are uncertain, selection coefficients (described in Section 4.5) may be calculated to estimate the overall selection against foreign populations (Davies & Snaydon 1976; Lovett Doust 1981b; McGraw & Antonovics 1983; Joshi et al. 2001). A few have even combined the reciprocal transplant approach with estimates of selection differentials or gradients (van Tienderen & van der Toorn 1991b; Bennington & McGraw 1995; Griffith & Watson 2005; Byars et al. 2007). Sometimes, factors presumed to be important to local adaptation such as soil type or chemistry, herbivory, or competition are purposefully manipulated in the transplant gardens (e.g., Kindell et al. 1996; Bischoff et al. 2006; Crémieux et al. 2008; Grøndahl & Ehlers 2008; Ortegón-Campos et al. 2012; Lovell & Menges 2013). This section provides a few illustrations of the diversity of approaches used in reciprocal transplant experiments. Additional examples are provided in later chapters that consider specific categories of selection agent.

Note that many reciprocal transplant experiments do not have straightforward outcomes that invariably demonstrate local adaptation in all populations examined. The ecological factors and statistical issues important in the analysis and interpretation of transplant experiments are explored by Blanquart et al. (2013). The demonstration of local adaptation depends on

- Which habitats (sites) are chosen. The more environmentally different they are, the more likely the populations will show local adaptation. Environmental differences are expected to be more pronounced in sites that are far apart, but this is not always the case.
- What is transplanted (seeds, seedlings, or ramets). Early stages of life history that could be important to local adaptation are missed, for example, when juveniles or adults are transplanted.
- Whether any environmental factors are purposefully manipulated. Recent variations on the classic reciprocal transplant design have sometimes included experimental manipulation of key abiotic or biotic factors.
- Which types of demographic parameters are measured. Common variables include the proportion of plants surviving and/or flowering, measures of growth, and fecundity.
- How many years the plants are monitored. Many experiments last only one or a few years, although there are a few exceptions (see Section 4.4.4).

A demographic approach was used with the reciprocal transplant design to investigate local adaptation of eight populations of the winter annual *Phlox drummondii* in central Texas (Schmidt & Levin 1985). This study is notable for the completeness of its design, the large number of seeds transplanted (more than 14,000 in the first year), and the fact that survivorship and fecundity were monitored for two growing seasons. The finite rate of increase (λ) was also estimated for each population at each planting site. After seed transplanting, the natural vegetation was not manipulated at a site and the *P. drummondii* plants emerging from the seeds were therefore in competition with the surrounding, mostly annual, plant community (Schmidt & Levin 1985). The mean fitness of aliens, whether assessed as survivorship or fecundity, was less than that of plants native to a site (Fig. 4.3), although there were several examples of alien plants outperforming natives at some sites in one or both years (i.e., the fitness of aliens relative to natives was more than 1.0). The mean λ in both years was much less for populations alien to a site relative to populations that were native (Fig. 4.3). This study provided solid evidence that populations of *P. drummondii* were

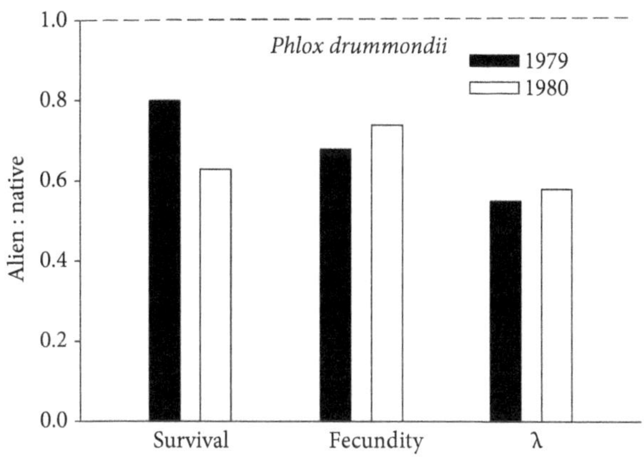

FIGURE 4.3
Performance of alien relative to native populations for reciprocal transplants of the annual *Phlox drummondii* in 1979 (seven sites) and 1980 (eight sites). λ is the finite rate of population increase. Data from Schmidt and Levin (1985).

mostly adapted to their local sites, although the specific environmental factors responsible remained elusive.

In two reciprocal transplant experiments with the perennial *Plantago lanceolata*, demographic data (survival, growth, flowering, and seed yield) were collected for several years (van Tienderen & van der Toorn 1991a). Three contrasting habitats in the Netherlands were used for population sources and planting sites: a late-mown, mesic hayfield; an early-mown, dry hayfield; and a pasture grazed regularly by cattle. The first experiment involved the reciprocal transplanting of 2-week-old seedlings, but there were no significant differences in survival between populations at two sites. Only in the pasture did the native population show significantly greater survival. In the second experiment, cuttings from 19 clones from each population were reciprocally transplanted. Survival of the native population was greatest at two sites, and the greatest proportion of flowering plants was mostly found for the native population at each site. Seed yield was also greatest for native populations, leading the authors to conclude that the evidence for local adaptation in this species was "very strong" (van Tienderen & van der Toorn 1991a, p. 39).

4.4.1 Manipulating the Planting Site

Reciprocal transplantation of the seeds of three grassland species from populations in the Czech Republic, Switzerland, and the United Kingdom was conducted by Bischoff et al. (2006). Plots with *Holcus lanatus*, *Lotus corniculatus*, and *Plantago lanceolata* were established at each site. In half of the plots, seeds of 10 plant species characteristic of grasslands were sown to provide a surrounding, competitive plant community. Note that this type of experimental manipulation of the planting habitat is all too rare in reciprocal transplant experiments (Donohue et al. 2001; Rice & Knapp 2008; Ariza & Tielbörger 2011), but has the potential to provide additional information on the selection agents relevant to local adaptation (in this case, competition with other herbs). Local adaptation based on fitness-related traits such as survival and reproduction was detected in *H. lanatus* and *P. lanceolata*. However, this result depended somewhat on the competition treatment. In *H. lanatus*, for example, the evidence for local adaptation was weaker under competition with the local plant community (Bischoff et al. 2006). In contrast, in *P. lanceolata*, the magnitude of the home-site advantage was greater in the competition treatment (e.g., the number of spikes was, on average, 22.3% greater for the local populations under competition, but only 8.2% greater without competition). This research shows that the home-site advantage of a local population depends on the habitat conditions established by the researchers, similar to the outcomes of common garden experiments described in Chapter 3.

In another experiment that involved manipulation of habitat conditions, Turkington (1989) reciprocally transplanted ramets of the clonal perennial *Trifolium repens* (white clover) collected from sites dominated by different perennial grass species. At each planting site, two types of plot were established: (1) native *T. repens* removed and (2) both native *T. repens* and the native dominant grass removed. Home-site advantage to replanted *T. repens* was found in terms of greater survival and biomass compared with ramets transplanted into a site from elsewhere, but only when the competitive grasses were *not* removed. This interesting result suggested that there had been past coevolution of the interacting species in these pasture communities such that competitive effects were minimized—a process driven by natural selection (Turkington 1989; Turkington & Mehrhoff 1990). However, for other species, the detection of local adaptation via reciprocal transplants does not depend on the presence of surrounding vegetation (Platenkamp & Foin 1990; Kindell et al. 1996; Lovell & Menges 2013). The evolutionary ecology of competitive interactions between plants is explored further in Chapter 7.

4.4.2 Comparing Planting Site Conditions

It has long been recognized that some of the most striking examples of local adaptation in plants come from the comparison of populations in sites exposed to anthropogenic pollution (e.g., localized heavy metal contamination of soils) with those that have not been so exposed (Jain & Bradshaw 1966; McNeilly & Bradshaw 1968; Dechamps et al. 2008; reviewed in Briggs 2009; see Chapter 6). This man-made situation is likely to impose selection pressures that are much more severe than what might be expected in unpolluted environments. Thus, a greater level of local adaptation may be detected among populations from environments that differ markedly (Rice & Mack 1991; Raabová et al. 2007; Hereford 2009).

Using six populations of the annual *Diodia teres* in northern Florida, Hereford and Winn (2008) explored the hypothesis that local adaptation would be more likely between populations from different habitat types than between populations from similar habitat types. Two populations were chosen from each contrasting habitat type: dunes, sand hills, and inland. The dune populations were on the coastal plain and about 120–140 km from the inland populations; the sand hills populations were about midway between them. Seedlings were grown to maturity from field-collected seeds in a common greenhouse environment to control for possible maternal effects. Seeds from these greenhouse plants were used to obtain seedlings transplanted into the field sites in a reciprocal design conducted over two years.

Environmental differences between all sites were calculated based on features of the vegetation and soil texture (Hereford & Winn 2008). As expected, sites within

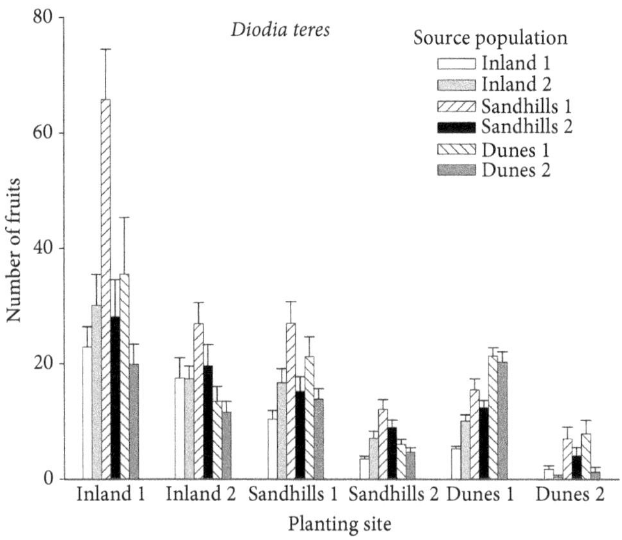

FIGURE 4.4
Mean ± standard error of the number of fruits produced by reciprocal transplants of the annual *Diodia teres* at six planting sites in 2003. Data from Hereford and Winn (2008).

a habitat type were more similar than those from different habitat types, with the exception of the two sand hills sites, which were substantially different from one another. Complex patterns in the per-capita number of fruits produced, used as an estimate of fitness, are evident in Figure 4.4 for source populations planted into the different sites and habitat types in the second year of the study (2003). The effects of planting site and source population on fruit number were highly significant ($p < 0.0001$). However, evidence for local adaptation to home sites was equivocal and few populations showed a distinct home-site advantage (Fig. 4.4). In several instances, alien populations produced significantly more fruits than the native population at a site. Nonetheless, the frequency of local adaptation was generally greater when populations were from different habitat types (although this was only statistically significant in 2003). The authors concluded that limitations to the evolution of local adaptation may be more widespread than expected because past reciprocal transplant experiments have typically been designed to study adaptation to very different environments (Hereford & Winn 2008).

4.4.3 Modification and Expansion of Reciprocal Transplant Designs

Reciprocal transplant experiments have sometimes been modified or expanded to examine more closely questions relating to which specific aspects of the immediate environment a population is adapted and to what extent the range limits for a species might act as constraints on local adaptation. In the previous section we saw how the prevailing environmental conditions at a site, such as the presence or absence of competing vegetation, could have an influence on the ability to detect local adaptation. Schoen et al. (1986) built on this approach by transplanting the study species (*Impatiens capensis* and *Impatiens pallida*) and also neighbors that were from the same or a different habitat as that of the study species. For example, a seedling replanted into its native, home habitat was surrounded by three neighboring seedlings 6 cm away that were either from the same (home) or a different (away) site. Thus, this experiment allowed the researchers to distinguish between effects on target plants resulting from abiotic and biotic conditions at a site, and effects resulting from the type of neighbors (i.e., native vs. alien). The latter effect was important to demonstrating a home-site advantage; for example, individuals surrounded by conspecific neighbors from the same population produced more fruits than individuals surrounded by neighbors from a different population (and site). Analogous to the studies of Turkington (1989) and Bischoff et al. (2006) described earlier, in which perennial competitors of the study species affected its relative home-site advantage, populations of *I. capensis* and *I. pallida* (both annual herbs) may have evolved a way to reduce the deleterious effects of competition in dense, conspecific stands and thereby retain a fitness advantage in their home sites. Platenkamp and Foin (1990) provide an additional example of a reciprocal transplant study in which the target species (*Anthoxanthum odoratum*) and the neighboring plants at a site were both transplanted.

In addition to their use in investigating biotic factors such as competition (and herbivory; see Section 9.2), modified reciprocal transplant designs have been used to investigate abiotic factors such as general soil type by transplanting both plants *and* soil collected from the study sites (Marañon & Bartolome 1993; Macel et al. 2007; Ortegón-Campos et al. 2012). Specific biotic factors associated with soil have also been examined in a similar "reciprocal" manipulation of the relevant factors (Ehlers & Thompson 2004; Grøndahl & Ehlers 2008; Sherrard & Maherali 2012).

To examine the specific role soil might play in the evolution of local adaptation in the perennial herb *Ruellia nudiflora*, two sites 55 km apart were chosen in Yucatan, Mexico (Ortegón-Campos et al. 2012). Seeds collected from the sites provided seedlings that were first grown for 18 months in a nursery under homogeneous conditions. Seeds from these plants were then used to generate seedlings transplanted to the field. At each site, soil samples were collected and mixed (but not sterilized). Before planting the seedlings, small pits (diameter, 10 cm; depth, 20 cm) were established in a grid at each site and pots were filled with soil of either the home or away site. Thus, the soils underwent reciprocal transplanting in addition to the plants. This design allows one to separate the gross effects of soil specific to a site from that of other environmental factors such as climate, which also differ between the sites (see Macel et al. [2007] for another example of this approach).

Based on survival, local adaptation was only detected at one site whereas, based on fruit production, there was no evidence for a home-site advantage at either site (Ortegón-Campos et al. 2012). At only one site was fruit production significantly greater for plants growing in their home soil (compared with the away soil at the same site). Thus, there was only limited evidence that *Ruellia nudiflora* had undergone adaptive evolution with respect to edaphic characteristics.

Reciprocal transplant experiments have been used not only to examine adaptation to local sites but also have been expanded to explore the expected lack of adaptation to sites beyond the usual range of the study species. Some of the transplant experiments do not incorporate a reciprocal component and have more in common with common garden studies, in which the gardens are placed outside of the normal species' range (Prince & Carter 1985; Griffith & Watson 2006; Samis & Eckert 2009).

Nonetheless, the shared goal of such studies is an exploration of the potential limits to adaptation.

By way of illustration, we examine a study in which two species of perennial monkeyflowers (*Mimulus cardinalis* and *Mimulus lewisii*) were transplanted both within and beyond their current elevational ranges in the Sierra Nevada Mountains of California (Angert & Schemske 2005). *M. cardinalis* normally ranges from sea level to 2,400 m, whereas *M. lewisii* normally ranges from 1,200 to 3,100 m. Six populations of each species were used, ranging from 590 to 2,750 m in elevation. Eight greenhouse-grown plants raised from field-collected seeds of each population were intercrossed to create a "genetically variable, outcrossed seed pool" (Angert & Schemske 2005, p. 1672). Experimental gardens at the elevations noted on the *x*-axis in Figure 4.5 were planted with seedlings (3 weeks old) from seeds of the intercrossed greenhouse plants. Survival, growth, and reproduction (number of flowers) were monitored, and fitness was estimated as the number of flowers summed over two growing seasons.

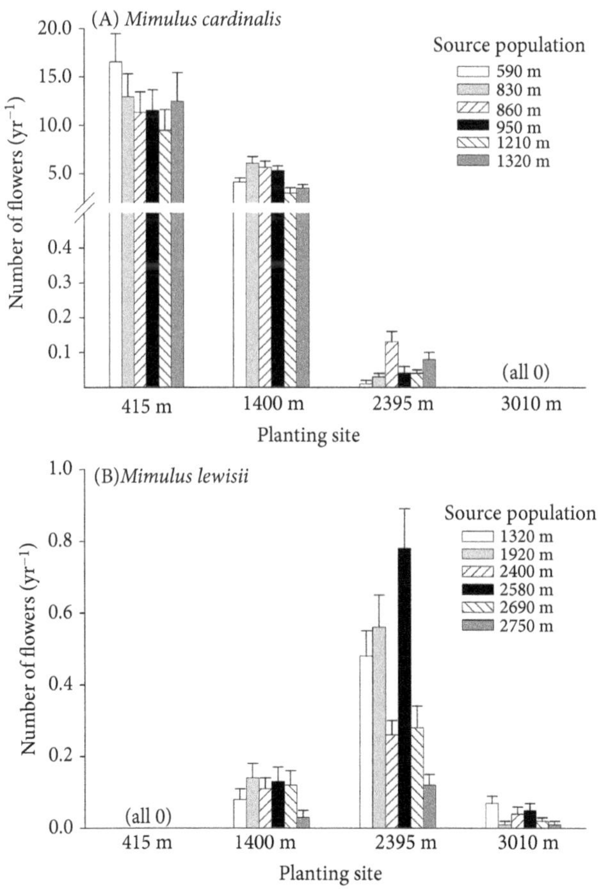

FIGURE 4.5

Mean ± standard error of the flowers produced by six populations of two species of perennial monkeyflowers transplanted into four gardens at different elevations in the Sierra Nevada Mountains, California. These gardens are both within and beyond the current elevational ranges for these species (see text). Data from Angert and Schemske (2005).

Mean annual fitness (flowers per year) is shown for the six source populations of the two *Mimulus* species in the four gardens (Fig. 4.5). For *M. cardinalis*, populations showed the greatest reproduction at the lowest elevation site, which was below the elevation of the various source populations but approximately near the center of this species' normal range. At the two highest elevations (above the range limit of this species), plants performed very poorly. For *M. lewisii* planted at 415 m, beyond its lower range limit, few individuals survived and there was no reproduction (Fig. 4.5B). Populations reproduced best near the center of the species' usual elevational range at 2,395 m. Thus, this expanded reciprocal transplant experiment provided good support for the idea that these species showed the best adaptation to sites nearest to their range centers and were not adapted to the conditions prevailing in sites beyond their range limits (Angert & Schemske 2005). Other studies in which plants have been transplanted outside their usual range have also indicated that adaptation to environmental conditions beyond the range limits is a likely prerequisite to geographic expansion (Prince & Carter 1985; Griffith & Watson 2006; Samis & Eckert 2009).

4.4.4 Long-Term Experiments

Most reciprocal transplant studies, even when conducted with potentially long-lived perennial species, seldom last more than a few years. However, slow-growing perennials (e.g., shrubs or trees; perennial herbs in stressful habitats) may not reveal local adaptation to particular sites for quite some time (Boshier & Billingham 2000; Miglia et al. 2005; Wright 2007). Fortunately, there are now long-term results from several reciprocal transplant experiments of perennial species from a variety of habitats.

Big sagebrush (*Artemisia tridentata*) is a relatively common shrub occurring as two subspecies in western North America that can live for several decades. Note that the "basin" subspecies occurs below 1,790 m in elevation whereas the "mountain" subspecies occurs at elevations above 1,850 m; hybrids occupy an elevational band between the parental subspecies at ~1,790 to 1,830 m (Miglia et al. 2005). Seedlings of two subspecies of big sagebrush and their hybrid were reciprocally transplanted among five sites in Salt Creek Canyon, Utah, in 1993 (Wang et al. 1997). After two years, a relative composite fitness metric was computed that took into account survivorship, proportion flowering, number of seeds produced, and seed germination rate. Local adaptation of the two subspecies and their hybrids was indicated by the fitness metric reported for gardens in the basin, hybrid, and mountain habitats (Wang et al. 1997). In other words, after two years each subspecies or hybrid performed best in its native habitat relative to the transplants from elsewhere.

The big sagebrush transplant sites of Wang et al. (1997) were revisited in 2002, nine years after they were originally established (Miglia et al. 2005). Although relative composite fitness was still greatest for basin and mountain subspecies in their native sites, in the hybrid zone, hybrid plants no longer had the greatest mean fitness; instead, mountain big sagebrush now showed the greatest mean fitness in the hybrid zone. The authors speculated that because density had increased during the nine years, the growth rate of hybrids may have diminished more than in the mountain subspecies (which was perhaps better adapted to greater densities). In addition, cooler, wetter weather conditions in later years of the study could have been more favorable for mountain big sagebrush compared with the hybrids and basin subspecies in the hybrid zone (Miglia et al. 2005). Regardless of the reasons, it is clear from this later study (but perhaps, not surprising) that results from reciprocal transplant experiments of a long-lived perennial can change in unexpected ways as both abiotic and biotic conditions vary continually over space and time.

Reciprocal transplant studies with two long-lived arctic perennials (*Dryas octopetala* and *Eriophorum vaginatum*) originally begun in 1979 and 1980, respectively, were revisited several times during the next three decades (Bennington et al. 2012). In Alaska, three transplant gardens had originally been established for two subspecies of *D. octopetala* along an altitudinal gradient (McGraw & Antonovics 1983) whereas six transplant gardens had been established for the perennial sedge *E. vaginatum* along a latitudinal gradient The most recent census recorded survivorship and flower production in 2010 (Bennington et al. 2012).

After 31 years, the original home-site advantage noted for *Dryas octopetala* (McGraw & Antonovics 1983) became even more striking with the complete demise of the subspecies that was foreign at two sites. Survival of *Eriophorum vaginatum* tussocks from 1980 to 2010 was remarkably high (>90%), and home populations at five of the six sites showed 100% survival over 30 years (Bennington et al. 2012). Local adaptation was also shown by the greater per-capita flower production in 2010 of *E. vaginatum* tussocks in their home sites. Furthermore, differences among populations in stomatal density and conductance could still be related after 30 years to the growing season temperatures that normally occur at the population's latitude of origin (Peterson et al. 2012). The strong, persistent evidence of local adaptation as revealed by reciprocal transplant experiments may indicate that this is a relatively common feature of arctic plant species (McGraw & Antonovics 1983; Galen et al. 1991; Byars et al. 2007; Gonzalo-Turpin & Hazard 2009).

The importance of long-term measurements to the demonstration of local adaptation in perennial plant species is nicely exemplified by 36 years of size data collected in a reciprocal transplant study of Ponderosa pine (*Pinus ponderosa*) in California (Wright 2007). This long-lived, outcrossing tree can be found on both serpentine and nonserpentine soils. Serpentine soils are nutrient poor, very dry, and high in magnesium and heavy metals. They have long been a favorite in studies of plant adaptation to soils (Kruckeberg 1954; Brady et al. 2005; Sambatti & Rice 2006; Wright et al. 2006; see Chapter 6). In the Ponderosa pine study, one-year-old seedlings were reciprocally transplanted between serpentine and nonserpentine sites in 1970, and measurements of height and basal area were taken periodically between 1971 and 2006 (Wright 2007). Significant differences in basal area between plants from serpentine (local) versus on-serpentine (foreign) populations growing at the serpentine site became apparent and statistically significant only after 15 years. Significant differences in height became apparent only after 20 years! In both instances, plants from the serpentine population were larger and taller at the home serpentine site than plants from the nonserpentine population. These differences gradually became greater from the point when they became significant on to the end of the experiment. At the nonserpentine site there were no size differences found between plants from serpentine versus nonserpentine populations for the 36 years of data collection. These results show that a very long-term reciprocal transplant experiment may be needed to demonstrate local adaptation in tree populations.

4.5 SELECTION COEFFICIENTS AND SELECTION GRADIENTS

Some researchers summarize the results of reciprocal transplant experiments by calculating selection coefficients. In the earlier literature, selection coefficients were typically expressed relative to the home population as

$$1 - \frac{\text{Performance of alien population}}{\text{Performance of home population}}.$$

"Performance" is any measure of survival, growth, or reproduction (or a composite index) averaged over all individuals in the population at any one transplant site. Note that values can be less than zero if the alien population has a greater mean fitness than the home population, and a value at zero would represent no home-site advantage. The coefficient ranges upward to a maximum of 1.0 (in which the alien population has zero fitness; e.g., no individuals survive). In essence, the performance ratio compares the mean fitness of alien populations with the home population; many researchers simply report these types of fitness measures without presenting the selection coefficients (which can be calculated by the reader, if desired).

Selection coefficients can be quite high, especially at environmentally extreme sites such as mine spoils, where coefficients were 0.95 or greater in two grass species (Jain & Bradshaw 1966). Less extreme sites typically show lower coefficients that depend greatly on which measures of performance are used. On plots in the Park Grass Experiment that varied in soil fertility, selection coefficients against the alien populations of *Anthoxanthum odoratum* based on dry mass ranged from 0.20 to 0.60, whereas those based on final survival varied from 0.09 to 0.77 (Davies & Snaydon 1976). Selection coefficients were greater on limed plots (0.44) relative to unlimed plots (0.21). In the reciprocal transplant study of *Ranunculus repens* described earlier in Section 4.3 (Fig. 4.2B), selection coefficients at the woodland site ranged from −0.25 to 0.79 (mean, 0.12), depending on which of the nine morphological measures were used to assess "performance" (Lovett Doust 1981b). At the grassland site, they ranged from −0.01 to 0.78 (mean, 0.39).

For their reciprocal transplant study of *Dryas octopetala*, McGraw and Antonovics (1983) calculated a slightly different selection coefficient in which the home population was no longer assumed to show the best performance. It was quantified as

$$1 - \frac{\text{Performance of any population}}{\text{Performance of the highest performing population}}.$$

for any transplant site. When calculated this way, the selection coefficient can only range between zero and one, and negative values cannot occur. Clearly, the most successful population at a site in terms of the performance measure used will, by definition, have a selection coefficient of zero, regardless of whether it happens to be the home population. Recent studies that report selection coefficients have used this formulation mostly to summarize results from reciprocal transplant experiments (Joshi et al. 2001; Santamaria et al. 2003; Bischoff et al. 2006).

An example of selection coefficients quantified using the formulation of McGraw and Antonovics (1983) is provided by Joshi et al. (2001). Three common perennial herb species were reciprocally transplanted as seedlings In a large-scale investigation of local adaptation across six European sites that included five countries. Data were obtained on survival and fecundity, estimated as the number of inflorescences per plant, for two years. The finite rate of population increase (λ) was calculated for each planted population at a site and was used as the performance measure for the selection coefficient: $1-\frac{\lambda}{\lambda_{max}}$ (the population with the greatest rate of increase at a site is denoted λ_{max}). At each site, selection coefficients were reported for the home and all alien populations separately (Joshi et al. 2001). If there is adaptation to local habitats, then home populations should show selection coefficients that are at, or close to, zero.

At four of the six sites, home populations of *Dactylis glomerata* had the expected selection coefficient of zero, indicating a home-site advantage; but, in Germany and Portugal, the home population was not the highest performing one (Table 4.1). Nonetheless, averaged across all sites, the selection coefficient against alien populations was significantly greater than that of home populations ($t = 3.16$, $p = 0.01$) in this widespread perennial grass.

For *Trifolium pratense* (red clover), the selection coefficient for home populations was zero at half of the sites, indicating a home-site advantage (Table 4.1). There was considerable selection against the alien populations at all sites (range of selection coefficients, 0.29–0.69) and the mean selection coefficient against alien populations was significantly greater than that against home populations ($t = 3.38$, $p < 0.01$).

In contrast to the other two species, there was no compelling evidence for home-site advantage in *Plantago lanceolata*. There was measureable selection against the home population at all six sites (Table 4.1), and the mean selection coefficient against alien populations was not significantly greater than that against home populations ($t = 1.66$, $p = 0.13$). Clearly, these three species vary somewhat in the extent to which their differentiated populations show evidence of adaptation to local conditions. As with so many reciprocal transplant studies, the selection pressures that generated locally adapted populations could not be identified explicitly; however, climatic factors were considered to be of minor importance because they did not explain much of the variance in selection coefficients among sites (Joshi et al. 2001).

Several researchers have combined the standard reciprocal transplant approach with multivariate methods (Lande & Arnold 1983) for the computation of selection gradients (Section 2.2.3). In a detailed investigation of population differentiation and natural selection in the annual *Impatiens pallida*, Bennington and McGraw (1995) calculated standardized directional selection differentials (S) and gradients (β) for two populations reciprocally transplanted between two sites in West Virginia. Significant positive S values occurred for several size measures (e.g., height, leaf area) for home and alien populations at both sites. Differences were apparent in selection gradients for the date of cleistogamous (CL) and chasmogamous (CH) flowering, as well as leaf area, between the two sites. On the drier hillside site, selection favored earlier CL flowering in home and alien populations; at the mesic floodplain site, selection favored later CH flowering in the two populations (Table 4.2 [Bennington & McGraw 1995]). Selection gradients were positive for leaf area in the two populations on the floodplain, but did not differ significantly from zero at the hillside site (Table 4.2). Although there were marked differences between the sites in the relative magnitude of selection for some phenotypic traits, and home-site advantage was demonstrated, major differences in S and β values between home and alien populations at the same site were not apparent. In other words, the relationship of these

Table 4.1 Selection coefficients for three perennial plant species reciprocally transplanted between six field sites across Europe.

Site	*Dactylis glomerata*		*Trifolium pratense*		*Plantago lanceolata*	
	Home	*Aliens*	*Home*	*Aliens*	*Home*	*Aliens*
Silwood, UK	0	0.382 ± 0.044	0.402	0.424 ± 0.139	0.036	0.112 ± 0.052
Sheffield, UK	0	0.279 ± 0.048	0.172	0.320 ± 0.112	0.102	0.127 ± 0.031
Germany	0.347	0.265 ± 0.055	0	0.436 ± 0.118	0.145	0.203 ± 0.050
Switzerland	0	0.445 ± 0.081	0	0.399 ± 0.131	0.194	0.307 ± 0.083
Sweden	0	0.851 ± 0.097	0.201	0.287 ± 0.121	Unavailable	0.188 ± 0.045
Portugal	0.310	0.654 ± 0.146	0	0.686 ± 0.199	0.178	0.302 ± 0.122
Mean ± SE	0.109 ± 0.069	0.479 ± 0.094*	0.129 ± 0.066	0.425 ± 0.057*	0.131 ± 0.028	0.207 ± 0.034

Coefficients were based on the finite rate of population increase. They are shown for the single home population and the mean± standard error (SE) of the seven (for *D. glomerata*) or six (for *T. pratense* and *P. lanceolata*) alien populations at each site. *Note*: There were several sites (e.g., Ireland and Greece for *D. glomerata*) from which populations were transplanted to another site where they were included as alien but were not used as a transplant garden site.

*$p < 0.01$, *t*-test comparing mean selection coefficients for home versus alien populations of each species. Data from Joshi et al. (2001).

Table 4.2 Standardized directional selection gradients at floodplain and hillside sites containing home and away populations of *Impatiens pallida* in a reciprocal transplant experiment in West Virginia.

Variable	*Floodplain*		*Hillside*	
	Home	*Away*	*Home*	*Away*
Date of CL flowering	−0.07	−0.03	−0.68*	−0.40*
Date of CH flowering	0.08*	0.15*	−0.03	0.11
Height in June	−0.04	−0.01	−0.07	0.04
Leaf area in June	0.34*	0.51*	0.33	0.16

Note that *I. pallida* produces both cleistogamous (CL) and chasmogamous (CH) flowers. Asterisks denote gradients significantly different from zero ($p < 0.05$). Data from Bennington and McGraw (1995).

traits to fitness, estimated as the sum of seeds in CH plus CL flowers, was mostly a function of the environment in which the populations were growing. Fitness in the floodplain was improved by greater leaf areas and later production of CH flowers, whereas at the hillside fitness was improved by early production of CL flowers. The authors speculated that earlier CL flowering on the hillside may be adaptive in a local environment that is stressful and imposes a high risk of early mortality (Bennington & McGraw 1995).

Several other reciprocal transplant studies have reported selection differentials and gradients for phenotypic traits at various transplant sites, but have not reported the values separately for each population as Bennington and McGraw did (van Tienderen & van der Toorn 1991b; Griffith & Watson 2005; Byars et al. 2007). Perhaps this reflects a predominant interest in estimating the collective strength of selection pressure that characterizes each transplant site and no particular expectation with regard to whether population origin would make a difference to these estimates.

At three transplant sites used for *Plantago lanceolata* in the Netherlands, selection favored larger plant size (β range, 0.29–0.41) based on a principle components analysis that included several vegetative traits (van Tienderen & van der Toorn 1991b). Selection differentials for flowering time were significantly negative at two sites (β = −0.36 and −0.39), indicating an adaptive advantage to early flowering in grasslands that are regularly mown or grazed by cattle.

At three transplant sites used for the annual *Xanthium strumarium* (cocklebur) in the midwestern United States, significant selection for early flowering (β = −0.66) was detected only in the northernmost garden at the edge of the species range in Michigan (Griffith & Watson 2005). In the garden near the central portion of the species' range in Indiana, later flowering increased fitness (β = 0.14). Path analysis for each garden site corroborated these analyses, and the authors concluded that earlier reproduction was a stress-avoidance adaptation for *X. strumarium* growing in a cool climate with a short growing season (Griffith & Watson 2005).

Vegetative clones and seedlings of the alpine grass *Poa hiemata* were reciprocally transplanted into sites at low and high altitudes in Alpine National Park, Victoria, Australia (Byars et al. 2007). Based on logistic regression analyses (Janzen & Stern 1998) and using seedling survival data to assess fitness, at the high-altitude sites, selection favored plants with small leaves (β range, −0.10 to −0.12) and large size (β range, 0.14–0.27). In contrast, at the low-altitude sites, selection favored larger leaves (β range, 0.19–0.27) and smaller size (β range, −0.13 to −0.25). Note that size was recorded as the circumference of each individual in this caespitose grass species. Thus, along this altitudinal gradient, contrasting patterns of selection were occurring among the different transplant gardens. Knowledge of the magnitude of selection experienced by plants along environmental gradients and the available genetic variation in ecologically relevant traits is clearly important when assessing potential responses to global climate changes (Geber & Dawson 1993; Potvin & Tousignant 1996; Etterson 2004b; Salmela 2014).

4.6 REASONS FOR THE LACK OF LOCAL ADAPTATION

Although there is little doubt that reciprocal transplant studies have provided solid evidence for local adaptation in a diversity of plant species, what about cases in which adaptation to local habitats was not demonstrated? Despite the recognized difficulty in publishing "negative" results in any biological discipline, there are studies in which reciprocal transplants did not reveal evidence of local adaptation (Table 4.3). Recall from Section 4.3 that, in a stringent comparison of local adaptation (Leimu & Fischer 2008) in which the performance of pairs of local and alien populations were compared at two sites, only 45% of 1,032 compared population pairs showed local adaptation (i.e., better performance of the home population) at *both* sites. The authors concluded that "local adaptation is less common in plant populations than generally assumed" (Leimu & Fischer 2008, p. 1).

What are the reasons that reciprocal transplant experiments are, at times, unable to document local adaptation of populations? Several possible explanations can be offered. Some of the reasons provided in studies in which local adaptation was not detected are given in Table 4.3 and are expanded on in the bulleted points that follow.

- Insufficient time has elapsed since populations were originally established. Adaptive evolution of a population in response to the unique selection pressures of a particular habitat can take generations. This may especially be true in highly disturbed, early-successional habitats where colonizing annual populations establish from seed, remain viable for a few years, but are then replaced by larger, long-lived perennials as succession proceeds. For example, the annual grass *Amphicarpum purshii* persists at disturbed sites in the Pinelands of southern New Jersey for only a limited number of years after establishment. The three sites chosen for a reciprocal transplant experiment had been disturbed by mining activities only 7 to 10 years earlier and there may not have been sufficient time for the evolution of local adaptation (Cheplick 1988). Insufficient time needed for adaptive evolution was also offered as a possible explanation for the lack of local adaptation in several perennial herb species (Rapson & Wilson 1988; Jakobsson & Dinnertz 2005; Zhao et al. 2013). Long-lived perennials have lengthy generation times, and it might be predicted that local adaptation does not evolve as quickly as in shorter-lived species (although Leimu & Fischer's [2008] analysis did not find this to be the case).
- The habitats chosen as transplant sites may be similar in regard to environmental conditions and the dominant selection pressures (Fig. 4.6A). This may especially be true when habitats are in close proximity, as in the adjacent fields used in the transplant experiments of Billington et al. (1990). This may also have been

Table 4.3 Reasons given for the lack of adaptation found for plant species used in 12 reciprocal transplant experiments.

Species	*Life habit*	*Number of sites*	*Distances between sites*	*Reason(s) for lack of adaptation*	*Reference*
Ageratina adenophora	Perennial herb	5	110–550 km	Insufficient time, high plasticity, gene flow	Zhao et al. (2013)
Agrostis capillaris	Perennial grass	5	12–114 km	Insufficient time, high plasticity, similar microhabitats	Rapson and Wilson (1988)
Amphicarpum purshii	Annual grass	3	480—950 m	Insufficient time, high plasticity, similar habitats (short distances)	Cheplick (1988)
Bromus tectorum	Annual grass	7	>10 km	High plasticity, genetic drift	Rice and Mack (1991)
Buddleja davidii	Shrub	3	316–540 km	Insufficient time, high plasticity	Ebeling et al. (2011)
Carlina vulgaris	Perennial herb	12	5–40 km	Insufficient time	Jakobsson and Dinnetz (2005)
Deschampsia cespitosa and four other species[a]	Perennial grass	2	40 m	High plasticity, short distance (?),	Seliskar (1985)
Holcus lanatus	Perennial grass	2	Adjacent fields	High plasticity, similar microhabitats	Billington et al. (1990)
Lupinus guadalupensis	Annual herb	3	2–10 km	Genetic drift (small, isolated populations)	Helenurm (1998)
Plantago lanceolata	Perennial herb	6	Adjacent–13 km	High plasticity	Antonovics and Primack (1982)
Ranunculus adoneus	Perennial herb	2	~100 m	Gene flow (seed dispersal)	Stanton and Galen (1997)
Vaccinium elliottii	Shrub	4	~1–4 km	Gene flow, small populations	Anderson and Geber (2010)

Species are listed in alphabetical order by genus.

[a] *Distichlis spicata, Grindelia integrifolia, Jaumea carnosa,* and *Salicornia virginica*.

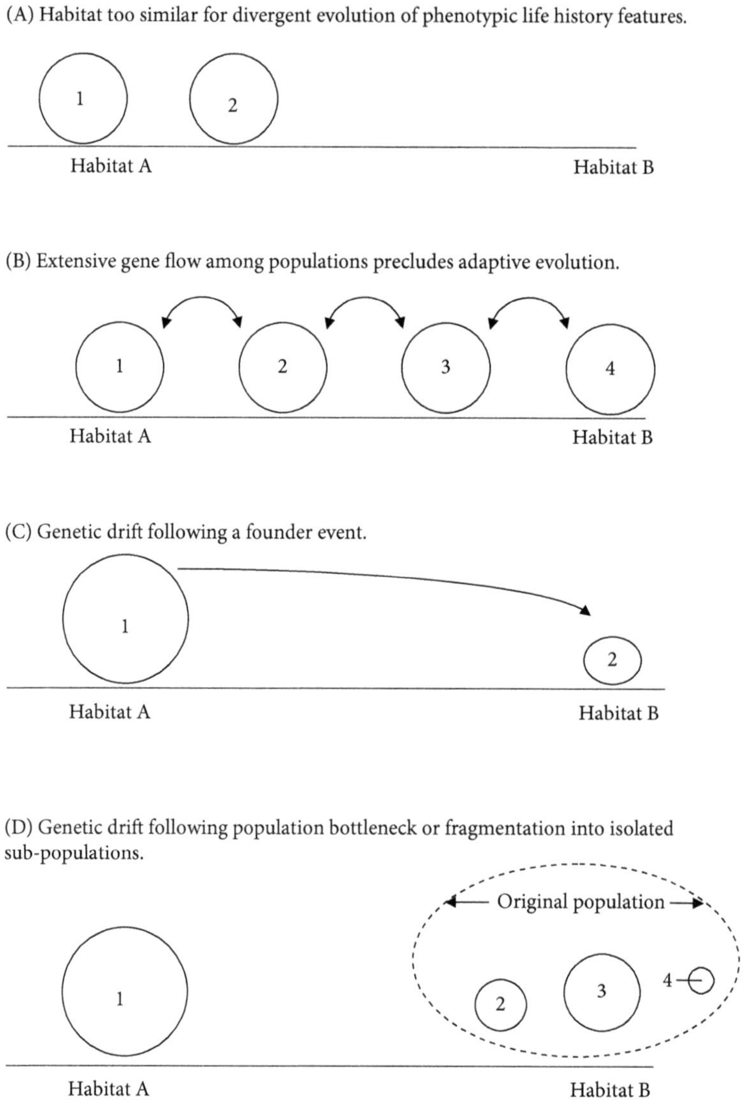

FIGURE 4.6
(A–D) Diagrammatic illustration showing some of the reasons local adaptation is sometimes not detected in reciprocal transplant experiments. Each circle corresponds to a population in habitat A or B.

a factor in a study of *Amphicarpum purshii* in which the transplant sites were separated only by 480 to 950 m (Cheplick 1988). However, the meta-analysis of Leimu and Fischer (2008) did not reveal any relationship between the geographic distance separating the compared sites and the strength of local adaptation. Billington et al. (1990, p. 11) speculated that the scale of "specialization" of individual *Holcus lanatus* clones was at the level of the "microhabitats within each field" that was used for the reciprocal transplant study. Thus, they were unable to demonstrate any home-site advantage of the population of clones at the scale of the 10-ha adjacent fields in Wales used as the transplant sites. In a similar interpretation, Rapson and Wilson (1988) maintained that the environmental dissimilarity between their transplant sites may have been masked by the *similarity of microhabitats within each site* at the spatial scale of individual clones. It was thought that, at this very fine scale, the agents of natural selection were probably most influential.

- There is high phenotypic plasticity of the study species. A perusal of the cases in Table 4.3 reveals this is a common explanation for the inability of some reciprocal transplant experiments to provide evidence of local adaptation. Note also that many of the species in the table are annual or perennial herbs of successional habitats that are likely to exhibit high levels of phenotypic plasticity as they adjust to local environmental conditions (Bazzaz 1996). The plasticity may be extensive enough that, when a population is transplanted into a new (alien) site, the individuals can grow and reproduce as well as the individuals native to the site. In other words, the alien plants may be buffered against selective elimination in the site (Rice & Mack 1991) as a result of a pronounced ability to adjust physiologically and morphologically to the novel environment. Such plasticity was clearly demonstrated in Seliskar's (1985) reciprocal transplant study of five saltmarsh species between upper and lower marshes in Oregon. Highly plastic phenotypic responses to local environmental conditions may override and mask genetic differences among individuals or populations (Antonovics & Primack 1982).
- There is extensive gene flow among populations (Fig. 4.6B). Obviously this can only be a factor when transplant sites are relatively close together; for example, it could have been a factor in the study of Billington et al. (1990) of a wind-pollinated grass in adjacent fields. Substantial gene flow among populations is recognized as a process that can preclude adaptive evolution by natural selection and can limit local adaptation (Barton 2001a; Lenormand 2002; Ellstrand 2014). Despite major differences in the habitats chosen for the reciprocal transplanting of seedlings and adults of the snow buttercup (*Ranunculus adoneus*), Stanton and Galen (1997) found that across a 150 × 200-m snow bed in the Rocky Mountains of Colorado, there was no evidence of local adaptation. However, there was significant directional gene flow via seed dispersal from early- to late-melting sites. Heavier seeds made by plants in the early-melting site consistently produced seedlings with a fitness advantage in all areas of the snow bed. In a blueberry shrub (*Vaccinium elliottii*) reciprocally transplanted between forest sites separated by a few kilometers in South Carolina, there was also no effect of habitat of origin on survival and growth over several years (Anderson & Geber 2010). Microsatellite markers were used to examine potential gene flow between upland and bottomland sites. Results suggested that asymmetric gene flow involved pollen movement from upland to bottomland populations, and this reduced the ability of the latter to adapt to their local habitat (Anderson & Geber 2010).
- Genetic drift may occur in small populations after founder events, bottlenecks, or habitat fragmentation (Fig. 4.6C, D). Any ecological process that results in small, isolated populations can lead to genetic drift—that is, random, unpredictable, and nondirectional changes in gene pools. Drift alone does not result in adaptive evolution, so small populations can be differentiated genetically, but not in the direction expected when natural selection shapes the gene pool. In addition, the loss of genetic diversity and the potentially high level of inbreeding associated with small populations reduce the opportunity for adaptive evolution. Several of the studies in Table 4.3 offered genetic drift as a possible reason for the lack

of adaptation in the populations examined. In the annual endemic *Lupinus guadalupensis*, restricted to San Clemente Island off the coast of California, there was no evidence of local adaptation in a reciprocal transplant experiment across three sites (Helenurm 1998). Limited seed dispersal and self-fertilization may have accentuated the effect of genetic drift in the populations studied. In their meta-analysis of local adaptation studies, Leimu and Fischer (2008) reported that large plant populations (>1,000 individuals) were much more likely to show local adaptation (i.e., home-site advantage) relative to smaller populations. In small populations, it may be difficult to obtain genotypes with the appropriate combinations of phenotypic traits necessary to evolve adaptation in response to natural selection (Antonovics 1976).

- There is a lack of sufficient additive genetic variance. This ties in to the previous explanation, as genetic drift and inbreeding in small populations result in the erosion of genetic variation (Sherwin & Moritz 2000). Bradshaw (1991, p. 298) used the term "genostasis" to denote the condition in which "evolution is limited by the lack of appropriate genetic variation." It has yet to be demonstrated convincingly, however, that genostasis has minimized the extent of local adaptation in any plant populations used for a reciprocal transplant experiment.

4.7 RECIPROCAL TRANSPLANT EXPERIMENTS: WHERE DO WE GO FROM HERE?

There can be little doubt that reciprocal transplant approaches to understanding adaptation have been important to the maturation of plant evolutionary ecology. Although this chapter has emphasized basic research, there are also applied reasons for conducting reciprocal transplant experiments. In restoration ecology, species may be deliberately translocated and introduced into a new region; transplant studies can provide insight into the extent to which populations are likely to be adapted to specific ecological conditions and to succeed in the new habitat (Hufford & Mazer 2003; McKay et al. 2005). The geographic source (provenance) of plants to be used for restoration purposes can be crucial to the success of restoration efforts (Bischoff et al. 2010). Provenance trials that assess the performance of transplanted populations are useful to predicting responses of plants to future climate change and have been encouraged by some researchers (Geber & Dawson 1993; Savolainen et al. 2007; Briggs 2009). Also, disentangling how environmental and genetic factors determine the success and spread of invasive species generally requires the establishment of transplant gardens in both the native and invasive ranges of the study species (Maron et al. 2004; Williams et al. 2008; Moloney et al. 2009). Reciprocal transplant experiments can be used to investigate species that might be used for phytoremediation of polluted environments (Lovett Doust et al. 1994) and to develop appropriate conservation measures for rare species (Raabová et al. 2007). There is clearly no shortage of applied reasons to use a transplant approach.

In their review of genetic differentiation in plant populations, Mazer and LeBuhn (1999) noted that reciprocal transplant experiments typically did not reveal the specific environmental factors and. therefore, the major selection pressures responsible for local adaptation. Experimental demonstration of a "pattern consistent with local adaptation . . . tells us little about the underlying [evolutionary] processes themselves"

(Kawecki & Ebert 2004, p. 1233). Identification of the critical agents of natural selection responsible for adaptive evolution is a difficult, but important, objective for the next generation of reciprocal transplant studies. Experimental manipulation of the agent(s) of selection hypothesized to be primarily responsible for the adaptation of populations to their local (home) habitats is a necessary expansion of the standard reciprocal transplant design. Several examples of reciprocal transplant experiments in which a biotic or abiotic factor was manipulated were presented in Section 4.4.1. Here I present a hypothetical scenario of what might be predicted in terms of average population performance in two habitats if one abiotic factor—soil moisture—differed greatly between sites and was thought to be the predominant selection agent responsible for local adaptation (Fig. 4.7).

In this example there are two conditions to which both home and alien populations are exposed in each habitat: an unmanipulated control (ambient conditions) and a treatment in which soil moisture is increased artificially (in the normally dry habitat) or in which soil moisture is reduced artificially (in the normally moist habitat). Comparison of the bars in the unmanipulated controls in each habitat reveals the classic pattern for local adaptation; fitness of each population was greatest in its home habitat (Fig. 4.7). In addition, when soil moisture was increased in the dry habitat, the fitness of the population from the moist habitat was greater than that of the population from the dry (home) habitat. In contrast, when soil moisture was reduced (perhaps by using rain exclosures) in the moist habitat, the population from the dry habitat had greater fitness than the population from the moist (home) habitat (Fig. 4.7). If the results of an actual experiment came out in such a neat way, the interpretation that soil moisture was the primary factor to which these populations are adapted would have solid support.

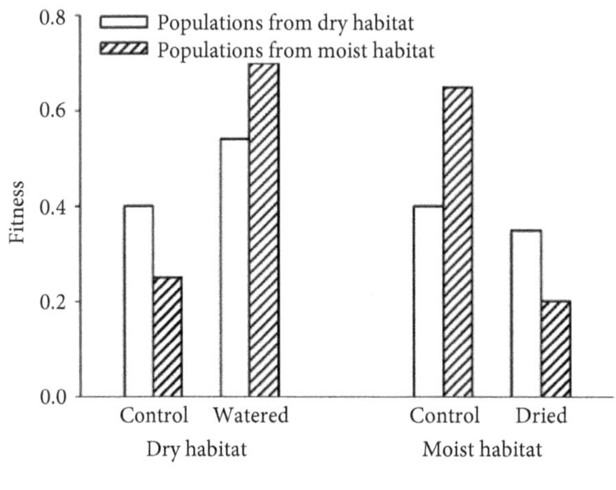

FIGURE 4.7

Hypothetical expansion of a reciprocal transplant experiment that includes experimental manipulation of an abiotic factor—soil moisture—thought to be the primary selection agent responsible for local adaptation of two populations to their respective home sites. "Control" shows population performance in an unmanipulated, naturally dry or moist habitat, and fitness measures reveal the predicted results of a standard reciprocal transplant experiment. In the dry habitat there was a treatment in which soil moisture was increased by supplemental watering; in the moist habitat there was a treatment in which soil moisture was reduced by means of a rain exclosure.

In addition to expansions that manipulate the key, suspected selection agents experimentally, more reciprocal transplant studies are needed in which multivariate analysis of selection gradients are conducted separately by population and site (e.g., Bennington & McGraw 1995). This would help identify the key traits important to fitness and how particular types of selection (directional, stabilizing) might be changing the mean and variance in the phenotypic values of these traits in transplanted populations.

Many types of questions remain with regard to adaptive evolution in plant populations. Should steeper directional selection gradients be expected for populations that are in a foreign site relative to those in their home sites? Would stabilizing selection be more likely for populations in their home sites? Does the extent of genetic variation within a population limit its ability to become adapted to a local environment? What are the particular QTLs that are involved in the expression of phenotypic traits that show local adaptation? These and many other questions can only be addressed by modifications of the reciprocal transplant design that incorporate detailed multivariate analysis of natural selection, experimental manipulation of suspected selection agents, or exploration of the molecular genetic basis for adaptive phenotypic traits (Section 5.3.2; Verhoeven et al. 2004; Knight et al. 2006; Latta 2009; Anderson et al. 2011).

5 Molecular Approaches

5.1 INTRODUCTION: WHAT IS MOLECULAR ECOLOGY?

It almost sounds like an oxymoron: *molecular ecology*. When one thinks of traditional ecology, one probably recalls many studies of whole organisms in populations and communities in natural settings. Yet, the use of molecular techniques to address issues in ecological genetics and evolutionary ecology has been increasing at a phenomenal pace. This should not be surprising because evolutionary processes between and within populations, which have long been of interest to ecologists, involve *gene* pools and changes in the frequencies of alleles or genotypes. Thus, molecular genetics approaches could potentially provide valuable insight into many questions posed by evolutionary ecologists.

As Burke et al. (1992, p. 1) noted in their editorial for the inaugural issue of *Molecular Ecology*, the principal aim of the new journal was to bring together "those who have predominantly molecular biological experience but who are interested in applying their expertise towards problems of the natural environment, and those with mainly ecological interests who wish to use molecular techniques to answer questions that are intractable to other approaches." Some of the research categories in which articles could be published included population and evolutionary genetics, population ecology and gene flow, genetic differentiation, and "molecular adaptation" (Burke et al. 1992). In their editorial report on the first year of *Molecular Ecology*, the editors further noted that the subject represented a "new discipline emerging, with hybrid vigour, from the fertilization of ecology with molecular biology" (Smith et al. 1993, p. 1). The remarkable growth in the discipline can be seen when comparing the journal in its first full year (1993), in which there were six issues (406 pp.); with 1997, when the journal expanded to 12 issues (1,201 pp.); and finally to

Table 5.1 A compilation of some of the general reviews of molecular genetics approaches used in plant evolutionary ecology.

Topic	*Reference*	*Title*
Genetic diversity and structure	Hamrick and Godt (1989)[a]	"Allozyme diversity in plant species"
	Clegg (1989)[a]	"Molecular diversity in plant populations"
	Karp et al. (1996)[a]	"Molecular techniques in the assessment of botanical diversity"
	Hamrick and Godt (1997)[a]	"Effects of life history traits on genetic diversity in plant species"
	McRoberts et al. (1999)[a]	"Assessing the ecological significance of molecular diversity data in natural plant populations"
	Buckler and Thornsberry (2002)[a]	"Plant molecular diversity and applications to genomics"
	Nybom (2004)[a]	"Comparison of different nuclear DNA markers for estimating intraspecific genetic diversity in plants"
	Vekemans and Hardy (2004)[a]	"New insights from fine-scale spatial genetic structure analyses in plant populations"
	Petit et al. (2005)[a]	"Comparative organization of chloroplast, mitochondrial and nuclear diversity in plant populations"
	Whitehead and Crawford (2006)	"Variation within and among species in gene expression: raw material for evolution"
	Arif et al. (2010)[a]	"A brief review of molecular techniques to assess plant diversity"
Molecular markers	Williams et al. (1990)	"DNA polymorphisms amplified by arbitrary primers are useful as genetic markers [RAPDs]"
	Bachmann (1994)[a]	"Molecular markers in plant ecology"
	Zietkiewicz et al. (1994)	"Genome fingerprinting by simple sequence repeats (SSR)-anchored PCR amplification"
	Vos et al. (1995)	"AFLP: a new technique for DNA fingerprinting"
	Ouborg et al. (1999)[a]	"Population genetics, molecular markers and the study of dispersal in plants"
	Baker (2000)	"Molecular methods in ecology [edited collection of papers]"
	Nybom and Bartish (2000)[a]	"Effects of life history traits and sampling strategies on genetic diversity estimates obtained with RAPD markers in plants"
	Merilä and Crnokrak (2001)	"Comparison of genetic differentiation at marker loci and quantitative traits"
	McKay and Latta (2002)	"Adaptive population divergence: markers, QTL and traits"
	Steinger et al. (2002)	"Does natural selection promote population divergence? A comparative analysis of population structure using amplified fragment length polymorphism markers and quantitative traits"

	Zane et al. (2002)	“Strategies for microsatellite isolation: a review”
	Garant and Kruuk (2005)	“How to use molecular marker data to measure evolutionary parameters in wild populations”
	Wright and Gaut (2005)[a]	“Molecular population genetics and the search for adaptive evolution in plants”
	Leinonen et al. (2008)	“Comparative studies of quantitative trait and neutral marker divergence: a meta-analysis”
	Helyar et al. (2011)	“Application of SNPs for population genetics of nonmodel organisms: new opportunities and challenges”
	Kirk and Freeland (2011)	“Applications and implications of neutral versus non-neutral markers in molecular ecology”
	Haasl and Payseur (2012)	“Microsatellites as targets of natural selection:
Genomics	Gibson (2002)	“Microarrays in ecology and evolution: a preview”
	Storz (2005)	“Using genome scans of DNA polymorphism to infer adaptive population divergence”
	Ehrenreich and Purugganan (2006)[a]	“The molecular genetic basis of plant adaptation”
	González-Martínez et al. (2006)[a]	“Forest–tree population genomics and adaptive evolution”
	Ouborg and Vriezen (2007)	“An ecologist’s guide to ecogenomics”
	Stinchcombe and Hoekstra (2008)	“Combining population genomics and quantitative genetics: finding the genes underlying ecologically important traits”
	Gossmann et al. (2010)[a]	“Genome wide analyses reveal little evidence for adaptive evolution in many plant species”
	Hohenlohe et al. (2010)	“Using population genomics to detect selection in natural populations: key concepts and methodological considerations”
	Siol et al. (2010)[a]	“The population genomics of plant adaptation”
	Grover et al. (2012)[a]	“Target sequence capture as a powerful tool for evolutionary analysis”
	Strasburg et al. (2012)[a]	“What can patterns of differentiation across plant genomes tell us about adaptation and speciation?”
	Des Marais et al. (2013)[a]	“Genotype-by-environment interaction and plasticity: exploring genomic responses of plants to the abiotic environment:
	Hough et al. (2013)[a]	“Patterns of selection in plant genomes”

Note that the literature on these topics is vast and updated continually. Within a topic, references are listed in order of publication date. [a]Denotes a specific focus on plants.

2007, when it expanded to 24 issues (5,340 pp.). Although *Molecular Ecology* covers all types of organisms and is not specific to plants, there has clearly been an explosive increase in the use of molecular tools in basic modern ecology, as well as in the more applied disciplines of conservation and invasive species biology.

As the field of molecular ecology rapidly expanded, it became clear that some articles were tightly focused on technical information for the development of molecular markers in select species. These articles were not necessarily germane to the general questions posed by most research studies in evolutionary ecology, in which molecular markers were used as tools for the investigation. Thus, a companion journal—*Molecular Ecology Notes*—was established in 2001 for the publication of technical protocols for molecular marker development in specific species. In 2008, the journal was renamed *Molecular Ecology Resources* and currently publishes six issues per year (~1,200 pages). In a similar way, the *American Journal of Botany* launched an online-only companion journal, *Applications in Plant Sciences*, in 2013 for the publication of "novel protocols and primers, software notes, reviews, and application and genomic resource articles" (Culley 2013, p. 1). The journal contains many articles on molecular characterization in a diversity of plant taxa. It developed from a prior online-only section of the *American Journal of Botany* called "AJB Primer Notes & Protocols in the Plant Sciences," established in 2010. Any prospective student or researcher planning to use molecular markers for a study of the evolutionary ecology of a nonmodel species would do well to consult the technical literature available in these (and other) journals to determine whether protocols already exist for their species.

The focus in this chapter is on the many uses of molecular information in the investigation of a diverse set of topics in plant evolutionary ecology. Details of molecular genetic techniques should not be expected! What types of research topics have been explored using molecular approaches? A synthesis of the literature reveals that molecular methods in ecology have mainly been used in four broad categories of investigation:

1. To quantify the patterns in genetic variation within and among populations of a species (Nybom 2004; Freeland et al. 2011; Weigel 2012). These studies have examined the spatial or geographic structure (architecture) of molecular genetic variation, and sometimes related the patterns to life history features of the species studied. Many plant systematics studies would fall into this category; multiple populations of the taxa studied are normally examined in an effort to work out their putative evolutionary relationships. Molecular genetic surveys typically use markers that are neutral with regard to natural selection (see Section 5.3), not the QTLs for phenotypic traits that are the targets of selection (McKay & Latta 2002). Thus, the population differentiation in phenotypic traits that are most of interest to evolutionary ecologists may not be correlated with differentiation in molecular markers.
2. To characterize kinship—that is, the genetic relatedness of individuals—within populations (Ennos 2001; Blouin 2003; Garant & Kruuk 2005; Waples & Waples 2011). This characterization is important to diverse issues in plant parentage, kin selection, sibling competition, hybridization, clonal plant biology, and mating system evolution (e.g., selfing vs. outcrossing).
3. To investigate the basic processes of microevolution: natural selection, genetic drift, and gene flow (Garant & Kruuk 2005; Petit et al. 2005; Dlugosch & Parker 2008; Hohenlohe et al. 2010; Ellstrand 2014). Neutral molecular markers may

vary among populations solely as a result of genetic drift in small populations (Willi et al. 2006), whereas markers linked to key QTLs may provide evidence of adaptive population differentiation molded by past selection.

4. To identify specific gene loci responsible for ecological traits important to adaptation (Borevitz 2004; Stinchcombe & Hoekstra 2008; Dalziel et al. 2009; Freeland et al. 2010; Colautti et al. 2012). These loci are commonly called *candidate genes* in the molecular ecology literature and their identification is an important first step in revealing the molecular genetic basis of adaptation (Stinchcombe & Hoekstra 2008; Brock et al. 2010).

As a guide to the types of molecular analyses used in ecology and evolution, Table 5.1 presents a compilation of some of the many reviews of topics to be considered in this chapter as studies of specific species are described. The reviews are grouped loosely into three categories: genetic diversity, molecular markers, and genomics. Of course, there is considerable overlap among these categories. Some reviews are focused specifically on plants (asterisks in Table 5.1), whereas others are more generally applicable to all types of organisms. It should be recognized that the literature on molecular genetics is vast and rapidly produced; thus, updated reviews of the topics in Table 5.1 can be expected. General overviews of the molecular techniques useful to ecology and evolution studies can be found in textbooks on the subject (Beebee & Rowe 2008; Freeland et al. 2011) and in the papers collected in Baker (2000). Textbooks also are an important source of information on many useful websites and software programs. In addition, a tabulation of 30 software programs for various uses in population genetic analysis, such as genetic structure and outlier locus detection, is provided by Helyar et al. (2011). A useful online information source is www.molecularecologist.com.

5.2 MOLECULAR GENETIC VARIATION WITHIN AND BETWEEN POPULATIONS

One of the basic premises necessary for Darwinian natural selection and adaptive evolution is that populations contain individuals with heritable variation (Darwin 1859; Mayr 1982). In other words, for a population to evolve it must have a variable gene pool (Lewontin 1974). This is true whether the pool is composed of selectively neutral alleles that can be changed randomly by genetic drift or whether the pool is composed of alleles that affect fitness and can be changed in a specific way by natural selection (or, again, changed randomly by drift). Of course, any population gene pool is going to contain alleles of both types. Surveys of molecular genetic variation within and between populations mostly examine presumably neutral markers, and the adaptive significance of the substantial variation that is normally found is typically unknown. As Bradshaw (1991, p. 295) pointed out, the presence of large amounts of genetic variation in a population "does not necessarily mean that the population has large amounts of variation available to increase the adaptation to the environmental conditions that it faces." As seen in Section 5.3, uncovering the molecular genetic basis for adaptive evolution is an ongoing process in evolutionary ecology (Storz 2005; Stinchcombe & Hoekstra 2008; Siol et al. 2010).

Historically, the study of variation has been central to the development of evolutionary biology (Mayr 1982; Briggs & Walters 1997; Bowler 2005). Genetic variation in nature is often conceptualized as having a hierarchical structure of organization.

FIGURE 5.1
Maritime pine (*Pinus pinaster*), a species used in several studies noted in this chapter for molecular approaches to the investigation of population genetic diversity and adaptive variation. Photo attributed to Manuel M. Ramos.

For example, genetic variation occurs (1) between individuals within populations; (2) between populations within a geographic region; (3) between geographic regions (including variation between ecotypes—clusters of populations adapted to the environmental conditions of a particular region), (4) between races, varieties, subspecies, or another taxonomic entity below the species level; and (5) between species (and higher taxa). Because this text is on evolutionary ecology, the primary focus of this chapter is mostly below the species level, emphasizing microevolutionary processes occurring within and between populations. Molecular genetic variation between species (and higher taxa) and macroevolutionary processes such as patterns of speciation are primarily the domain of systematics and are not considered here.

As a result of the plethora of molecular genetic studies on the diversity of plant species, only a small sampling are presented here to illustrate the topics at hand. For continuity, several of the approaches used to examine molecular genetic variation are presented for a single, well-studied tree species: the maritime pine (*Pinus pisaster*; Fig. 5.1). This species occurs in the western Mediterranean region from Portugal to Italy in Europe and from Morocco to Tunisia in North Africa. It is economically important as a source of timber and for secondary compounds extracted from its bark that apparently have medicinal value (Maimoona et al. 2011).

5.2.1 Allozymes

The first molecular markers used to characterize plant population gene pools were allozymes distinguished and visualized by gel electrophoresis (Gottlieb 1971). Allozymes are different forms of the same enzyme coded for by different alleles at a particular gene locus. (The older term *isozyme* is also used generally for different forms of an enzyme, but is sometimes restricted to those coded for by *different* gene loci). Allozymes are separated by the differences in their net electrical charges and size, which affect their mobility in an electrical field set up across a starch or polyacrylamide gel (Gottlieb 1971). Although developed in the mid 1950s, the use of protein electrophoresis to characterize genetic diversity in natural populations did not become widespread until the 1970s and beyond. The early studies, mostly on species of fruit flies (*Drosophila*), revealed that relatively high levels of genetic variation could be found both within and between populations (Lewontin 1974). In his insightful book on the genetic basis of evolutionary change, Lewontin (1974, p. 120) cited only a few plant studies available at the time and noted that "in proportion to their potential, plants have been greatly neglected as material for studies of genetic variation." That situation quickly changed! For example, only 16 years later, in a review of allozyme diversity in plants, Hamrick and Godt (1989) included 449 species (using data from articles published between 1968 and 1988).

The separation of allozymes by electrophoresis allowed the identification of polymorphic gene loci (and their frequency), and revealed the extent of heterozygosity found in natural populations. It also allowed the calculation of allele frequencies, so critical to basic models of population genetics, as long as sample sizes were large enough. Estimates of total genetic diversity (H_T) based on allele frequency data could be partitioned into that occurring within populations (H_S) and that remaining, which was distributed among populations (D_{ST}), as outlined by Nei (1973). These parameters could be averaged over all loci examined to characterize genetic structure. The proportion of the total diversity found among populations $\left(G_{ST} = \frac{D_{ST}}{H_T}\right)$ became an important estimate of the extent to which populations were differentiated genetically (Loveless & Hamrick 1984; Hamrick & Godt 1989). Although most molecular ecology studies today use DNA-based markers, the data initially provided by allozyme electrophoresis permitted some of the first analyses of population genetic structure and represented an important milestone in the history of molecular ecology (Beebee & Rowe 2008).

As an example, we can consider eight allozyme loci that were examined in six populations of the maritime pine (*Pinus pisaster*) as part of a larger investigation of genetic differentiation in this important Mediterranean conifer (Petit et al. 1995; Fig. 5.1). Starch gel electrophoresis was used to separate the variant proteins for each of the polymorphic loci. Populations were sampled across a wide geographic range and included Spain, Portugal, France, and Italy. Mean H_T across all populations and loci was 0.211, a relatively high level of genetic diversity that is expected for long-lived, outcrossing trees (Hamrick & Godt 1989). The mean genetic differentiation (G_{ST}) was 0.161, a relatively high value that the authors speculated was a result of the highly fragmented range of this species in the western Mediterranean region (Petit et al. 1995). A subsequent study of 18 allozyme loci in 12 populations restricted to Spain and Portugal reported a mean H_T of 0.153 and G_{ST} of 0.076 (Salvador et al. 2000). The relatively low level of population differentiation is identical to the mean value (G_{ST} = 0.076) reported for 131 species of woody perennials by Hamrick and Godt (1989).

For several decades, especially during the 1970s and 1980s, allozyme electrophoresis was used extensively to characterize population (and species) gene pools by evolutionary biologists and systematists. As a result of the popularity of this technique, Hamrick and Godt (1989) were able to include 653 studies that involved 449 plant species in their extensive review of allozyme diversity. However, in the same volume in which their review occurred, other authors were beginning to consider the promise of DNA-based molecular markers for future research (Clegg 1989; Stuber 1989).

Freeland et al. (2011) noted that the use of allozymes as molecular markers has now been supplanted largely by DNA-based markers. There are drawbacks to using allozymes; the leaf material has to be stored in liquid nitrogen in the field and then stored at 80°C in the lab, whereas when getting samples for microsatellites, one can simply collect leaf sections from each plant and store them in bags with silica beads for up to a few weeks or longer until used for molecular analysis. However, allozymes may be valuable for some types of application. Because they are codominant markers, whereas some commonly used DNA markers (e.g., amplified fragment length polymorphisms [AFLPs]) are interpreted as dominant, allozymes are very amenable to allele frequency estimation and comparison with Hardy-Weinberg expectations (J.L. Hamrick, pers. comm., July 5, 2013). In the same communication, Hamrick also noted how allozyme studies typically overlap in 80% to 85% of the loci examined; this nonrandom sampling of allozyme loci facilitates comparisons among species examined by different investigators. This is not true for DNA markers such as randomly amplified polymorphic DNA (RAPD), AFLPs, and microsatellites in which random sequences of DNA from throughout the genome are examined (after their amplification via polymerase chain reaction). Last, allozymes can be useful for determining precise multilocus genotypes in polyploid plants because allele dosage can be inferred accurately from high-quality allozyme electrophoresis (Trapnell et al. 2011).

5.2.2 DNA Markers

DNA-based molecular markers are small parts of the genome that are chosen by researchers in the hope that they are representative of much larger segments of DNA (Beebee & Rowe 2008). Some of the types of DNA markers that have been used in plant biology can be gleaned from a perusal of the reviews compiled in Table 5.1. These markers may involve a single locus or multiple loci, and the DNA may be extracted from the nucleus or organelles, such as the chloroplast and mitochondrion. As with allozymes, gel electrophoresis is used; but, for nuclear DNA fragments, separation is based mostly on their size (Conner & Hartl 2004). Actual DNA sequences may serve as markers in organelle genomes. Because the focus is at the level of individual genes, DNA fragments, or nucleotides, DNA-based techniques have the potential to reveal far more genetic variation within and between plant populations than allozyme electrophoresis. Table 5.2 briefly describes the meaning of some commonly used abbreviations for various DNA markers. Note that RAPD is no longer used extensively and some journals (e.g., *Molecular Ecology*) prohibit publishing results based solely on its use.

To continue with the example of maritime pine (*Pinus pinaster*), chloroplast microsatellites have been used to characterize genetic diversity and its distribution across the species' Mediterranean range (Vendramin et al. 1998; Ribeiro et al. 2001; Bucci et al. 2007). Microsatellites are repetitive sequences of DNA that are relatively short (less than 10 nucleotides long) and are often referred to as *simple sequence repeats* (SSR). Genetic variation in the haploid genotypes (i.e., haplotypes) of chloroplasts (cp) from different individuals can be detected by analysis of SSRs. Measures of within- and between-population genetic variation, analogous to those calculated in allozyme studies, can be obtained from DNA marker data (Nybom 2004). A sampling of 30 trees, each from 10 natural maritime pine populations (from France, Italy, Morocco, and Portugal), revealed greater levels of genetic differentiation based on cpSSRs (23.5% of the haplotype variation was distributed between populations [Vendramin et al. 1998]) compared with those based on allozymes (16% of the allozyme variation was between populations [Petit et al. 1995]). The genetic diversity of haplotypes within populations was relatively high; about 75% of the molecular variation was found within populations. Thirty-four different haplotypes were observed, but 25 of them were limited to a single population (Vendramin et al. 1998).

In a subsequent study, six polymorphic cpSSR loci were examined in 12 maritime pine populations in Portugal (Ribeiro et al. 2001). Almost no differentiation was detected among populations, perhaps because of extensive gene flow, and about 87% of the genetic diversity was within populations. A more inclusive investigation of 48 maritime pine populations across its range using chloroplast microsatellite markers detected 103 distinct haplotypes and 8 major genetic clusters ("gene zones") that could be related to the evolutionary history of this species (Bucci et al. 2007).

Genetic diversity has also been examined within and among populations of maritime pine with nuclear microsatellite markers and AFLPs (Mariette et al. 2001; Ribeiro et al. 2002). Using the AFLP technique, DNA is cut by restriction enzymes (endonucleases) to generate DNA fragments that vary in length. Adapters with attached sequences to which primers can anneal are then ligated to the digested DNA fragments. Many copies of these fragments can be synthesized ("amplified") simultaneously by polymerase chain reaction using those primers that anneal to the adapters; the DNA fragments are then separated by gel electrophoresis to provide a specific genome fingerprint (Vos et al. 1995). Thus, a large number of DNA markers can be generated and used to estimate population genetic parameters (Ribeiro et al. 2002), although dominant markers like AFLPs cannot normally be used to calculate allele frequencies. For 23 populations of maritime pine, the average level of differentiation among populations was 10.2% for AFLP markers and 11.1% for three nuclear microsatellites (Mariette et al. 2001). However, diversity within populations was much greater for the microsatellite markers. Ribeiro et al. (2002) compared the genetic variation for 24 populations (12 in France and 12 in Portugal), determined using AFLPs or cpSSR loci. For six cpSSR loci, 108 different haplotypes were found and total diversity was very high ($H_T = 0.944$). Based on 100 AFLP loci, diversity was much lower ($H_T = 0.179$). However, the levels of population differentiation were similar for both type of molecular marker (although the comparison is complicated by the fact that two distinct genomes—nucleus and chloroplast—are being used in the comparison).

It should be apparent from this brief presentation of molecular diversity studies of the maritime pine that different markers can give genetic diversity estimates that are quite disparate. For example, measures of population differentiation (e.g., G_{ST}) have been shown to be lower for nuclear DNA markers compared with markers from organelles in a data set of 183 plant species (Petit et al. 2005). However, in a compilation of 307 studies using only nuclear DNA markers from plants, Nybom (2004) found reasonably good agreement among genetic diversity estimates for a number of dominantly inherited markers, including AFLPs, RAPD, and intersimple sequence repeats (ISSRs; Table 5.2). The estimates for within- (H_S) and between-population genetic diversity (G_{ST}), compiled for many plant species (Hamrick & Godt 1989; Nybom 2004), are plotted for allozymes and for DNA markers in Figure 5.2. For within-population diversity, allozymes, RAPD, AFLPs and ISSRs produced similar estimates despite the vastly different number of studies used to generate each mean (see legend to Fig. 5.2 for sample sizes). Sequence-tagged microsatellite site (STMS) markers were much more variable within populations (Nybom 2004). Microsatellites are typically highly polymorphic, with much allelic variation resulting from variable numbers of DNA repeats (Beebee & Rowe 2008), and both homozygotes and heterozygotes can be distinguished (codominance).

Table 5.2 Navigating the alphabet soup of DNA-based marker abbreviations.

Marker	*Meaning*	*Description*
AFLP	Amplified fragment length polymorphism	Variable lengths of DNA pieces (= fragments) from different genotypes after digestion by the same restriction enzyme and subsequent multiplication (= amplification) using polymerase chain reaction
EST	Expressed sequence tag	Segment of nucleotide bases at the ends of an active (= expressed) gene that marks (= tags) that gene
ISSR	Intersimple sequence repeat	Short, repeating sequence of nucleotide base pairs (= microsatellite)
QTLs	Quantitative trait loci	Multiple gene loci that affect phenotypic characteristics collectively (such as the traits commonly measured by organismal ecologists)
RAPD	Randomly amplified polymorphic DNA	Variable lengths of DNA pieces from different genotypes after multiplication (= amplification) of random portions of the genome ("chosen" by the same randomly generated primer) using polymerase chain reaction
RFLP	Restriction fragment length polymorphism	Variable lengths of DNA pieces (= fragments) from different genotypes following digestion by the same restriction enzyme
SNP	Single nucleotide polymorphism	Variation in individual (= single) base pairs at the same specific point in the DNA from different genotypes
SSR	Simple sequence repeats	Short, repeating sequence of 2–5 nucleotide base pairs (= microsatellite)
STMS	Sequence-tagged microsatellite site	Short, repeating sequence of 2–5 nucleotide base pairs (= microsatellite) marked (= tagged) by a known segment of DNA

An overview of molecular markers can be found in Arif et al. (2010). Synonymous terms are indicated parenthetically in the description.

Estimates of population differentiation based on DNA markers in many species ranged from $G_{ST} = 0.21$ (AFLPs) to $G_{ST} = 0.34$ (ISSRs), although note that the latter estimate was only based on six studies (Nybom 2004). For allozymes, mean G_{ST} was 0.22 (Hamrick & Godt 1989). Thus, there is general agreement among differentiation estimates for allozymes and DNA markers (Fig. 5.2B). It should be recognized, however, that these mean values mask much of the variation in diversity estimates among species, some of which can be explained by differences in life history attributes, as considered in the next section.

5.2.3 Life History Traits and Molecular Variation

It was not long after extensive allozyme data had become available that researchers began to review the potential relationships between variation patterns of allozymes and life history features in plants (Hamrick et al. 1979). Reviews of the effects of life

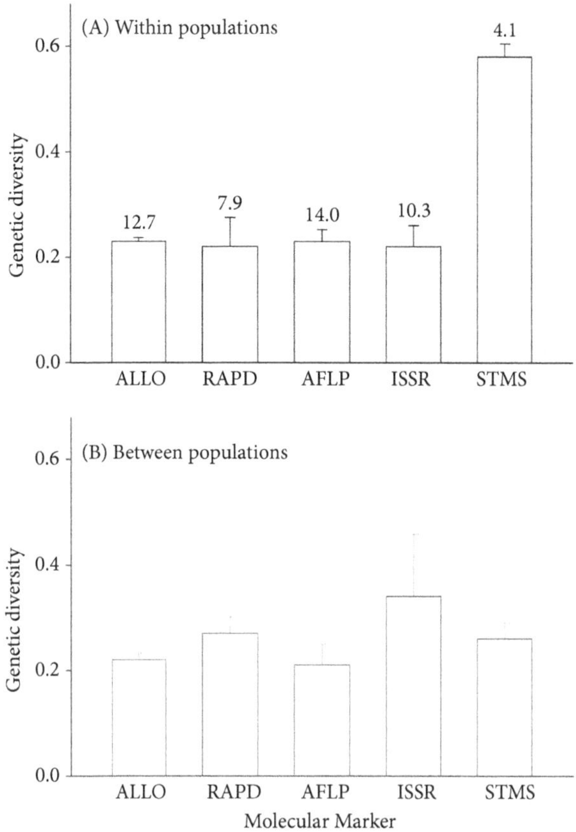

FIGURE 5.2
(A, B) Estimates of within-population (A) and between-population (B) genetic diversity assessed using allozymes (Hamrick & Godt 1989) or various DNA-based molecular markers (Nybom 2004). Numbers above the bars in (A) represent the mean number of populations examined per study. AFLP, amplified fragment length polymorphism; ALLO, allozymes; ISSR, intersimple sequence repeats; RAPD, random amplified polymorphic DNA; STMS, sequence-tagged microsatellite sites. The number of studies included in each estimate is 406, 60, 13, 4, and 80 (A) and 406, 46, 12, 6, and 33 (B) for ALLO, RAPD, AFLP, ISSR, and STMS, respectively.

history traits on allozyme diversity were subsequently updated to include hundreds of plant species (Hamrick & Godt 1989, 1997). Similar types of analyses were attempted with RAPD markers, which were popular throughout the 1990s (Nybom & Bartish 2000), but are no longer used much. The database on RAPD in plant studies was later expanded, and the relationship of the genetic diversity revealed by microsatellite (STMS) markers to life history traits was also explored by Nybom (2004).

All of the reviews have indicated that certain kinds of life history features, especially those associated with life habit, breeding system, or dispersal mode, have consistent effects on both within- and between-population genetic diversity in plant species. Some of these associations have become so widely accepted that they have almost become axiomatic in the evolutionary ecology literature. Through their effects on population genetic structure, these life history features have the potential to shape evolutionary processes and the rate of adaptive evolution (Bradshaw 1984; Bone & Farres 2001; Ennos 2001). Low levels of genetic diversity (e.g., resulting from extensive self-fertilization), for example, might limit the ability of some populations to respond to selection pressures (Pujol & Pannell 2008; Kirk & Freeland 2011).

Whether assessed by allozymes, RAPD, or STMS, annual species tend to have less genetic diversity within, and more between, populations compared with short-lived (herbaceous) perennials or long-lived (woody) perennials (Hamrick & Godt 1989; Nybom 2004). Much of this pattern may simply be the result of the association of life habit with different types of breeding system. Annuals, for example, are more likely to have a breeding system that is mostly self-fertilizing, whereas woody perennials are predominantly outcrossing. When considered without regard to life habit, selfing species generally show reduced levels of genetic diversity within, and greater levels between, populations than outcrossing species (Fig. 5.3). The patterns in population genetic structure appear to hold reasonably well regardless of the markers used. Hamrick and Godt (1997) analyzed life habit and breeding system together for allozyme studies; the genetic patterns within and between populations were maintained consistently for the three breeding systems shown in Figure 5.3 when compared within

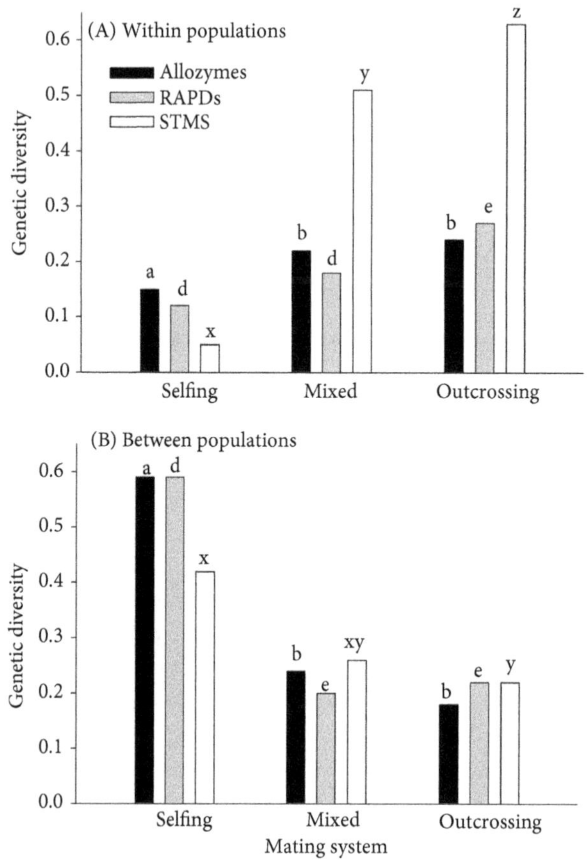

FIGURE 5.3

(A, B) Estimates of within-population (A) and between-population (B) genetic diversity assessed using allozymes (Hamrick & Godt 1989, 1997) or two types of DNA-based molecular markers (Nybom 2004) in relation to breeding systems. RAPD, random amplified polymorphic DNA; STMS, sequence-tagged microsatellite sites. Different letters above bars denote statistically significant differences in genetic diversity across breeding systems for a particular type of marker.

annual or short-lived perennial life habits. They highlighted a fivefold difference in G_{ST} between annual, selfing species versus long-lived, outcrossing perennial species.

Plants with mating systems that are a mixture of selfing and outcrossing showed higher within-population genetic diversity for allozymes and STMS markers compared with predominantly selfing species (Hamrick & Godt 1989, 1997; Nybom 2004). The proportion of genetic diversity found between populations in outcrossing plants versus plants with a mixed mating system was similar for all markers (Fig. 5.3B); however, within-population diversity of the latter was significantly less than that of outcrossing plants when based on RAPD or STMS markers (Fig. 5.3A [Nybom 2004]).

There are a lot of unknowns regarding plant species with a mixed mating system, not the least of which is the usual extent of reproduction by selfing versus outcrossing. Thus, species-specific patterns in population genetic structure can be expected. For example, species that produce both open-pollinated chasmogamous (CH) flowers and self-pollinated cleistogamous (CL) flowers are a classic case of a readily discernible mixed mating system (Darwin 1877; de Jong & Klinkhamer 2005; Culley & Klooster 2007). Sun (1999) investigated population genetic structure using allozymes and RAPD in the perennial herb *Scutellaria indica*, a species producing both CH and CL flowers. In this species, only 5% of CH flowers set fruits (possibly because of a scarcity of pollinators) compared with 96% of the CL flowers (Sun 1999). Thus, most of the sexual reproduction occurs by self-fertilization within CL flowers. As might be expected, within-population genetic diversity (H_S) was extremely low ($H_S = 0.008$, based on 30 allozyme loci in 20 populations; and $H_S = 0.027$, based on 95 RAPD loci in 10 populations). These values are actually far less than the average within-population levels of diversity for selfing species shown in Figure 5.3A. The proportion of genetic diversity found between populations (G_{ST}) was very high in *S. indica*: $G_{ST} = 0.926$ for allozymes and $G_{ST} = 0.809$ for RAPD (Sun 1999)—literally "off the chart" compared with the average values for selfing species shown in Figure 5.3B. Extremely high levels of population differentiation can also be found in asexual species such as the apomictic orchid *Zeuxine strateumatica* (Sun & Wong 2001). The general consequences of extreme population genetic structuring for the future evolutionary potential of plant species are not known.

The final life history factor to be considered here in terms of its effect on population genetic structure is seed dispersal mode. This is another feature that clearly interacts with both breeding system and life habit (Hamrick & Godt 1997). For allozymes, species with seeds dispersed by animals (after ingestion) had the greatest within-population diversity and was significantly greater than those species with no specific dispersal mechanism, often simply said to be gravity dispersed (Hamrick & Godt 1989). No differences among four dispersal categories used by Nybom (2004) were found for RAPD markers in within- or between-population genetic diversity; however, microsatellite (STMS)-based studies revealed the greatest within-population diversity for species with animal-ingested seed dispersal (observed heterozygosity [H_O] = 0.72, $n = 24$). The lowest H_O was for plants dispersed by adhesion to animals ($H_O = 0.27$, $n = 6$), whereas H_O was intermediate for species dispersed by wind/water ($H_O = 0.54$, $n = 26$) or gravity ($H_O = 0.50$, $n = 14$ [Nybom 2004]).

Because species with gravity dispersal show reduced gene flow and would be more likely to exhibit mating between relatives (inbreeding) compared with species with more effective dispersal modes, it might be predicted that they would show the greatest levels of genetic diversity distributed between populations (G_{ST} [Loveless & Hamrick 1984]). For allozymes, Hamrick and Godt (1989) reported a G_{ST} of 0.28 (n = 161), which was greater (but not significantly so) than that for species with seeds dispersed by animal ingestion (G_{ST} = 0.22, n = 39) or wind (G_{ST} = 0.14, n = 121). When considering the interaction of seed dispersal mode with breeding system, Hamrick and Godt (1997) compared two extremes, noting that selfing species with gravity dispersal showed five times more population differentiation than outcrossing species with wind dispersal. For STMS markers, species with gravity dispersal also showed greater genetic diversity between populations than those species dispersed by wind or animal ingestion, although sample sizes were relatively small (Nybom 2004).

Surveys of the relationships between life history traits and population genetic structure in plants can provide clues regarding which specific features of species' life history are important to population microevolution. When beginning an investigation of genetic diversity in a plant species, it is clearly important to know its life habit, breeding system, and dispersal mode because of their general associations with population genetic structure. However, it should be remembered that molecular markers are presumed to be neutral, unless proved otherwise. Thus, the question arises: Is the selectively neutral genetic variation revealed by molecular markers related to variation in the QTLs that affect ecologically important phenotypes?

5.2.4 Comparisons of Population Differentiation: Molecular Markers Versus Quantitative Traits

For studies (or reviews) that compare measures of population differentiation based on molecular markers with those based on quantitative trait measurements, Wright's (1951) fixation index (F_{ST}), or an analogue such as G_{ST} (Nei 1987; Holsinger & Weir 2009) is typically compared with Q_{ST}, an analogous measure of differentiation in quantitative traits (McKay & Latta 2002; Whitlock 2008). Note that these indices typically involve a ratio of the genetic variance distributed between populations to the sum of the between- and within-population variance (Merilä & Crnokrak 2001; Freeland et al. 2011). Because molecular markers are presumed to be selectively neutral, any genetic differentiation among populations detected by F_{ST} represents the combined effects of genetic drift, gene flow, and mutation. If the level of population differentiation for a quantitative trait (as determined in a common garden trial, for example) is no different than that revealed by neutral molecular markers (i.e., $Q_{ST} = F_{ST}$), then any detectable genetic divergence is unlikely to have been caused by natural selection. Rather, the extent of differentiation in the trait is about what would be expected from random chance fluctuations (drift) in the population gene pools (Willi et al. 2006). However, if the estimate of Q_{ST} is significantly greater than F_{ST}, then different phenotypic values of the quantitative trait are being favored in the different populations, implying the influence of natural selection (McKay & Latta 2002; de Kort et al. 2013). Last, if Q_{ST} is significantly less than F_{ST}, stabilizing selection may be favoring the same phenotypic values in the different populations. Note that there has been some criticism regarding such comparisons of Q_{ST} and F_{ST} (Pujol et al. 2008; Pannell & Fields 2014). Whitlock (2008) provides a thorough overview of the assumptions involved in the interpretation and statistical analyses of Q_{ST} and F_{ST} estimates, which are not detailed here. Further information on the general uses of F_{ST} to detect selection is available elsewhere (Holsinger & Weir 2009; Pannell & Fields 2014).

Depending on the species surveyed and the markers used, correlations between molecular and quantitative measures of genetic variation have been shown to be relatively weak (Reed & Frankham 2001; McKay & Latta 2002) or relatively strong (Price et al. 1984; Merilä & Crnokrak 2001; Steinger et al. 2002)! However, when specific comparisons were made between estimates of molecular (F_{ST}, G_{ST}) and quantitative trait differentiation (Q_{ST}) for plant and animal species, surveys found that Q_{ST} often exceeded F_{ST}, indicating an important role for natural selection in generating the differentiation patterns of phenotypic traits (Merilä & Crnokrak 2001; McKay & Latta 2002; Leinonen et al. 2008; de Kort et al. 2013). This same result has been found in several studies of specific plant species using a variety of molecular markers and quantitative traits. In white spruce (*Picea glauca*), Q_{ST} for several traits (e.g., height, wood density) measured in provenance trials exceeded G_{ST} estimates, whether based on allozymes or expressed sequence tag polymorphisms (Jaramillo-Correa et al. 2001). For two quantitative traits (time to germination and age at flowering) in *Silene latifolia*, Q_{ST} exceeded F_{ST} estimates based on microsatellites, although formal statistical tests were not performed (Jolivet & Bernasconi 2007). In the common ragweed (*Ambrosia artemisiifolia*), Q_{ST} for reproductive allocation was significantly greater than F_{ST} based on microsatellites, but did not differ significantly for other traits such as height and biomass (Chun et al. 2011). For eight measured traits, multiple populations of the aggressive invader purple loosestrife (*Lythrum salicaria*) showed significant differentiation in a common garden in Iowa; yet, Q_{ST} for any trait was never significantly greater than estimates of F_{ST} based on AFLP markers (Chun et al. 2009). Thus, in this species other evolutionary processes such as genetic drift, migration, or mutation (instead of natural selection) may be responsible for population differentiation.

5.3 MOLECULAR APPROACHES TO STUDYING SELECTION AND ADAPTATION

We saw in the previous section how, when a differentiation index based on neutral molecular markers such as Wright's (1951) F_{ST} is significantly less than that based on a quantitative trait (Q_{ST}), it is generally inferred that divergent selection has favored different phenotypic values for the trait in different populations/environments. The value of comparing Q_{ST} with F_{ST} is that it can rule out random genetic drift as an explanation for divergence in phenotypic traits among populations (Whitlock 2008). Unfortunately, such analyses still contain a big black box in terms of what particular gene loci or molecular genetic variants are responsible for adaptive differentiation in quantitative traits and which selection agents might be responsible (Barrett & Hoekstra 2011). However, molecular approaches are now available to begin the process of identifying the specific regions of the genome under selection (Nielsen 2005; Storz 2005; González-Martinez et al. 2006; Hohenlohe et al. 2010; Kirk & Freeland 2011; Hough et al. 2013) and to identify the QTLs or candidate genes that underlie ecologically relevant adaptive traits (Borevitz 2004; Stinchcombe & Hoekstra 2008). Furthermore, molecular markers can be used to estimate selection gradients in natural populations (Garant & Kruuk 2005) and to provide evidence of local (molecular) adaptation to specific environments (Namroud et al. 2008; Verhoeven et al. 2008; Coop et al. 2010; Siol et al. 2010; Grivet et al. 2011; Keller et al. 2012).

5.3.1 Correlation of Molecular Markers with Environmental Variables

Analogous to the classic indirect way of analyzing selection by correlating variation in phenotypic traits with environmental variables (Section 2.2.3), correlations between allele frequencies based on molecular markers and environmental variables can identify regions of the genome influenced by selection (Kirk & Freeland 2011). In Chapter 2 on natural selection, reference was made to a study of cork oak (*Quercus suber*) in which the frequency of a few alleles correlated strongly with MAT (Fig. 2.4 [Ramírez-Valiente et al. 2010]). What was not mentioned then was that these alleles were nuclear microsatellite markers that were most likely linked to genes coding for several quantitative traits (e.g., leaf size).

For the alpine perennial *Arabis alpina*, a large genome scan was conducted that involved 825 polymorphic AFLPs for plants sampled from 99 locations in the French Alps and 109 locations in the Swiss Alps (Poncet et. al. 2010). This type of screening approach uses the polymorphism common to many types of DNA marker to identify the "signature" of natural selection (Storz 2005). Loci showing a greater differentiation among populations than expected based on neutral models can be detected statistically as outliers, presumably linked to genes that are under positive selection (Nielsen 2005; Stinchcombe & Hoekstra 2008; Hohenlohe et al. 2010). Overall, the French and Swiss populations of *A. alpina* showed significant differentiation ($F_{ST} = 0.16$, $p < 0.0001$). Of the 825 AFLP loci screened, 78 (9.4%) were found to exhibit correlations with environmental variables, especially mean annual minimum temperature and precipitation (Poncet et. al. 2010). A later analysis of the data set used outlier analysis to identify a locus that consistently had the highest frequencies in moist habitats (Buehler et al. 2013). The locus was sequenced, compared with available gene bank data for the evolutionary relative *Arabidopsis*, and found to match with several phosphatase-associated family proteins. Unfortunately, the specific function of this gene family is not known; however, this candidate gene locus would be worthy of further evaluation in *A. alpina* because it may provide a molecular signature of habitat-mediated selection (Buehler et al. 2013). Clearly, identifying the specific genes that the DNA markers under selection are actually marking and what they code for is important to the analysis of the molecular genetic basis of adaptation.

Grivet et al. (2011) examined associations between variation in candidate genes and climatic variables for two species of pine: *Pinus pinaster* (maritime pine) and *Pinus halepensis* (Aleppo pine). The maritime pine (Fig. 5.1) is the genetically variable conifer, native to the Mediterranean region that we used earlier to illustrate molecular genetic diversity research (Sections 5.2.1 and 5.2.2). The Aleppo pine is another widespread tree species of the Mediterranean region and is closely related to the maritime pine. Twelve populations of *P. pinaster* and nine populations of *P. halepensis* were sampled (77–122 trees per population) from the complete ranges of both species. Previously established primers for three well-known multigene families were used to amplify regions of the genome that corresponded to these putative candidate genes. These genes were already known to be potentially important to drought tolerance in the two pine species (details and references in Grivet et al. [2011]). Chloroplast microsatellites were also examined and provided information on haplotype diversity. Many distinguishable haplotypes and single nucleotide polymorphisms (SNPs) were detected in the two species. After controlling statistically for neutral processes that might also generate clines in frequencies, several significant correlations remained between specific temperature indices and the frequency of a particular haplotype or SNP. For example, in *P. pinaster* one gene identified as a possible target of natural selection was a member of the dehydrin family, known to play an important role in protecting cells against desiccation (Grivet et al. 2011).

Thousands of SNP markers were used to identify gene loci that might be tied to habitat aridity in loblolly pine (*Pinus taeda*) across its range in the southeastern United States (Eckert et al. 2010b). Five loci were found to be associated significantly with gradients in aridity, and they were matched to sequences in *Arabidopsis* that are known to be responsive to two abiotic stress hormones: abscisic acid and jasmonic acid.

Evidence of local adaptation at the molecular level was found in the Balsam poplar (*Populus balsamifera*), once again using SNP genotyping (Keller et al. 2012). Outlier analysis was used to identify the molecular targets of natural selection. Then, covariance between the frequencies of the candidate SNPs and environmental variables important to climate and photoperiod were explored in detail. Particular SNPs were matched to specific candidate genes; in all, 335 SNPs from 443 trees were distributed among 31 populations in northern Canada and Alaska. Several strong associations of candidate SNPs and environmental variables were detected. Several SNPs of a circadian clock gene were correlated with latitude and longitude of populations plus MAT and precipitation. Other SNPs of an abscisic acid gene correlated with maximum temperature of the warmest month. Additional targets of selection appeared to be loci important to flowering (Keller et al. 2012).

Identifying the signature of natural selection at the molecular level typically begins with a search for the markers that have failed the test of neutrality (Siol et al. 2010). These markers show up as outlier loci, showing greater differentiation among populations than expected (based on neutrality models; see example in Fig. 5.4). One recognizes that these DNA markers are not the actual loci responsible for the outlier effect, but rather are likely to be linked to the genes showing adaptive differentiation (Stinchcombe & Hoekstra 2008). Statistical correlations between the frequencies of these markers and specific environmental variables (Coop et al. 2010) further strengthen the evidence for natural selection at the molecular level. In addition, such correlations can suggest which types of selection agents are relevant to adaptive evolution in the species studied.

5.3.2 The Molecular Genetic Basis of Adaptation

We noted in the previous section how significant differentiation among populations for a particular marker (relative to other markers) implies that the marker is not neutral and instead may be linked (and hitchhike with) a particular gene locus that is the molecular target of natural selection. Of course, target genes can be selected only if they affect phenotypic traits that influence fitness. However, it must be recognized that, like reciprocal transplant experiments (Chapter 4), most studies that use molecular genetics to reveal the "signature of natural selection" (Nielsen 2005) are sampling and studying the *products* of adaptive evolution, not the selection process itself. Therefore, an important goal of molecular evolutionary ecology has been to delineate variants of the candidate genes that have been selected for, or against, in populations that are already differentiated. Note that molecular ecologists typically refer to selection for a particular variant as *positive selection*, and selection against a particular variant as *negative* or *purifying selection* (Freeland et al. 2011).

Besides searching for correlations between marker frequencies and environmental variables to identify putative loci that have been shaped by selection, one can also

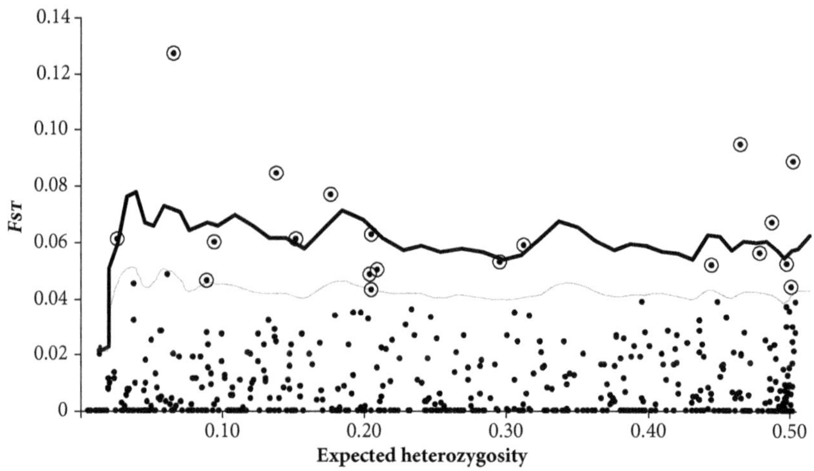

FIGURE 5.4
Plot of Wright's fixation index (F_{ST}) as a function of expected heterozygosity to detect outliers in an analysis of hundreds of single nucleotide polymorphisms (SNPs) in white spruce (*Picea glauca*). The solid line estimates the 99% and the thin gray line estimates the 95%, upper and lower confidence levels. Twenty significant outlier SNPs are denoted by the circled points. Figure used with kind permission of John Wiley & Sons, Inc. Source: From Namroud, M.C., Beaulieu, J., Juge, N., Laroche, J. & Bousquet, J. 2008. Scanning the genome for gene single nucleotide polymorphisms involved in adaptive population differentiation in white spruce. Mol. Ecol. 17: 3599–3613.

examine the statistical association of phenotypic variation with marker frequencies (Stinchcombe & Hoekstra 2008; Kirk & Freeland 2011). This allows one to determine whether the markers are designating genetic loci that affect the phenotypic traits exposed to selection. Formal analysis of QTLs can be conducted, typically using controlled crosses, and specific loci that influence (control?) the expression of ecologically relevant phenotypic traits can be singled out for further study (Colautti et al. 2012). This is a promising aspect of molecular approaches to evolutionary ecology; it helps to forge that all-important link between genotype and phenotype (Dalziel et al. 2009) that is needed to develop a realistic framework of adaptive evolution (Barrett & Hoekstra 2011).

By way of illustration, several examples of molecular genetic approaches to the study of adaptation are presented. Note that some studies have supplemented such an approach with a reciprocal transplant experiment (Verhoeven et al. 2004; Knight et al. 2006; Latta 2009).

A genomewide scan for SNPs differentiated among populations was conducted for white spruce (*Picea glauca*) sampled in Quebec, Canada (Namroud et al. 2008). Expressed sequence tags were used to mark active gene loci, and published gene data banks were used to identify particular functions of the genes important to adaptive population differentiation. Detection of the genes targeted by selection was accomplished by outlier analysis in which SNPs that showed significantly ($p < 0.05$) greater population differentiation (measured by F_{ST}) than expected showed up as outliers in a graph of F_{ST} as a function of expected heterozygosity (Fig. 5.4). These types of graphs are commonly seen in molecular ecology papers reporting outlier analysis to locate putative adaptive gene loci. Twenty SNPs were outliers presumed to be important to adaptive differences among populations (circled points in Fig. 5.4). One gene that showed a high level of differentiation belonged to a family of loci known to be important to reproductive success and flowering phenology. Another gene belonged to a group associated with the control of nitrogen uptake (Namroud et al. 2008). In general, the expressed genes showing evidence of adaptive differentiation in the white spruce populations belonged to gene families known, from comparison with model plant databases, to code for enzymes, regulatory proteins, and hormones important to growth, reproduction, and stress tolerance.

A study by Knight et al. (2006) is notable in that it used three techniques to investigate local adaptation in the short-lived perennial *Boechera* (formerly *Arabis*) *holboellii*, a relative of *Arabidopsis*. First, a reciprocal transplant experiment was conducted across two sites (26 km apart) in Idaho that differed greatly in water availability. Adaptation to the local habitat was indicated based on survival and WUE, which was greatest for plants from the site with low soil moisture. Second, a dry-down experiment was conducted in a growth chamber using genotypes from both field sites. Plants from the dry site showed significantly greater WUE and less transpiration relative to those from the moist site (Fig. 5.5). Furthermore, plants from the

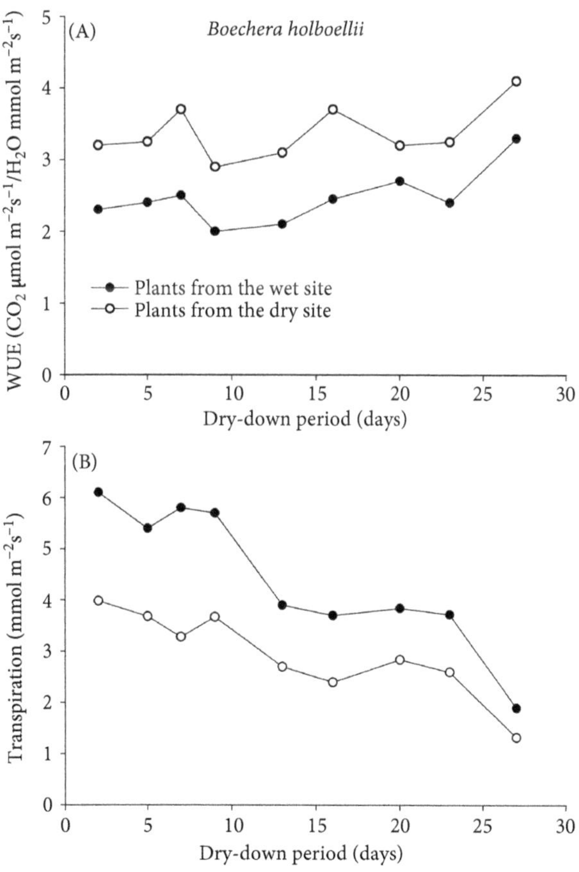

FIGURE 5.5

(A, B) Water-use efficiency (WUE) (A) and transpiration (B) of *Boechera holboellii* plants from sites with low or high water availabilities grown in a 29-day, dry-down experiment. Plants from the dry site had significantly greater WUE and lower transpiration relative to plants from the wet site. Figure redrawn from data in Knight et al. (2006) and used with kind permission of John Wiley & Sons, Inc.

dry site had significantly greater leaf mass per unit area and a greater root-to-shoot ratio, traits that also suggest adaptation to dry soils. Note that this was essentially a common garden experiment in which one experimental variable—soil moisture—was manipulated.

The third technique used to explore local adaptation in *Boechera holboellii* was gene expression profiling with AFLPs and complimentary DNA (cDNA), which permitted identification of expressed candidate genes in plants in dry versus moist environments (Knight et al. 2006). The cDNA is obtained from reverse transcription of messenger RNA synthesized by expressed gene loci. Comparisons of base sequences with the well-known genome of *Arabidopsis* revealed a number of candidate genes in *B. holboellii* that were induced differentially by drought, including homologs of dehydrin, several transcription factors responsive to water stress, and a protein responsive to the stress hormone abscisic acid. It is intriguing that some of these proteins are of the same general types described for two pine species in the previous section (Section 5.3.1) (Eckert et al. 2010a; Grivet et al. 2011).

A QTL mapping approach was used to investigate the molecular genetic basis of adaptive population differentiation in wild barley, *Hordeum spontaneum* (Verhoeven et al. 2004). This annual, selfing grass species (Fig. 5.6) was already known to exhibit local adaptation, as evidenced by reciprocal transplant experiments. Crosses were performed between a coastal Mediterranean and an inland steppe population in Israel and F_3 progeny (obtained by selfing F_2 families) were reciprocally planted into the two field sites. In addition, a common garden trial was established in a greenhouse in which potted plants received a low- or high-nutrient treatment (it was suspected that soil fertility could be important to population adaptation). QTL analysis of field and garden plants was conducted using AFLP markers (details in Verhoeven et al. [2004]). Data on several seed traits tended to show greater performance of genotypes in their native habitat, especially at the more fertile, productive site. For example, there was substantial crossing of the reaction norms for seed number in the F_3 lines across the two field sites. In the common garden trial, plants from the more fertile site had greater reproductive fitness in the high-nutrient treatment, with seed number showing a significant population origin-by-nutrient treatment interaction. QTLs affecting seed number and mass (and additional traits important to fitness) were detected in both field and greenhouse plants. Although some QTLs had an effect at only one field site, in general alleles tended to confer a greater fitness advantage at their native site (Verhoeven et al. 2004). A linkage map included in the

FIGURE 5.6
Wild barley (*Hordeum spontaneum*) inflorescences in a population in Aleppo, Syria. This annual grass has been used in several studies of the molecular genetic basis of adaptive population differentiation. Photo used with permission from Brian Steffensen, University of Minnesota, St. Paul, Minnesota.

article shows that some QTLs explained more than 25% of the variance in a particular fitness trait. Such mapping of QTLs can aid in identifying genomic regions that are important to the expression of ecologically important phenotypic traits (Slate 2005; Freeland et al. 2011; Grillo et al. 2013).

Selection gradients (β) were also examined for the wild barley populations and were reported in a subsequent article (Verhoeven et al. 2008). At both sites there was negative selection for flowering time and positive selection for seed weight, implying fitness benefits to flowering early and producing heavy seeds. In addition, a QTL associated with variation in flowering time was identified. The locus showed close correspondence to a gene known to play a role in controlling the time of flowering in cultivated barley. This QTL appeared to be a target of divergent selection because it was significantly associated only with greater fitness in the steppe population, where early reproduction might be advantageous in the short, unpredictable growing season there (Verhoeven et al. 2008). This is consistent with traditional selection gradient analysis (Lande & Arnold 1983) which showed $\beta = -0.13$ in the steppe population compared with $\beta = -0.04$ in the coastal population.

The final example of the use of molecular tools to detect local adaptation to specific environments involves a dominant seagrass, *Zostera marina*, which is widely distributed in shallow, coastal waters of the northern hemisphere (Oetjen et al. 2010). This ecologically important species has been well studied and shows considerable genotypic variation in morphological and physiological traits (e.g., Hughes et al. 2009). There were two habitat-specific phenotypes investigated by Oetjen et al. (2010): a subtidal form that is completely submerged and reproduces asexually by rhizomes or sexually by flower and seed production, and an intertidal form that is subjected to fluctuations in water level and mostly reproduces sexually. Genome scanning of three subtidal and three intertidal populations was conducted using microsatellite and SNP markers (284 individuals and 46 markers in all). Six outlier loci were detected that could be related to the two habitats. Some were linked to candidate genes that code for proteins involved in osmoregulation at the cellular level and one was linked to a candidate gene that codes for a protein used for seed maturation (Oetjen et al. 2010). As with other studies, the function of the genes was assigned based on published homologous loci available for *Arabidopsis thaliana* and *Oryza sativa* (rice).

The sample of studies reviewed here shows how molecular genomics using DNA markers such as microsatellites or SNPs can inform our understanding of plant adaptation at the molecular level (González-Martínez et al. 2006; Siol et al. 2010; Helyar et al. 2011; Strasburg et al. 2012). Although some methods have not provided strong evidence for adaptive evolution in plants (e.g., Gossmann et al. 2010), population differentiation in molecular markers as revealed by outlier analysis after genome scanning has been demonstrated convincingly in a variety of plant species (Strasburg et al. 2012). These markers have helped to identify candidate genes, and sometimes the probable function of these genes has been determined (Ehrenreich & Purugganan 2006). In addition, the frequencies of marker loci have been associated with specific environmental factors in some species. QTLs have been detected in other species that explain significant proportions of the variation in ecologically important phenotypic traits (Colautti et al. 2012). The study of adaptation in plant populations has clearly gained much from the merging of molecular approaches with more traditional approaches (Anderson et al. 2011).

5.4 OTHER USES FOR MOLECULAR MARKERS

In addition to their use in detecting the genetic signature of natural selection and adaptive evolution, molecular markers have been used extensively to investigate other aspects of the evolutionary ecology of plant populations. These aspects include microevolutionary processes such as gene flow and hybridization as well as genetic drift. Relevant population attributes such as effective population sizes and fine-scale genetic structure based on kinship have also been estimated using the tools of molecular genetics. This is a vast topic and only a smattering of examples are culled from the literature to illustrate some of the basic questions in evolutionary ecology that can be addressed with these methods.

5.4.1 Gene Flow

Gene flow is usually described as the movement of alleles among populations, and it occurs in plants as pollen or seeds (or sometimes vegetative fragments) are transported (Ellstrand 1992, 2014). A pollen grain carries a haploid genome whereas a seed carries a diploid genome (the haploid genomes of the paternal and maternal parents) in an embryo surrounded by various maternal tissues. In a broader sense, one might also visualize gene flow as occurring *within* a population (although this can be distinguished from true gene flow as "gene dispersal" [e.g., Hardy et al. 2006; Born et al. 2008]) when seeds are dispersed by wind or animal vectors and fall among the maternal plants on which they matured. Likewise, gene dispersal occurs as the haploid genome of a pollen grain is transported from one flower to another, between individuals. These individuals are often considered to reside in different populations when using the traditional notion of gene flow.

In general, gene flow tends to reduce genetic structuring within and between populations (Ennos 2001). Gene pool distinctiveness is reduced by extensive gene flow between populations and it may be difficult for two populations to evolve adaptations to their local habitat (Lenormand 2002). The relevance of gene flow to understanding the basic evolutionary ecology of plant populations has been long recognized, but it had always been very difficult to keep track physically of seeds or pollen as they were transported. For example, early researchers had to estimate pollen dispersal by measuring pollinator foraging distances or by using pollen marked with chemicals or dyes. Later adoption of molecular markers aided greatly the quantification of gene flow in a range of plant species (Ellstrand 2014).

In the absence of selection, genetic drift can still cause population gene pools to differentiate, especially when they are small and spatially isolated. However, a relatively low level of gene flow between two populations can counteract this effect of drift, as formalized a long time ago by Sewall Wright (1951) in his often-reprinted equation:

$$F_{ST} = \frac{1}{4N_e m + 1}$$

Here, F_{ST} is the measure of population differentiation (i.e., the genetic diversity distributed among populations [Holsinger & Weir 2009]), m is the rate of migration (i.e., the fraction of the population gene pool that is represented by migrant alleles), and N_e is the effective population size (technically a reflection of the rate at which genetic diversity is lost as a result of drift, but often simply characterized as the size of the breeding population (see Silvertown [2001], Luikart et al. [2010], and Freeland et al. [2011] for further discussion). Although this equation has been flipped around to estimate gene flow rate (*m*) indirectly from empirically derived F_{ST} values (based on neutral molecular markers), there are a host of assumptions and problems with this technique (Ouborg et al. 1999; Whitlock & McCauley 1999; Holsinger & Weir 2009). However, an approximation for *m* can be obtained if the populations can be said to correspond roughly to the so-called island model of population structure. For example Bos et al. (1986) used the equation to estimate gene flow among subpopulations in a large population of the perennial herb *Plantago lanceolata* in the Netherlands. The eight subpopulations were separated by an average of 14 m and contained an average of 73.3 adult plants (used for N_e in the Wright equation). F_{ST} (= 0.04) was estimated as the average for eight allozyme-coding loci. This gives a value of 0.08 for the migration rate (*m*), implying that about six individuals ($N_e m = 73.3 \times 0.08 \approx 6$) enter a subpopulation each generation (Bos et al. 1986). Gene dispersal distances were calculated to be quite low (0.2–1.4 m) and the small-scale genetic structure (i.e., differentiation among subpopulations) was attributed to the "combined result of genetic drift and restricted gene flow" (Bos et al. 1986, p. 51).

Like the studies that first characterized population genetic structure, the neutral molecular markers first applied to the problem of estimating gene flow were allozymes (Adams et al. 1992; Ouborg et al. 1999). For example, Schaal (1980) established an experimental population of the outcrossing annual legume *Lupinus texensis* (Fig. 5.7) in which plants were homozygous for different alleles at a locus coding for phosphoglucose isomerase. Three allozymes were distinguishable by their electrophoretic mobility and were used as genetic markers to quantify gene flow distances via pollen transport by bees. The foraging distances of pollinating bees were also determined. Gene flow estimated by the markers was quite limited, averaging only 1.8 m, although some pollen was transported over 4 m. However, this was greater than the mean pollinator flight distance (~1 m). Measured seed dispersal was also low, averaging only 0.6 m. Schaal (1980) concluded that gene flow was quite restricted, with most pollen transport being to neighboring plants, and most seeds moving only a short distance. This conclusion meshed with the zeitgeist of the time—that is, that gene flow, whether by pollen or seeds, is likely to be quite limited in most plants (Levin & Kerster 1974; Levin 1981).

Other studies using allozymes showed that gene flow for some plants could be substantial. The wild radish *Raphanus sativus* is an annual, outcrossing (insect-pollinated) weed known to be polymorphic for a number of allozyme loci and was used during the 1980s as a model system to investigate gene flow via pollen under

FIGURE 5.7
Variable population of the annual legume *Lupinus texensis* in Montgomery, Texas. This species was included in early studies of gene flow that used allozymes as molecular markers. Photo attributed to Lisa Henry.

natural conditions (Ellstrand & Marshall 1985; Ellstrand et al. 1989). Between-population gene flow was estimated by identifying immigrant pollen using three allozyme markers. For three populations in southern California, the proportion of seeds sired by immigrant pollen varied from 8.2% to 17.9% per generation (Ellstrand & Marshall 1985). Each population was isolated from other flowering *R. sativus* plants by ~600 m to 1 km. In a later study, three experimental populations of *R. sativus* (between 255 m and 400 m apart) were established that were monomorphic for specific allozyme marker loci (Ellstrand et al. 1989). Gene flow via pollen into the experimental populations ranged from 0.2% to 4.5% (expressed as the proportion of between-population matings detected). In addition, apparent gene flow into seven natural populations assessed by paternity exclusion analysis ranged from 3.2% to 18.0% (Ellstrand et al. 1989). These populations were separated by 100 m to 1 km.

Parentage analysis based on 11 polymorphic allozyme loci in a population of the perennial lily *Chamaelirium luteum* showed a mean pollen dispersal distance of 10.4 m, with some pollen dispersed as far as 30 m (Meagher & Thompson 1987). More recent studies of pollen-mediated gene flow have mostly used DNA-based markers such as microsatellites. They have revealed much greater distances (>100 m) for gene flow by pollen, especially in trees that are mostly wind pollinated (Dow & Ashley 1998; Latta et al. 1998; Burczyk et al. 2006; Savolainen et al. 2007; Born et al. 2008; Ashley 2010). For example, the average pollination distance within a stand of bur oak (*Quercus macrocarpa*) was 75 m, based on analysis of four microsatellite loci, and more than half the trees were pollinated by trees outside the stand (Dow & Ashley 1998). Across other tree species, the average distance of pollen movement varies widely: 17 m in *Quercus alba*, 65 to 121 m in *Quercus lobata*, 136 m in *Pinus sylvestris*, more than 300 m in *Quercus robur*, and 743 m to more than 1 km in *Sorbus torminalis* (references in Savolainen et al. [2007]). Ashley (2010) summarized the results of parentage assignment studies based on microsatellites in a diversity of plant species from around the world and cites studies that commonly report gene flow distances that exceed 100 m.

These types of studies using allozyme or DNA markers were instrumental in challenging earlier notions that gene flow in plants was quite restricted (e.g., Levin & Kerster 1974). Gene flow via pollen is now considered to be a significant process in plant microevolution and to occur at far greater distances than previously suspected (Ellstrand 2003, 2014).

We have mostly examined gene flow resulting from pollen transport, but what about seed dispersal? Although it is more challenging to identify the parents of seeds at the molecular level, both allozymes and DNA markers have been used successfully to estimate seed dispersal (Broquet & Petit 2009; Hamrick & Trapnell 2011). Parentage analysis (aka parentage assignment) is a direct method that uses molecular markers to identify the most likely parents of seeds or seedlings (Dow & Ashley 1996; González-Martínez et al. 2002; Burczyk et al. 2006; Hamrick & Trapnell 2011), or sometimes the maternal parent when maternally inherited tissues such as the pericarp are used for DNA extraction (Grivet et al. 2005).

The parentage of 100 saplings in a stand of bur oak (*Quercus macrocarpa*) was resolved by Dow and Ashley (1996) using four microsatellite loci that were highly variable. The population of 62 adult trees in northern Illinois was mapped, and distance between saplings and their maternal parents was determined. Only a relatively small fraction of the trees (6.5%) were the seed parents of the majority of the saplings.

Mean sapling distance from maternal source trees was 23.8 m, and 69% of the saplings were within 15 m of their maternal source tree, although some saplings showed evidence of long-distance (>90 m) dispersal of acorns (Dow & Ashley 1996).

Gene flow via seeds has been documented using microsatellites for other plant species (see examples cited in Ashley [2010] and Hamrick and Trapnell [2011]). For example, in the maritime pine (*Pinus pinaster*), considered earlier in this chapter as an example of the use of molecular markers to describe population genetic diversity (Section 5.2), parentage analysis has been applied to seedlings and adults of a native stand in central Spain (González-Martínez et al. 2002). Average distance between seed parents and offspring was 26.5 m, reflecting the relatively modest dispersal that might be expected from a pine with heavy seeds that are dispersed by wind. In the oak *Quercus lobata*, dispersed by acorn woodpeckers, and studied in California, seed movement was also relatively limited (mostly <50 m), as determined by a maternity analysis (Grivet et al. 2005).

5.4.2 Fine-Scale Genetic Structure

Within plant populations, many studies using molecular markers have revealed that genetic structure can occur at very small spatial scales, generally as a result of the clustering of genetically related individuals (kin). Although a variety of ecological factors, including environmental heterogeneity and selection, can contribute to spatial genetic structure (SGS), pollen and seed dispersal (as well as ramet spread via clonal growth), have long been recognized as being particularly important (Epperson 1989; Heywood 1991; Ennos 2001; Epperson 2007). The theoretical foundation for SGS was developed by Wright (1943) in the concept of isolation by distance. Neighboring individuals (or subpopulations) are more likely to share alleles, resulting from the physical limitations of pollen and seed movement, than those separated by greater distances. Thus, the genetic similarity between individuals for neutral molecular markers should decrease with increasing separation distance. This pattern is well supported by many molecular ecology studies of SGS in plant populations (e.g., Campbell & Dooley 1992; Williams 1994; Fenster et al. 2003; Torres et al. 2003; Dutech et al. 2005).

Most studies of SGS begin with a precise mapping of the individuals in the population(s) followed by tissue sampling for genetic analysis, typically using microsatellites. Then, the genetic data are subjected to spatial autocorrelation analysis (Heywood 1991). Pairwise kinship coefficients, which quantify the genetic relatedness of pairs of individuals, are computed and plotted against the distance between the individuals being compared . This generates a spatial autocorrelogram like that depicted in Figure 5.8. It can be determined whether individuals separated by some specified distance show a kinship coefficient that is significantly greater than zero (shown by asterisks in Fig. 5.8), indicating the individuals are genetic relatives. Typically, this occurs in the shortest distance classes (Vekemans & Hardy 2004) because of the limited dispersal of seeds from their maternal parents; poorly dispersed seeds from the same mother contain embryos genetically related as siblings (Cheplick 1993b). Limited seed dispersal is commonly invoked as an explanation for highly significant SGS in many plant species (e.g., Williams 1994; Latta et al. 1998; González-Martínez et al. 2002; Torres et al. 2003; Hardy et al. 2006; Mathiasen & Premoli 2013). In contrast, very little SGS may be evident in species in which dispersal is especially well developed, as in the tropical tree *Guaiacum sanctum*, which can show long-distance seed dispersal by birds (Fuchs & Hamrick 2010).

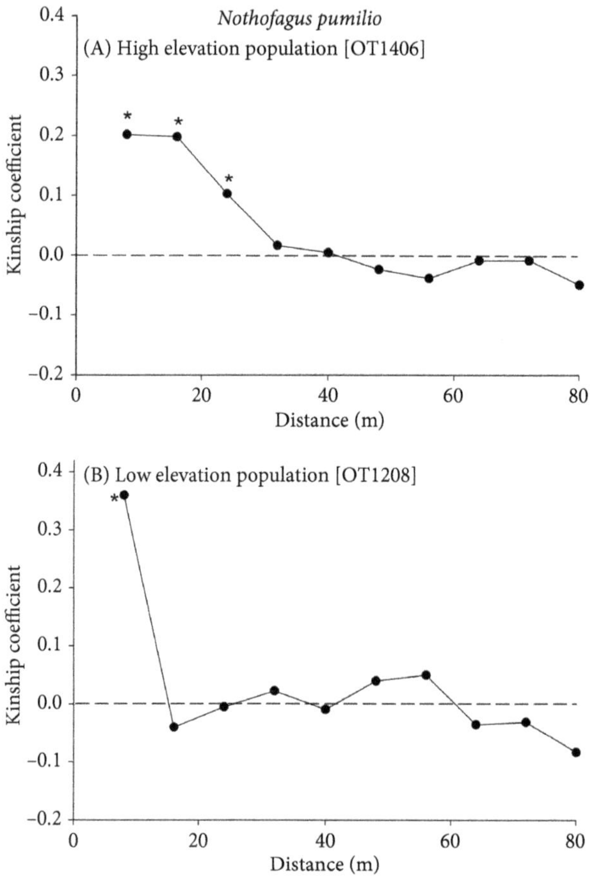

FIGURE 5.8

Spatial autocorrelograms for two populations of *Nothofagus pumilio* in mountain ranges of Argentina. Kinship coefficients are based on combined information from five allozyme and five microsatellite loci. Asterisks indicate coefficients that are significantly ($p < 0.05$) different from zero (dashed horizontal line). (A) One of three sites at high elevation. (B) One of three sites at low elevation. Figure redrawn from data in Mathiasen and Premoli (2013) and used with kind permission from Springer Science and Business Media.

In addition to dispersibility, other aspects of a plant species' biology can be important determinants of SGS. Furthermore, SGS can change over time (Hossaert-McKey et al. 1996; Kalisz et al. 2001). Mating system can clearly affect SGS. In species that are predominantly self-fertilizing and/or mostly inbred as a result of mating among kin, SGS can be more pronounced than in species that are mostly outbreeding (Williams 1994; Hossaert-McKey et al. 1996; Vekemans & Hardy 2004; Epperson 2007). This is because selfing and mating among genetic relatives increases the potential for isolation by distance by reducing pollen transport distances (Heywood 1991). Clonal growth by the vegetative production of ramets is common in many plant species and affects SGS as a result of the spatial clumping of ramets made by each clone (Heywood 1991; Reusch et al. 1999; Mizuki et al. 2010) and inbreeding among the clumped individuals (Handel 1985; Hossaert-McKey et al. 1996; Charpentier 2001). In the sprouting tree *Notholithocarpus densiflorus* (tanoak) in California, spatial autocorrelograms at six sites revealed significant SGS at the shortest distance class

(5 m), reflecting the effect of clonal growth of genets (Dodd et al. 2013). Last, population density has also been shown to affect SGS; it tends to be more pronounced at low density (Heywood 1991; Hamrick et al. 1993; Williams 1994; Vekemans & Hardy 2004).

We next examine a study of a widespread South American tree (*Nothofagus pumilio*) as an example of the use of molecular markers to explore fine-scale genetic structure in multiple populations (Mathiasen & Premoli 2013). The researchers examined six sites, three at high elevations and three at low elevations, in the mountain ranges of Patagonia, Argentina. Five polymorphic allozyme loci and five microsatellite loci were analyzed for 50 individuals per site. Tree populations in all sites showed significant SGS, as revealed in spatial autocorrelograms that plot pairwise kinship coefficients as a function of distance, using 8-m intervals (Fig. 5.8). Most of the SGS occurred at small scales (<40 m) at all sites, and kinship coefficients were significantly ($p < 0.05$) greater than zero for some short-distance intervals (Fig. 5.8). Using the spatial statistic (S_p) developed by Vekemans and Hardy (2004) to quantify SGS for comparison among sites, Mathiasen and Premoli (2013) found that SGS was much greater in the high-elevation stands of *N. pumilio* ($S_p = 0.043$) than in the low-elevation stands ($S_p = 0.004$). These high-elevation stands consisted mostly of relatively even-age trees, possibly indicative of sporadic episodes of seedling establishment after restricted seed dispersal. At low elevations, plants are of multiple ages, indicating continuous seedling establishment, but competition can limit available microsites for establishment and potentially weaken the genetic structure of some low-elevation stands (Mathiasen & Premoli 2013).

It is clear from this brief overview that fine-scale genetic structure is common within the populations of many plant species. Neutral molecular genetic markers have contributed much to the documentation of distinct patterns of SGS. Explanations for SGS vary depending on species' life history attributes, but often entail restricted gene dispersal, whether by seeds or pollen. SGS has implications for a number of microevolutionary processes in plant populations. These include natural selection resulting from G × E interactions (including intergenotypic competition), genetic drift resulting from reduced effective population size, and gene dispersal resulting from potential mating among adjacent kin.

5.4.3 Hybridization

Successful mating between genetically distinct groups in nature, whether differentiated populations, ecotypes, subspecies, or species, has long been acknowledged as an evolutionarily significant process (Anderson & Stebbins 1954; Arnold 1997; Rieseberg & Carney 1998; Barton 2001b). Although Arnold (1997, p. 10) maintained that "the populations from which the hybridizing individuals derive do not need to be assigned to particular taxonomic categories," often the term *hybrid* has been restricted to organisms formed by mating between separate species (Rieseberg & Carney 1998). Regardless of how one interprets hybridization, neutral genetic markers have been very useful to the documentation of hybrid species in a variety of plant groups (e.g., sunflowers [Rieseberg et al. 2007; Scascitelli et al. 2010], grasses [Paul et al. 2010; Jiang et al. 2013], trees [Moran et al. 2012; Hamilton & Aitken 2013], and many others). In addition, hybridization has been touted as a powerful tool for studying natural selection under field conditions (Lexer et al. 2003a). More recently, the incorporation of genomic approaches to research on hybridization has revealed possible candidate loci targeted by selection (Lexer et al. 2004; Hamilton et al. 2013).

Space precludes a thorough synopsis of the many investigations of plant hybridization that are relevant from the perspective of evolutionary ecology. It is clear that hybridization can contribute to adaptive evolution, as novel genetic combinations arise in hybrids that are then subjected to natural selection pressures that may favor some combination of hybrid traits (Lexer et al. 2003a; Rieseberg et al. 2007). Reviewing eight experimental hybridization studies of plant species from which 149 estimates of directional selection gradients (β) were culled, Lexer et al. (2003a) reported that 56% differed significantly from zero, with a mean ± SE of 0.12 ± 0.01. This is quite similar to the mean of 0.14 (based on 656 estimates) reported in the review by Kingsolver and Diamond (2011) described earlier in Section 2.3. In hybrid sunflower (*Helianthus*) species, differentiated with respect to their usual habitats, significant selection gradients acting on synthetic hybrids planted into the field tended to favor the phenotypic traits typically found in each natural hybrid species (Rieseberg et al. 2007). For example, leaf succulence showed significant positive selection (β = 0.17) in synthetic hybrids of *Helianthus annuus* × *Helianthus petiolaris* placed into the sandy dune habitat to which the natural hybrid species *Helianthus anomalus* is restricted. Selection for leaf succulence (β = 0.11) also occurred in the synthetic hybrids placed into salty desert marshes where the natural hybrid *Helianthus paradoxus* normally grows (Rieseberg et al. 2007). Several candidate genes associated with salt tolerance were identified in *H. paradoxus* (Lexer et al. 2004). Note that these natural hybrid species have been shown by molecular genetic comparisons to have originated from mating between the same two parental species (*H. annuus* and *H. petiolaris*).

A broad natural hybrid zone between Sitka spruce (*Picea sitchensis*) and white spruce (*Picea glauca*) has been investigated with neutral genetic markers along a climatic gradient in northwestern British Columbia, Canada (Hamilton & Aitken 2013). Based on nuclear microsatellite markers, gene flow appeared to be widespread between the two species. The climatic gradient of the hybrid zone spanned an ecotone from maritime conditions (nearer the ocean) inland to more continental conditions. There were clines in hybrid ancestry, reflecting the relative proportions of the two parental genomes, in relation to mean annual precipitation and temperature. The authors reasoned that precipitation was likely to be a strong selection agent in the drier parts of the hybrid zone (Hamilton & Aitken 2013).

The same researchers also investigated 384 SNPs in 29 populations across the *Picea sitchensis* × *P. glauca* hybrid zone in an effort to identify candidate genes using an F_{ST}–outlier detection method (Hamilton et al. 2013). Three candidate genes were revealed that might be important to adaptation to the precipitation and temperature gradient associated with distance from the ocean. Once again, precipitation appeared to be an especially strong selection agent as some of the SNPs were homologous to *Arabidopsis* genes important to drought response (Hamilton et al. 2013). In short, the same molecular approaches used to investigate the genetic basis of adaptation (Section 5.3.2) can be readily applied to research on the adaptive dynamics important to the evolution of plant hybrids in nature (Andrew et al. 2013).

5.5 WRAP-UP

Of all the approaches to plant evolutionary ecology considered in this book, those examined in this chapter could easily fill the pages of an entire volume of this size. Molecular ecology often incorporates evolutionary theory and is a vast subject.

The journal *Molecular Ecology* is a great place to witness the integration of ecology, evolution, and genetics in modern molecular studies. Plant species have been used to examine a diversity of key questions relating to genetic diversity and structure, population differentiation, natural selection and adaptation, genetic drift, gene flow, and hybridization. Although this chapter is drawing to a close, we are not done with molecular ecology/evolution, because many other examples of marker-based empirical research arise in upcoming chapters that explore abiotic and biotic agents of selection in greater detail.

Yet, clearly much remains to be learned. Andrew et al. (2013) have conveniently (for me) provided an excellent "road map for molecular ecology," outlining important future research priorities. This article, co-authored by 18 leading researchers in the field, includes ecological genomics and molecular adaptation, hybridization and speciation, and kinship/parentage among the key research areas. They describe one major objective of ecogenomics as "linking genotype to phenotype and ultimately fitness" (Andrew et al., 2013, p. 2613). There is still much to discover regarding the manner in which variation at the molecular genetic level translates into the phenotypic variation that is exposed to the agents of natural selection. Future work that uses molecular approaches should lead to a better understanding of the relative roles of gene flow (and hybridization), genetic drift, and natural selection in shaping plant microevolution.

6 Abiotic Agents of Selection

6.1 INTRODUCTION

More than 100 years ago, Clements (1905, p. 151), in his text *Research Methods in Ecology*, observed that "for the critical investigation of the origin of new forms, an exact knowledge of the features of the habitat, both physical and biotic, is imperative. . . ." He referred to these factors as the agents of selection and proposed a number of experimental techniques, including reciprocal transplanting and deliberate "modifications of the habitat" (Clements 1905, p. 156), for the investigation of adaptation. More recently, the microevolutionary process responsible for adaptation to a specific set of environmental conditions has been neatly embodied in such phrases as "habitat-mediated selection" (Buehler et al. 2013) or "habitat-specific selection" (Scott et al. 2010).

This chapter focuses on the abiotic agents of selection that affect all plant populations. They are predominantly edaphic and climatic factors (Table 2.1) and include natural and anthropogenic sources (Briggs 2009). There is a lengthy history of such factors being recognized as key players in the differentiation of plant populations and their evolutionary adaptation to specific environments (Clements 1905; Turesson 1922, 1925; Marsden-Jones & Turrill 1938b; Clausen et al. 1948; Böcher 1949; Heslop-Harrison 1964; Snaydon 1970; Nevo et al. 1986, 1988; Linhart & Grant 1996; Macel et al. 2007). Methodological approaches to their investigation in relation to microevolution include many of those described in earlier chapters: common gardens and reciprocal transplants, multivariate analysis of selection, and molecular genetics approaches to the study of adaptation.

6.2 EDAPHIC FACTORS

Soil is a complex material with physical features, such as the relative proportions of sand, silt, and clay or moisture content, and chemical features, such as the mineral ion composition of the soil solution (for a detailed, book-length overview of soil properties, see Tan [2009]). In addition, there is a biotic component to soil that includes a diverse set of microbes, especially bacteria and fungi, as well as invertebrate animals. The myriad of physical, chemical, and biotic factors affect plants collectively in complex ways, rendering it a challenge to determine which particular factors may be the critical agents of natural selection in any one population. Because the focus in this chapter is on abiotic agents of selection, the emphasis here is on chemical and physical factors that affect population microevolution and adaptation. Biotic soil factors that are agents of selection (Lau & Lennon 2011) are considered in Chapter 8.

Given the inherent difficulty of teasing apart the many edaphic factors that can vary in soils from different habitats, it is perhaps not surprising that plant ecologists have historically focused on soil types that differ in very distinct ways (Heslop-Harrison 1964). For example, many comparisons have been made between plant populations growing in serpentine versus nonserpentine soils or in metal-contaminated versus uncontaminated soils. These pairs of soil types differ in dramatic ways, and many plant species show adaptive population differentiation in relation to them (Kruckeberg 1954; Antonovics et al. 1971; Macnair 1987; Shaw 1990; Brady et al. 2005; O'Dell & Rajakaruna 2011). Several studies are detailed in this section to illuminate the diverse approaches used.

6.2.1 Population Responses to Distinct Soil Types

The ability of growth and reproductive traits in plant species to respond phenotypically to natural soil types was demonstrated early in the history of ecology by extensive transplant experiments conducted by the British Ecological Society during the late 1920s and 1930s (Marsden-Jones & Turrill 1930, 1938a, b). Samples of several distinct soil types were collected in the vicinity of Potterne, Wiltshire, UK: two were predominantly sand, two were predominantly clay, and another was sandy silt natural to the area at Potterne, where experimental plants were usually grown. The soils differed in available mineral nutrients such as nitrogen, potassium, and calcium (Marsden-Jones & Turrill 1930). Six perennial herb species were originally planted into the different soils in 1928, although additional species were used later (Marsden-Jones & Turrill 1938b). A number of morphological traits were recorded for each species in different years and the full data set on the transplants spanned 10 years. The means ± SD of two traits scored for *Plantago major* ramets of uniform genetic background growing in the five soil types (n = 13–14 ramets planted per soil type) are shown for two years in Figure 6.1. It is clear that soils that were mostly sand and relatively nutrient poor were less favorable for flowering stem production and height growth compared with more nutrient-rich clay soils (Marsden-Jones & Turrill 1938a). In their summary report, Marsden-Jones and Turrill (1938b, p. 388) forcefully concluded that "there is no doubt that different soils have, with certain species, a selective effect."

Of course, these types of early transplant experiments only indicated that edaphic features affected plant growth and reproduction in specific ways and were

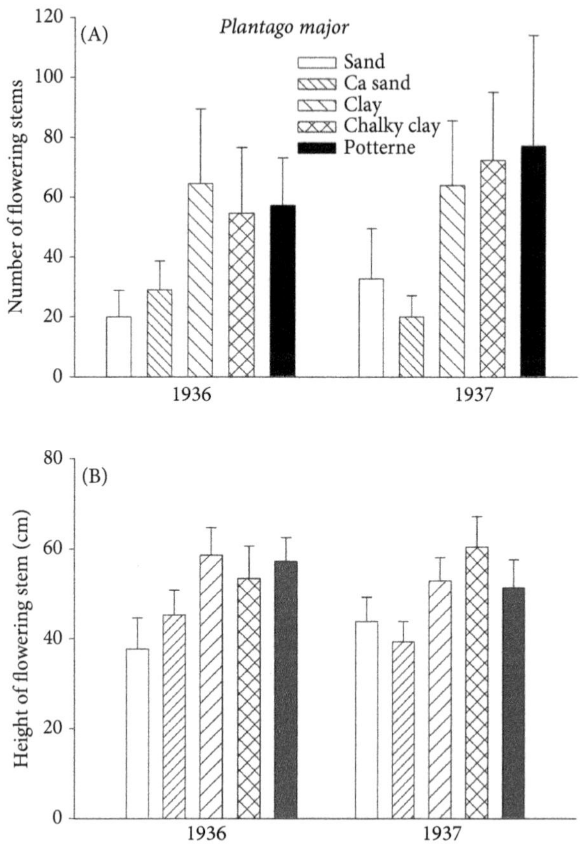

FIGURE 6.1

(A, B) Mean number ± standard deviation of flowering stems (A) and height of the flowering stem (B) in *Plantago major* ramets of uniform genetic background planted into five distinct soil types and grown at Potterne, Wiltshire, UK, for several years. "Ca sand" is a sandy soil rich in calcium, "Chalky clay" is soil obtained from a chalk (limestone) pit and is also rich in calcium, and "Potterne" is a sandy silt soil obtained locally from where experimental plants were grown at Potterne (Marsden-Jones & Turrill 1930). Drawn from data in Marsden-Jones and Turrill (1938a).

likely to be significant agents of selection in natural populations. Other approaches were necessary to demonstrate that selection had resulted in adaptive differentiation of populations in response to distinct soil types.

The long-running Park Grass Experiment in Rothamsted, UK, in which grassland plots were subjected to periodic fertilizer and liming treatments for many years (Silvertown et al. 2006), has provided a unique opportunity to investigate microevolution in plant populations in relation to distinct, quantifiable soil conditions (Briggs 2009). Much of the research has focused on the outcrossing perennial grass *Anthoxanthum odoratum* growing for years in limed or unlimed field plots. Limed plots contain calcareous soils with a higher pH (lower acidity) than the more acidic soils of unlimed plots. Snaydon (1970) collected tillers of *A. odoratum* from limed and unlimed plots and grew plants on local soils that were either calcareous or acidic. Data were expressed as the biomass yield on calcareous soil relative to that on acidic soil and this ratio was compared with the soil pH of the original Park Grass source plots. The ratio showed a highly significant increase with increasing pH of the source plots from which plants were collected, indicating that each population (limed vs. unlimed) performed best on soils that matched the original source plots (i.e., populations from limed plots [pH 7] grew best on calcareous soils whereas populations from unlimed plots [pH 4] grew best on acidic soils [Snaydon 1970]). Because these treatments had been conducted for 40 years at the same site, Snaydon (1970, p. 265) concluded that differences between the populations in response to soil features "have evolved . . . over distances of less than 30 m in less than 40 years."

As briefly noted in Section 4.5, reciprocal transplants of *Anthoxanthum odoratum* among the Park Grass plots revealed substantial selection against alien populations in both limed and unlimed plots (Davies & Snaydon 1976), which suggested that local adaptation to distinct soil types had evolved in this species over a relatively short time. Indeed, a later comparison of *A. odoratum* plants collected from subplots that had been limed (or not limed) for six years only and grown in a spaced-plant trial showed that populations had already diverged for a multitude of phenotypic traits in response to the altered soil environment (Snaydon & Davies 1982).

6.2.1.1 Serpentine Soils

Other well documented examples of plant adaptation to natural soils with distinctive characteristics can be found in the many studies of species that grow in serpentine soils (Brady et al. 2005; O'Dell & Rajakaruna 2011; Rajakaruna & Boyd 2014). These soils, formed by the weathering of ultramafic rocks, typically show higher pH, higher magnesium and lower calcium levels, elevated levels of heavy metals such as nickel, and lower levels of essential nutrients and moisture relative to nonserpentine soils (Brady et al. 2005). The data in Figure 6.2 compare some of these edaphic features that typically differ between serpentine and nonserpentine soils. These soils, analyzed at three sites in Scandinavia, were in pairs separated by 2 to 50 km (Nyberg Berglund et al. 2004). With the distinctive profiles in mineral ion composition and abundance associated with serpentine soils (Fig. 6.2), it is perhaps not surprising that there is a long history of ecologists documenting the unique flora of serpentine sites and, within species, the distinct "edaphic ecotypes" (Kruckeberg 1951, p. 418) that can tolerate and grow on serpentine soil (Kruckeberg 1954; reviewed by Anacker 2014).

Of 21 plant species in California flora examined by Kruckeberg (1954, p. 269), "12 showed definite differentiation into serpentine and non-serpentine races." His methodology mostly entailed growing plants from both habitats in serpentine and nonserpentine soils (Kruckeberg 1951, 1954). He also did experiments in which he grew plants in soils exposed to different levels of calcium amendments (Kruckeberg 1954). He maintained that the capacity to obtain calcium at low levels was an important trait for serpentine-adapted populations. He also noted some morphological distinctions between serpentine and nonserpentine populations within species—differences that had been noticed by previous workers as well. For example, populations from serpentine soils tended to have a smaller stature and smaller, thicker leaves (xeromorphic features important to drought tolerance [Brady et al. 2005]). Kruckeberg (1954, p. 271) presumed that the striking differences between serpentine and nonserpentine populations were genetically based and had evolved as a result of the "selective action of the serpentine environment." Since that time, differentiation

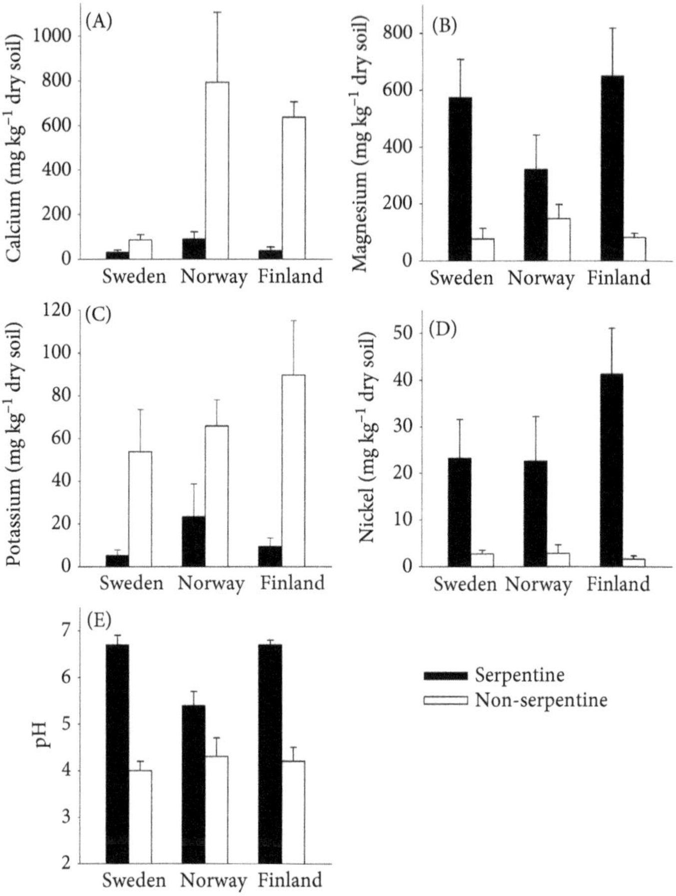

FIGURE 6.2

(A–E) Characteristics of three pairs of serpentine and nonserpentine soils from three Scandinavian countries. Values are mean plus the 95% confidence interval (n = 10). Drawn from data in Nyberg Berglund et al. (2004).

of a number of plant species in relation to serpentine soils has been documented using a variety of experimental and comparative approaches, including reciprocal transplants (Rajakaruna & Bohm 1999; Nyberg Berglund et al. 2004; Sambatti & Rice 2006; Wright et al. 2006; Wright 2007; Wright & Stanton 2011).

Two edaphic races of the outcrossing annual *Lasthenia californica* were detected along transects through a large serpentine outcrop in the Santa Cruz Mountains in California (Rajakaruna & Bohm 1999). One race was normally found on soils with characteristics that were typical of serpentine, with higher pH and a lower calcium-to-magnesium ratio, than soils of the other race. In addition, plants of both races were grown in a greenhouse on the two soil types, as well as in a potting soil mixture. Plants from the nonserpentine soil showed poor growth and did not flower when grown in the serpentine soil. The plants from serpentine soils showed similar growth and flowering in both soil types (Rajakaruna & Bohm 1999). Note that these two races (or ecotypes, to use a more widespread term) occupied well-demarcated positions on the transects along which soil chemical properties (e.g., pH, calcium-to-magnesium ratio, and sodium levels) changed in a gradual, clinal manner with distance. Thus, variation in edaphic factors along these transects, including soil water, which was investigated in a second study (Rajakaruna et al. 2003), is likely to have had a primary role in the evolution of between-population differences in *L. californica* and differences between this species and the closely related species *Lasthenia gracilis* (Yost et al. 2012).

The reciprocal transplant technique has been used to demonstrate local adaptation to serpentine soils in the annual sunflower *Helianthus exilis* (Sambatti & Rice 2006), in two cryptic species of annual *Lasthenia* (Yost et al. 2012), and in the selfing annual herb *Collinsia sparsiflora* (Wright et al. 2006). Populations of the latter species were reciprocally transplanted, as seeds from experimental crosses, between three pairs of serpentine and nonserpentine sites in California. The number of fruits produced by transplants in 2002 and 2003 shows the classic pattern expected if plants are locally adapted to their home (native) soil type (Fig. 6.3). Wright et al. (2006) noted that the distances between soil types were as short as 75 m and no farther than 1 km. Thus, *C. sparsiflora* apparently is differentiated into two distinct edaphic ecotypes adapted locally to either serpentine or nonserpentine soil. Furthermore,

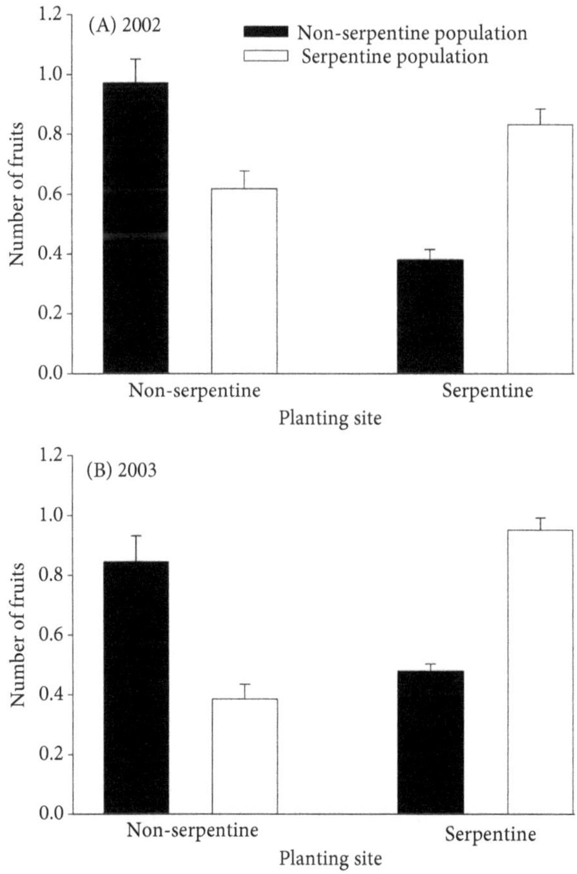

FIGURE 6.3
(A, B) Local adaptation in plants from reciprocally planted seeds of the annual *Collinsia sparsiflora* between nonserpentine and serpentine sites in California. Data are mean fruit number ± standard error in 2002 (A) and 2003 (B). Redrawn from Wright et al. (2006).

hybrids between the two ecotypes have been shown to be relatively sterile, and thus barriers to hybridization may be evolving within this species (Moyle et al. 2012).

Adaptation in relation to serpentine soil was also found for Ponderosa pine (*Pinus ponderosa*) in a long-term reciprocal transplant experiment that was described earlier (Section 4.4.2). Wright and Stanton (2011) reviewed the use of reciprocal transplant experiments to study plant adaptation to serpentine soils and concluded that such experiments, used in conjunction with molecular techniques, would continue to be relevant investigative tools.

Molecular genetics approaches have also begun to be used to more closely analyze the mechanistic basis of genetic adaptation to serpentine soils (Brady et al. 2005; Turner et al. 2010; von Wettberg & Wright 2011). As might be expected, most screening for candidate genes important to serpentine adaptation has involved study of *Arabidopsis* species. In *Arabidopsis thaliana*, survival in solutions with a low calcium-to-magnesium ratio (and other serpentine characteristics) depended on a loss-of-function mutation at a locus called *CAX*1, which normally codes for an antiporter in the vacuole membrane that moves calcium from the cytoplasm into the vacuole (Bradshaw 2005). In wild-type plants growing on serpentine soil, cytoplasmic calcium levels may become very low as a result of the antiporter performing its normal function; also, magnesium levels can become high enough to be toxic as a result of a second membrane cation channel that opens in response to reduced calcium. Plants with the *CAX*1 mutation are able to maintain sufficient calcium levels within the cytoplasm, and accumulation of magnesium does not occur (Bradshaw 2005). This is an interesting analysis because allelic variation at only a single gene locus appeared to be the key player in adaptation to serpentine soils.

In *Arabidopsis lyrata*, genomic analysis was performed on DNA from plants of serpentine and nonserpentine soils (Turner et al. 2010). Millions of polymorphisms were detected after DNA sequencing and 96 of them showed allele frequency differences greater than 80% across the soil types. Several candidate mutations for serpentine adaptation were identified, mostly involving calcium and magnesium transporters or proteins important to detoxification of heavy metals (Turner et al. 2010).

QTL analysis has also been used to identify potential genes that may be important to adaptation to serpentine soils. Bratteler et al. (2006) performed such an analysis with the bladder campion (*Silene vulgaris*) in Switzerland, and linkage maps were constructed using AFLP markers. The magnitude of QTL effects for particular marker loci on several quantitative traits was expressed as the percent of the phenotypic variance explained (Table 6.1). Although the specific candidate genes labeled by these markers were not identified, several of them explained a remarkably high fraction of the variation in some traits known to be critical to serpentine adaptation. These included several markers for leaf succulence and a marker for nickel tolerance that explained 61.7% of the variance in this trait (Table 6.1). The authors summed up by stating that results of the QTL analysis fit the hypothesis that divergence in leaf succulence and nickel tolerance had occurred between serpentine and nonserpentine ecotypes of *S. vulgaris* as a result of consistent selection over a long time (Bratteler et al. 2006).

6.2.1.2 Other Soil Types

In addition to serpentine soil, reciprocal transplant experiments using other soil types collected from different sites have also revealed local adaptation to soils in some

Table 6.1 QTL analysis of AFLP marker loci potentially important to serpentine adaptation in *Silene vulgaris*.

Quantitative trait	*AFLP marker*	*Variance explained (%)*	*Direction of effect*
No. of flowers	AT3–290	45.5	Negative
Leaf succulence	FT9–146	24.8	Positive
	FT7–102	37.1	Positive
	FT7–146	41.7	Negative
	AT3–290	50.9	Positive
Leaf area	FM6–150	24.5	Negative
Nickel tolerance	DM5–391	61.7	Positive

The percent of the phenotypic variance explained by each marker for several quantitative traits is indicated. AFLP, amplified fragment length polymorphism; QTL, quantitative trait loci.

Source: Bratteler, M., Lexer, C. & Widmer, A. 2006. Genetic architecture of traits associated with serpentine adaptation of *Silene vulgaris*. *J. Evol. Biol.* 19: 1149–1156.

plant species (Baltzer et al. 2005; Ellis & Weis 2006; Pregitzer et al. 2010; Ortegón-Campos et al. 2012; Smith et al. 2012). For example, two species of perennial dwarf succulents in the genus *Argyroderma* in southern Africa showed a home-site advantage in terms of survival when reciprocally transplanted (as seeds) between sites 10 to 1,000 m apart that differed mostly in percent quartz cover and stone content (Ellis & Weis 2006).

A transplant experiment in which the invasive annual grass *Taeniatherum caput-medusae* was grown in a greenhouse in four types of soil (two from the invaded range in California and two from the native range in southern France) showed differential growth in native versus introduced soils (Blank & Sforza 2007). Much of the variation (more than 80%) in aboveground mass was explained by the amount of phosphorus, iron, and manganese in the soils, all of which were highest in the soils of the invaded range. However, local adaptation was not indicated because all plants performed best in the more nutrient-rich soils of the invaded range regardless of their population origin (France or the United States [Blank & Sforza 2007]). Note that the hypothesis of local adaptation to soils has also not been upheld in several other investigations of herbaceous species (Abdala-Roberts & Marquis 2007; Macel et al. 2007; Göransson et al. 2009).

A fine-scale level of local adaptation to variable soils was found in seedlings of the riparian forest tree *Populus angustifolia* (narrowleaf cottonwood) planted into soils collected beneath adults of the same species along a river in Utah (Smith et al. 2012). The soils were from within 1 m of the trunk of five adult trees separated by 5 to 20 km. Analysis of the soils showed they differed in physical, chemical, and biotic properties (especially microbial biomass and a fungi-to-bacteria ratio). Seedling survival was about 2.5 times greater when growing in local versus nonlocal soils. Growth of seedlings based on height, and the number and area of leaves were also greatest in local soil. It is not clear which soil properties were specifically responsible for local adaptation, although results of this study (Smith et al. 2012) and a prior study of the same species (Pregitzer et al. 2010) suggested that biotic properties could be especially important. In any case, research has shown that natural variation in soils within a single ecosystem type (a river drainage system) can still act as a subtle agent of natural selection even when the soils cannot be categorized readily into highly distinctive types.

6.2.2 Metalliferous Soils

Some of the earliest examples of plant microevolutionary responses to distinctive soils come from the many studies of tolerance of populations to heavy metals, whether natural or anthropogenic in origin (Bradshaw 1952; Gregory & Bradshaw 1965; Jain & Bradshaw 1966; McNeilly 1968; McNeilly & Bradshaw 1968). In an early review of natural selection commemorating the centennial of Darwin's (1859) *Origin of Species*, Morley (1959) used tolerance of lead contamination in some plant species (e.g., Bradshaw 1952) as one example of how adaptation can arise in populations exposed to variable habitats. In his review of pollution as it affects plant microevolution, Briggs (2009, p. 209) noted that "hundreds of research papers have been produced, together with many important reviews" on the subject of heavy-metal tolerance in plants. It may indeed represent a model evolutionary system (Macnair 1987). In their review of local genetic differentiation in plants, Linhart and Grant (1996, p. 244) referred to studies of metal tolerance as "among the most detailed and elegant in evolutionary biology."

6.2.2.1 Metal Terminology

In sampling some of the extensive literature on heavy-metal tolerance in plants, I was struck by how few authors ever defined "heavy metals" in a general way (but see Antonovics et al. [1971] and Boyd and Rajakaruna [2013]). Looking at the earlier literature on plant adaptation to soils rich in heavy metals, it is clear that the term includes copper, lead, and zinc, which were commonly studied (Bradshaw 1952; Gregory & Bradshaw 1965; McNeilly 1968; McNeilly & Bradshaw 1968; Walley et al. 1974; Wu et al. 1975). In addition, it is evident that elevated concentrations of these metals in soil have detrimental effects on plant growth—that is, they are phytotoxic, at least to populations or species not adapted to such soils (Antonovics et al. 1971; Macnair 1993). Nonetheless, a technical report by the International Union of Pure and Applied Chemistry (Duffus 2002, p. 794) maintained there was no "authoritative definition to be found in the relevant literature" and that the term "heavy metals" was essentially meaningless! However, the report also noted that a common use of the term in biological research reflected the toxicity of metals to living things and included elements such as arsenic, cadmium, copper, lead, nickel, selenium, and zinc, although not all of these are "heavy" (i.e., of high density) or even entirely metallic (Duffus 2002). In any event, all these elements in soils have been studied by plant ecologists in relation to tolerance and adaptation, and the term *heavy metals* is still in use (e.g., Briggs 2009; Lal 2010). The research on plants in relation to heavy metals continues to involve evolutionary, genetic (quantitative and molecular), and physiological perspectives, depending on goals, and appears to have become increasingly reductionist in recent years.

The soils contaminated by heavy metals have been referred to as *metalliferous* (e.g., Gibson & Risser 1982; Dechamps et al. 2008) or *metallicolous* (e.g., Bert et al. 2000; Pauwels et al. 2005), with no apparent distinction between the two adjectives. Plant species that can grow on such soils are sometimes called *metallophytes* (Antonovics et al. 1971, p. 9), and species that can grow on both metalliferous and nonmetalliferous soils are *facultative metallophytes* (Schat et al. 1996), also called *pseudometallophytes* (Antonovics et al. 1971; Pauwels et al. 2005; Dechamps et al. 2008).

It should be recognized that the elevated levels of metal ions found in some soils can be the result of natural processes, such as the weathering of ultramafic rock, which yields serpentine soils rich in chromium and nickel (Briggs 2009). Plant adaptation to these natural soils was covered in the previous section (6.2.1). Here the focus is on microevolution and adaptation of plants in relation to pollution from metals released during anthropogenic activities such as mining or agriculture.

6.2.2.2 Adaptation to Metalliferous Soils

The work of A.D. Bradshaw and others during the 1950s and 1960s focused mostly on *Agrostis capillaris* (formerly *Agrostis tenuis*) and other perennial grasses (Bradshaw 1952; Jowett 1958; Gregory & Bradshaw 1965; Jain & Bradshaw 1966; McNeilly 1968; McNeilly & Bradshaw 1968). The investigation by Gregory and Bradshaw (1965) used techniques that are representative of the early research. They collected *A. capillaris* plants from 12 populations on copper, lead, or zinc mines in Wales and two populations on normal pasture soil. Chemical analyses revealed greatly elevated levels of different metals in soils from the various mining sites. For example, the copper level in the control soils was 50 ppm whereas in the mine soils it varied from 90 ppm to as high as 2,300 ppm (Gregory & Bradshaw 1965). Plant tillers were placed into solutions of metals at different concentrations and root growth was measured to generate an index of tolerance. This widely used index (Macnair 1993; Briggs 2009) was simply the ratio of the growth in a metal solution to the growth in a control solution, often expressed as a percentage. Root growth as expressed by the tolerance index was considered a sensitive metric for metal tolerance in plants (Wilkins 1957, 1978). Gregory and Bradshaw (1965) found a good match between the specific metal tolerances of the populations and the specific metals that were abundant at the mine sites from which the populations had been collected. For example, the indices of tolerance of populations to zinc correlated significantly with levels of zinc in the original soil in which the plants had been growing ($r = 0.84$, $p < 0.01$). This was also true for copper (but not for lead or nickel). They also reported that individual clones from the same population could vary substantially in their tolerance indices, suggesting genetic variation within populations for metal tolerance (Gregory & Bradshaw 1965). Several other species previously shown to be metal tolerant (including serpentine species) were also discussed by Gregory and Bradshaw (1965), who concluded that natural selection had been responsible for the evolution of metal tolerance in these herbaceous species. It was apparent that strong selection pressures exerted by metalliferous soils over a relatively short time span had led to the evolution of distinct populations separated by short distances (Jain & Bradshaw 1966; McNeilly & Bradshaw 1968).

The possibility exists that the alleles responsible for metal tolerance could spread as gene flow occurs from populations on metalliferous soils to populations on nearby nonmetalliferous pasture soils (McNeilly 1968). Karataglis (1980) found some support for this possibility in *Agrostis capillaris* sampled at different distances from a mine in North Wales and tested for tolerance to copper. In the direction of the prevailing wind, tolerance decreased gradually with increasing distance from the mine (Fig. 6.4A).

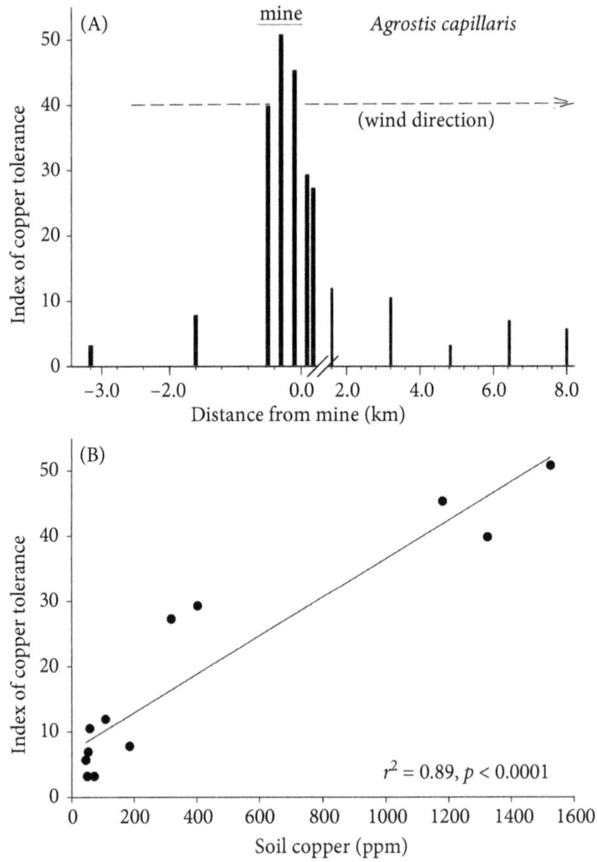

FIGURE 6.4
(A) Copper tolerance in populations of *Agrostis capillaris* collected from a mine in North Wales, UK, and from sites at various distances upwind and downwind from the mine. Note that the three populations at the mining site have been displaced slightly for ease of viewing. (B) Relationship for these same populations between copper tolerance (measured with a root growth assay) and the copper content of the soils from which the populations were collected. Redrawn from data in Karataglis (1980).

However, a plot of copper tolerance versus the copper content of the soil in these populations reveals a strong correlation ($r^2 = 0.89$, $p < 0.0001$; Fig. 6.4B). Thus, there is some doubt that distance from the mine and its relation to gene flow is the only factor correlated with copper tolerance, especially because plants tolerant of elevated soil copper can be found in populations growing in uncontaminated soil in several species, including *A. capillaris* (Gartside & McNeilly 1974). In the Karataglis (1980) study, the two populations downwind from the mine and closest to it had a relatively high tolerance to copper (Fig. 6.4A), suggesting gene flow from the mine populations into the two nearby populations. However, the levels of copper in the soils of these two populations were moderately high (300–400 ppm; see the two middle points above the regression line in Fig. 6.4B). Thus, selection acting on alleles that may have been contributed via gene flow from the mine populations probably contributed to the adaptive tolerance of elevated copper levels in the two nearby populations.

In terms of gene flow between metal-adapted and nonadapted populations, one might predict that there would be a fitness disadvantage for a plant adapted to metalliferous soil to produce seeds sired by nearby plants growing on nonmetalliferous soil. This is because the input of alleles that do not impart adaptation to metalliferous soils into the metal-adapted population after outcrossing could reduce genetic adaptation in the offspring of plants already adapted to metalliferous soil. In other words, it makes more sense from an evolutionary standpoint for metal-adapted plants to mate primarily with other metal-adapted plants within the metalliferous population to preserve the alleles that provide adaptation to metals. Indeed, barriers to gene flow across the two types of population apparently have evolved in the two grass species most thoroughly investigated in regard to metal tolerance: *Agrostis capillaris* and *Anthoxanthum odoratum*. The flowering times between metalliferous and nonmetalliferous populations in both species have diverged, with metalliferous plants flowering about a week earlier (McNeilly & Antonovics 1968). This genetically determined characteristic would tend to reduce or prevent gene flow between adjacent metalliferous and nonmetalliferous populations. Antonovics (2006) later reported that these phenological differences have persisted for more than 40 years in *A. odoratum*, maintaining a high degree of prezygotic isolation between tolerant and nontolerant populations. Other work has shown how metal-tolerant populations of both grass species also had a greater degree of self-fertility than adjacent nontolerant populations, providing an additional barrier to gene flow that was probably favored by natural selection (Antonovics 1968; Antonovics & Bradshaw 1970). Perhaps for the same reason, barriers to gene flow have also been reported to be evolving in the annual *Collinsia sparsiflora* between ecotypes adapted to either serpentine or nonserpentine soil (Moyle et al. 2012).

Given the pervasive evidence for plant adaptation to metalliferous soils, one may well ask how metal tolerance originates before the sorting of variable genotypes during natural selection. In textbook examples of natural selection, such as the evolution of antibiotic resistance in bacteria, in a genetically variable population with no prior history of exposure to a particular agent of selection, genetic variants usually exist that are predisposed to respond to the agent of selection when it does arise. Undergraduate students in my evolution class sometimes write as if the selection agent somehow *causes* the genetic variants to appear magically, but of course that is not how Darwinian selection functions. Thus, in facultative metallophytes, one would expect to find at least a small fraction of plants in a population at an uncontaminated site to, by chance, have the ability to tolerate specific metals. This is apparently the case in several plant species in which metal-tolerant individuals were identified in populations from uncontaminated normal soils (Gartside & McNeilly 1974; Walley et al. 1974; Meharg et al. 1993; Bert et al. 2000). Bradshaw (1991) maintained that the lack of appropriate genetic variation in some species was the reason they had not evolved copper-tolerant populations. He lists several species of facultative metallophytes with demonstrated copper tolerance, and all of them showed some small percentage of copper-tolerant individuals in normal populations, suggesting that the species have the requisite genetic variation needed to evolve copper tolerance. In contrast, none of the seven species that never occurred on mine wastes showed genetic variants tolerant of copper (Bradshaw 1991).

A striking example of how metal tolerance can be identified in populations from uncontaminated sites comes from detailed root growth tests conducted to examine tolerance to soil arsenate in populations of the grass *Holcus lanatus* from across Britain (Meharg et al. 1993). Seeds were collected from at least 50 individuals from 39 uncontaminated populations and 11 populations on arsenic mine spoils. For seedlings from uncontaminated sites (n = 4,657), an average of 45.3% were tolerant to the level of arsenate tested. The percentage of tolerant individuals in populations from arsenic mine soils (n = 1,276), as expected, showed much greater levels of arsenate tolerance (92%–100%). Meharg et al. (1993) considered that plants with root growth exceeding 10 cm could be classified as arsenate tolerant in the rooting test. Based on this criterion, the frequency distribution of root lengths for plants from uncontaminated sites shows how plants could be grouped readily into tolerant and nontolerant classes (Fig. 6.5A). The high frequencies of plants from contaminated sites showing arsenate tolerance is also evident in Figure 6.5B. Thus, in *H. lanatus* the high proportion of arsenate tolerance found in populations of uncontaminated sites provides the appropriate genetic variability necessary for a population to evolve tolerance when challenged by soil arsenate as an edaphic agent of natural selection.

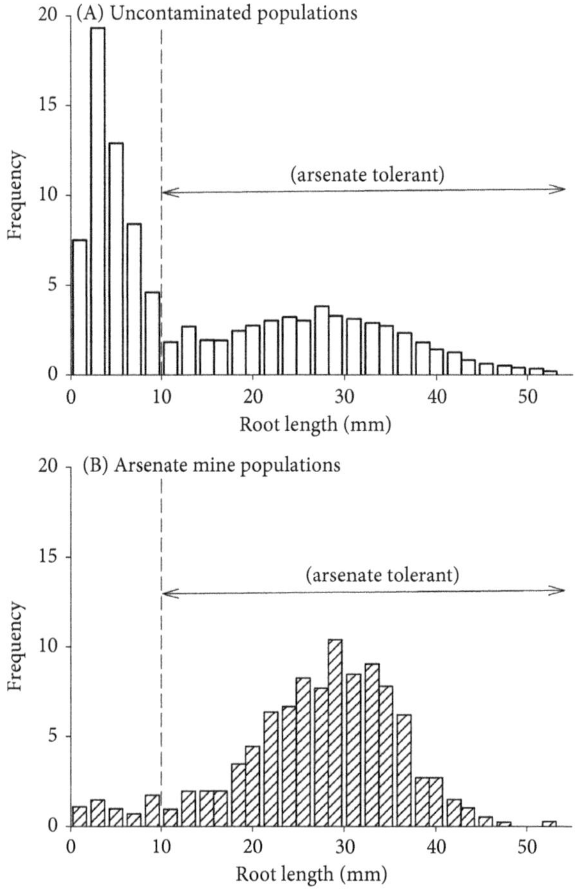

FIGURE 6.5
(A, B) Frequency distribution of root lengths of *Holcus lanatus* in a growth test for tolerance to arsenate in 39 uncontaminated populations (n = 4,657 seedlings) and 11 arsenic mine populations (n = 1,276 seedlings) from Britain. Plants with root lengths that exceeded 10 mm were classified as being arsenate tolerant. Redrawn from Meharg et al. (1993).

6.2.2.3 Genetic Aspects of Metal Tolerance

Early researchers surmised that tolerance to heavy metals had a genetic component, as evidenced from common garden experiments with perennial grasses (Gregory & Bradshaw 1965; McNeilly & Antonovics 1968; McNeilly & Bradshaw 1968; Antonovics & Bradshaw 1970). More recently, the reciprocal transplant approach has been used to demonstrate local adaptation of metalliferous populations of *Thlapsi caerulescens*, a European herb of the mustard family (Brassicaceae) with an annual or biennial life cycle (Dechamps et al. 2008). The soils examined were characterized by elevated levels of metals, especially copper, lead, and zinc. Reciprocal transplants between metal-contaminated and uncontaminated substrates were also used to demonstrate adaptation to elevated levels of copper in a copper smelter population of the perennial grass *Calamagrostis epigejos* (Lehmann & Rebele 2004). Interestingly, although metal-tolerant plants of *T. caerulescens* not in their home environment showed survival, growth, and reproduction comparable with nonadapted plants growing in their nonmetalliferous home environment, the metal-tolerant plants showed a greater susceptibility to leaf herbivory in that environment (Dechamps et al. 2008). Thus, the fitness advantage to metal-adapted plants was only apparent in their metalliferous home environment, supporting the hypothesis of local, genetic adaptation and suggesting they may not be as well adapted to growing in normal, uncontaminated soil.

It became clear from ecological transplant studies that metal tolerance in plants was under genetic control (Antonovics et al. 1971), and subsequent reviews described both genes with a major effect and genes of more modest effects that were key players in metal tolerance (e.g., Macnair 1993; Lal 2010). In five metal-tolerant populations of *Silene vulgaris* in Europe, there appeared to be only a few major gene loci responsible for tolerance to a number of metals (Schat et al. 1996). Analysis of molecular markers, including allozymes (Vekemans & Lefebvre 1997), chloroplast microsatellites (Mengoni et al. 2001), and restriction fragment length polymorphisms of chloroplast DNA (Pauwels et al. 2005) have generally revealed that metal-tolerant populations of herbaceous species have multiple, independent evolutionary origins from nearby, nontolerant populations.

6.3 CLIMATIC FACTORS

There are a host of factors that may broadly be considered "climatic" in nature and include such general features as precipitation, temperature, and light availability (Table 2.1). These show geographic spatial and temporal patterns, of course, but they can also vary at much smaller scales, including that of individual plants. Physical factors identifiable at this fine scale can be considered collectively to comprise a microclimate that affects both individuals and populations. These factors can influence plant fitness directly (e.g., high temperature causing reduced seed maturation) or indirectly (e.g., low precipitation causes soil moisture deficits, thereby reducing seed maturation). Thus, climatic factors, regardless of the scale at which they are considered, have long been recognized as major agents of natural selection for plants (Clements 1905; Böcher 1949; Heslop-Harrison 1964; Langlet 1971).

We examined the classic use of the common garden to demonstrate population differentiation along geographic clines that vary for a number of climatic factors in

Sections 3.2 and 3.3. Historically, these complex climatic gradients from which populations are collected are typically distributed along geographic clines that change gradually with latitude, longitude, or elevation (e.g., Clausen et al. 1948; McMillan 1956; Clausen & Heisey 1958; McMillan 1959; Quinn 1969). Research on phenotypic variation in forest trees examined extensively with provenance tests in common gardens has provided many examples of population differentiation with respect to climatic features (e.g., Mátyás 1996; Rehfeldt et al. 1999; Savolainen et al. 2007; Bower & Aitken 2008; Mimura & Aitken 2010).

The number of climatic factors that can be defined for a particular area is potentially enormous, but most studies have singled out specific variables related to precipitation, temperature, and growing season length as the major defining features of climate that matter to plant adaptation. This makes sense because, on a global scale, these climatic features, often expressed as long-term averages for a specific area, clearly influence the composition of plant communities (Woodward 1987; Box 1996). Extensive data on many climatic variables such as average maximum and minimum temperatures or average rainfall on a monthly or yearly basis for a particular site can be found online at sites such as WorldClimate (www.worldclimate.com) or Climate Europe (www.climatedata.eu). In the United States, much climatic data can be assessed for many locations through the National Climate Data Center (www.ncdc.noaa.gov) or the National Weather Service (www.nws.noaa.gov/climate). These data can then be used in statistical analyses to tease apart some of the long-term climatic variables most relevant to adaptive evolution at different sites.

6.3.1 Temperature

Experiments that entail the transplanting of populations into different natural environments or artificially constructed environments have yielded insights into some of the physical factors that may be most important to plant adaptation to climate (Billington & Pelham 1991; Howe et al. 2003; Stinchcombe et al. 2004; Mimura & Aitken 2010; Kim & Donohue 2013; Weinig et al. 2014). For example, the date when leaf buds opened was determined for populations of two birch species (*Betula pubescens* and *Betula pendula*) collected from 10 sites in Scotland and then grown in a field common garden (Billington & Pelham 1991). There was significant variation among populations (and among families within populations) for the date that buds opened in the garden, implying genetic variation for this important trait. Furthermore, realizing the importance of temperature as a selection agent for this trait, the authors projected that the predicted increases in annual temperature resulting from climate change would be too rapid for birch populations to evolve effectively in response (Billington & Pelham 1991).

Temperature-related climatic variables, such as the long-term MAT for an area, have commonly been found to be especially important to phenotypic and molecular adaptation in forest tree populations (Carter 1996; Rehfeldt et al. 1999; Howe et al. 2003; Jump et al. 2006; Aitken et al. 2008; Eckert et al. 2010a; Holliday et al. 2010; Ramírez-Valiente et al. 2014). Howe et al. (2003) summarized many of the quantitative and molecular genetic studies of the adaptation of trees to low temperature in temperate and boreal regions. They reported that tree populations are often well differentiated with respect to traits important for adaptation to seasonally cold conditions, such as the timing of growth cessation in buds (bud set) and dormancy.

Common garden studies have shown that bud set and frost hardiness are typically associated with the temperature regime found at the sites where populations were collected (Howe et al. 2003). Plants from colder regions tend to show an earlier cessation of growth and bud set than those from warmer regions, where the growth period is markedly longer.

Local adaptation has been studied intensively in Sitka spruce (*Picea sitchensis*) populations along the Pacific coast of North America in relation to climatic factors (Holliday et al. 2010; Mimura & Aitken 2010). These studies have used both traditional ecogenetic approaches, such as examination of phenotypic variation in common gardens, as well as modern molecular genetic approaches, such as association mapping (Neale & Savolainen 2004) of traits that are especially related to climate. Mimura and Aitken (2010) collected seeds from 17 populations distributed from Alaska in the north, to British Columbia, Canada, near the range center, to California in the south. Seedlings from the seeds were moved into three growth chambers set up with environmental conditions (temperature, photoperiod) designed to simulate the climate of the northern periphery, the range center, and the southern periphery. About 25 seedlings were used per population and climate condition. A variety of quantitative phenotypic traits were recorded, including days to bud set, height, biomass, and daily growth rate (based on height). For most traits, there were significant differences among populations and a striking clinal pattern of variation with distance along the coast (primarily a latitudinal transect from north to south).

The researchers reported highly significant correlations of MAT ($r^2 = 0.93$, $p < 0.001$) and number of days with temperatures more than 5°C ($r^2 = 0.92$, $p < 0.001$) with geographic distance—both declining as one proceeded northward (Mimura & Aitken 2010). Although the authors showed the clines in graphs of phenotypic trait values versus geographic distance, I chose to plot one of the traits—daily growth rate—more directly against the MAT at the population site of origin for the three simulated climates (Fig. 6.6). In the northern and central climates, populations from the north, where temperatures are colder, showed a greater growth rate than those from the south (northern climate: growth rate = 0.80–0.01 [MAT], $r^2 = 0.24$, $F = 4.67$, $p = 0.0473$; central climate: growth rate = 0.91–0.04 [MAT], $r^2 = 0.74$, $F = 43.74$, $p < 0.0001$). However, in the southern climate, growth rate was not related to MAT ($r^2 < 0.01$, $F < 0.01$, $p = 0.9943$; Fig. 6.6C). Thus, in the warmer, southern climate, all populations appear to do equally well in terms of their average daily growth rates. Other traits such as height, biomass, and bud set showed the greatest values for populations from the south in the southern climate, suggesting local adaptation of these Sitka spruce populations (Mimura & Aitken 2010). However, the same trend also occurred in the northern climate; plants from the southern populations showed the greatest values, clouding the issue of adaptation considerably!

Details of the genomic basis for the local adaptation of Sitka spruce to temperature-based climatic variables are beginning to become available as researchers "genotype" individual plants for hundreds of SNPs at hundreds of gene loci (Holliday et al. 2010). The researchers used a common garden in British Colombia, Canada, to grow 410 plants from 14 source populations spanning the species' range on the northwestern coast of North America. DNA was extracted from bud tissue and candidate genes, chosen on the basis of prior work on gene expression in Sitka spruce (Holliday et al. 2008), were amplified and SNPs were genotyped. An association analysis based on the candidate genes (Neale & Savolainen 2004) was used to identify allelic variants

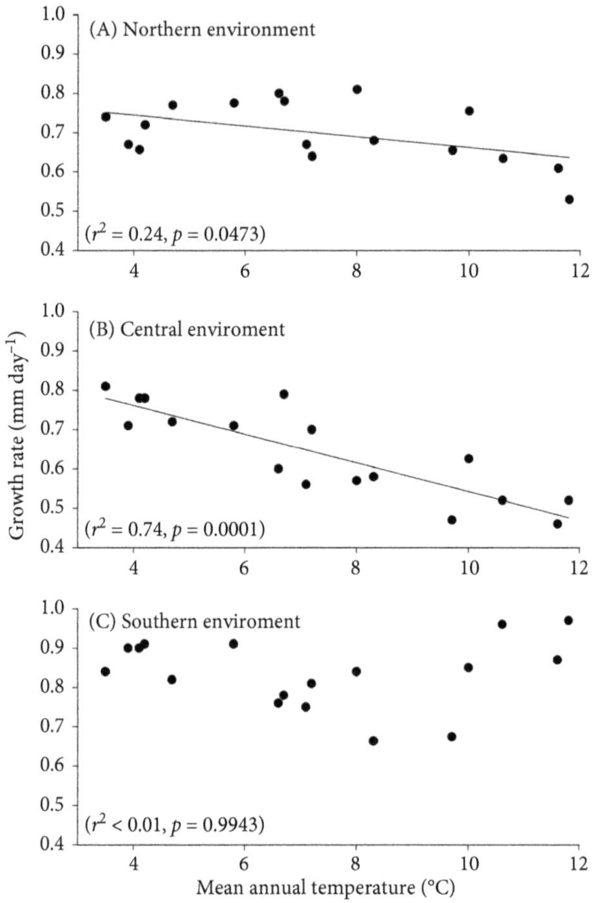

FIGURE 6.6
(A–C) Relation between daily growth rate of Sitka spruce populations and the mean annual temperature at the population site of origin in environments simulating northern (A), central (B), and southern (C) conditions. Drawn using data presented in Mimura and Aitken (2010).

that explained significant fractions of the phenotypic variance in bud set and cold hardiness. Correlations of SNP allele frequencies with several climatic variables were also explored.

Significant associations were found between 35 SNPs (distributed among 28 candidate gene loci) and one or both of the two phenotypic traits (bud set and cold hardiness). Putative identification of the candidate loci was made through comparison with published gene banks. Several of these loci, their products, and their possible functions, are listed in Table 6.2. Although individual SNPs explained only a small, but significant, proportion of the phenotypic variance in either trait, collectively they explained 34.4% and 28.1% of the variance in bud set and cold hardiness, respectively (Holliday et al. 2010). The allele frequencies of several SNPs showed significant correlations with climate-related variables (Table 6.2), especially MAT (but not annual precipitation). Other genomic studies have also found correlations between the frequencies of AFLP markers and MAT in European beech (*Fagus sylvatica* [Jump et al. 2006]), and between SNP markers and multivariate metrics of climate in loblolly

Table 6.2 Some of the SNP markers in Sitka spruce that had significant effects on phenotypic variation in bud set and/or cold hardiness, and the associated gene loci.

Locus	*Number of SNPs*	*Gene product*	*Putative function*	*Variance explained (%)*		*Correlated climatic variables*
				Bud set	*Cold hardiness*	
co	1	Regulatory protein (constans)	Promotion of flowering, season growth cessation	1.1	—	MAT, MCMT, DD
cry2	2	Flavoprotein (cryptochrome)	Blue light photoreceptor, inhibition of stem elongation	1.1	1.5	—
expl2	1	Protein (expansin)	Loosening of cell walls	—	2.3	—
hta3	2	Regulatory protein (histone)	Epigenetic modification of gene activity	1.0–1.3	1.4	MAT, MCMT, DD, MWMT
per6	1	Enzyme (peroxidase)	Assembly of cell walls	5.1	2.6	MWMT
xth1	4	Enzyme (xyloglucan endotransglucosylase/hydrolase)	Assembly of cell walls	1.4–4.3	1.1–5.4	MAT, MCMT, DD
All 28 loci (cumulative)	35	Various proteins	Various functions	34.4	28.1	Variable

The percent of the phenotypic variance explained by each marker is indicated (when significant) along with the climatic variables with which their allele frequencies correlated significantly. Adapted from Holliday et al. (2010). Putative functions are simplified from Taiz and Zeiger (2010). DD, degree-days warmer than 5° C; MAT, mean annual temperature; MCMT, mean temperature of coldest month; MWMT, mean temperature of warmest month; SNP, single nucleotide polymorphism.

pine (*Pinus taeda* [Eckert et al. 2010a]). And in the molecular workhorse *Arabidopsis thaliana*, genomewide scans of more than 200,000 SNPs have revealed many loci that reflect adaptation to a diversity of climatic variables across its native range (Hancock et al. 2011).

There is now good evidence that environmental variables that are part of what we call "climate" found at the different sites from which populations are collected can be correlated with the allele frequencies of a variety of molecular markers (Nevo et al. 1988; Gupta et al. 2002; Ramírez-Valiente et al. 2010). In addition to the examples presented here, several other studies were reviewed in Section 5.3.1 to illustrate one molecular approach used to study selection and adaptation, especially in tree species. Of course, not all climate-related correlations with molecular markers, or ecological studies of climate adaptation, involve temperature variables; precipitation and drought are also critical components of climate, and they are explored next.

6.3.2 Precipitation, Drought, and Soil Water

It can be expected that a plant species will be exposed to much variability in precipitation across its range, and populations are likely to be subjected to differences in the quantity of rainfall received and its timing. The likelihood of lengthy dry periods (droughts) and their potential consequences for plant survival, growth, and reproduction differ from place to place. Of course, patterns in the amount and timing of precipitation greatly influence soil water availability, as do landscape topography and edaphic characteristics (e.g., soil texture). Probably more than any other aspect of climate, precipitation affects the soil directly and indirectly, which in turn affects the dynamics of plant populations. It should be recognized that climatic factors such as precipitation and temperature (and the likelihood of drought) covary in complex ways, and it can be very difficult, if not impossible, to separate the particular variables responsible for adaptive differentiation.

6.3.2.1 Ecological Approaches

Plant ecologists have generally studied population responses to soil water availability by subjecting plants from different sites to experimentally induced drought conditions in the greenhouse or field. These are essentially common garden experiments with a manipulated environmental variable (Section 3.4). In the perennial grass *Danthonia sericea*, Quinn (1975) showed that populations from drier upland sites differed significantly in seed germination, growth, and survival from populations from wet bog sites when exposed to different moisture levels in the greenhouse. Plants from drier sites were relatively intolerant of wet conditions. Significant differentiation of populations of *Eucalyptus viminalis* in Australia on dry granitic versus moist basaltic soils has been demonstrated when plants were exposed to experimental droughts (Ladiges 1976). The slower growing population from the granitic soil showed greater resistance to drought, which is presumed to be adaptive in its native habitat.

Two populations of the annual composite *Lasthenia californica* (Fig. 6.7) were collected from the dry upper part and moister lower part of a serpentine outcrop at Jasper Ridge Biological Preserve in California (Rajakaruna et al. 2003). Enough distinctive features between the two population types in allozymes and flavonoids, and fruit morphology, had been previously demonstrated so that two edaphic races were

FIGURE 6.7
Population of the annual composite *Lasthenia californica* at Jasper Ridge Biological Preserve, California. Photo courtesy of N. Rajakaruna.

distinguished in the species (Rajakaruna & Bohm 1999). Rajakaruna et al. (2003) collected seeds from ~100 plants of each race and, in a growth chamber, grew plants in a standard soil mixture and subjected them either to low, medium, or high watering treatments. A number of phenotypic traits were recorded, with the number of flowering heads used as a measure of fitness. There were highly significant effects of race ($F = 33.61$, $p < 0.001$) and treatment ($F = 53.33$, $p < 0.001$) on fitness. The interaction between race and treatment was also significant ($F = 11.61$, $p < 0.001$); the race from the drier area was able to maintain a greater reproductive output under low water availability relative to the race from the moister area (Fig. 6.8). However, differences in reproductive output between the races were not apparent in the other watering treatments (Fig. 6.8). Thus, the two races appeared to show genetic differentiation in relation to soil water conditions normally found at the field sites (Rajakaruna et al. 2003). Population differentiation in life history traits in relation to soil water conditions has also been shown in a greenhouse study of populations of *Plantago coronopus* from southern Spain (Hansen et al. 2013).

Ecological approaches, such as common gardens and reciprocal transplants, have been used to conduct investigations of population adaptation to soil moisture availability under field conditions. For example, Heschel et al. (2002) developed inbred lines of the annual *Impatiens capensis* derived from populations of wet and dry sites in Rhode Island and then planted them into two relatively dry sites (open and woodland). Selection analysis was used to determine selection differentials (S) for photosynthetic rate, stomatal conductance, and WUE. Relative fitness was based on the number of flowers and fruits produced. In both field gardens, selection favored increased photosynthetic rate ($S \pm SE = 0.12 \pm 0.04$ and 0.28 ± 0.15 for open

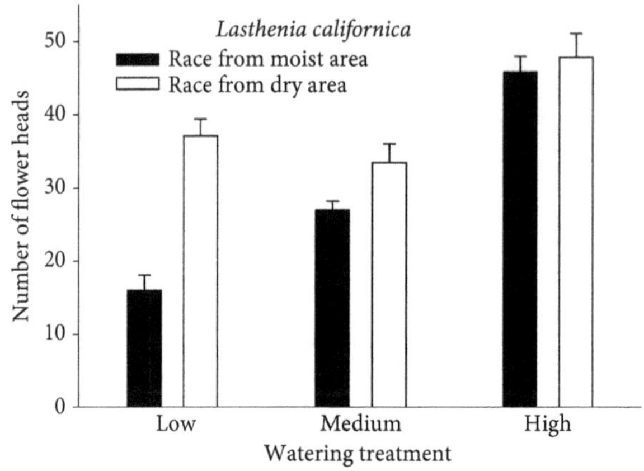

FIGURE 6.8
Mean number ± standard error of flower heads produced by two races of the annual composite *Lasthenia californica* when subjected to three experimental watering treatments. Redrawn from Rajakaruna et al. (2003).

and woodland, respectively) and greater WUE (0.08 ± 0.04 and 0.27 ± 0.15 for open and woodland, respectively). Mean WUE was greater for the population derived from plants of the dry site in both transplant gardens, suggesting adaptive population differentiation for more efficient water use under dry field conditions (Heschel et al. 2002).

In another common garden experiment, several ecotypes of four perennial grass species in Europe were planted under shelters in the field exposed to two precipitation and two temperature treatments (Beierkuhnlein et al. 2011). One precipitation treatment simulated extreme drought conditions. As with many recent studies of climate effects on plants, the objectives were couched in terms of predicting the future effects of climate change (discussed later in this chapter). Results showed the expected effect of drought and ecotype on biomass production over a four-month growing period (temperature effects were minor). More interesting were several interaction effects between ecotype and drought, indicating that the ability to regrow after drought showed ecotypic differentiation within species. Although it was hypothesized that ecotypes from southern sites would be better adapted to drought than local ecotypes near to the transplant garden (in Bayreuth, Germany), this was not the case. The authors concluded that, although there was population-level variation within species for responses to changed climatic conditions (especially drought), one could not predict performance based on the climate at a population's site of origin (Beierkuhnlein et al. 2011).

Surprisingly few ecological studies of plant adaptation to soil water availability per se have been conducted with reciprocal transplants among field sites. However, it should be recognized that many reciprocal transplant experiments are likely to have involved sites that differed in soil water content among other factors that may not have been identified or quantified. The following examples are studies with the explicit objective of examining growth responses to sites that differed in soil water conditions. The first example is the study of *Danthonia sericea* populations from dry upland and wet bog sites mentioned earlier in this section (Quinn 1975). In addition to the greenhouse experiment, in which moisture levels were manipulated, Quinn (1975) also conducted a reciprocal transplant experiment at an upland and a bog site in New Jersey. He presented evidence of local adaptation to dry soils, with two populations from dry upland sites showing much better survival at the upland site compared with two populations from the wet bog sites.

A second example of the reciprocal transplant approach used in relation to soil moisture variation involved three perennial grasses in four seasonal wetlands at Archbold Biological Station in Florida (Lovell & Menges 2013). All species were reciprocally transplanted among low-, middle-, and high-elevation zones (each dominated by one of the three species) that represented a gradient of decreasing soil moisture (from low to high zones). Dry-season drought causes a dramatic reduction in soil moisture in these wetlands (Lovell & Menges 2013). The species (*Panicum abscissum*) that normally dominates the driest, high-elevation zones showed the most effective phenotypic responses to soil moisture conditions and maintained greater leaf water content and root growth in all zones. However, the greatest growth did not occur in the native zone for each species, as greater soil moisture improved the growth of all species regardless of their zone of origin.

Reciprocal transplants of the bunchgrass *Festuca lenensis* along a topographic gradient in northern Mongolia provided evidence of local adaptation to soil water conditions (Liancourt et al. 2013). These two sites were separated only by 300 m, and thus climatic conditions could be presumed to be the same; one site was in the drier, upper part and the other was in the wetter, lower part of the gradient. Differential responses to supplemental watering and neighbor removal also occurred between the two populations, which the authors referred to as distinct ecotypes (Liancourt et al. 2013). As in the garden study by Beierkuhnlein et al. (2011), the importance of ecotypic variation to predicting species-specific effects of climate change was evident.

6.3.2.2 Molecular Approaches

A fair amount of research effort has been devoted to identification of the molecular targets of selection for drought tolerance (Juenger 2013), and several examples were provided in Chapter 5 to illustrate molecular approaches to the study of adaptation (Knight et al. 2006; Eckert et al. 2010b; Grivet et al. 2011). Only a few additional studies are described here.

In balsam poplar (*Populus balsamifera*), 46 SNPs from 14 candidate genes showed evidence of local adaptation to climate in trees sampled from 31 populations across its North American range (Keller et al. 2012). Covariance between 12 and 11 candidate SNP frequencies were associated significantly with MAT and precipitation, respectively. Candidate genes associated specifically with adaptation to dry conditions have been identified in several plant species (Stinchcombe et al. 2004; Knight et al. 2006; Eveno et al. 2008; Eckert et al. 2010b; Hancock et al. 2011). Although it is not always clear which physiological role the molecular products of these genes play in drought tolerance, there is solid evidence that selection has shaped the variation patterns found at the molecular genetic level.

Eveno et al. (2008) analyzed 11 polymorphic candidate genes in the pine *Pinus pinaster* that had previously been shown to be potentially important to drought tolerance. Ten populations were used, distributed across sites along the species' range in France, Spain, and northern Africa, that differed in mean annual precipitation, ranging from 348 to 1,343 mm. Neutrality tests based on F_{ST} distribution of SNPs followed by the detection of outlier loci were used to find the molecular signatures of natural selection (see Section 5.3.2 and Fig. 5.4). Two genes, with a role in drought response that is not known, showed greater differentiation (F_{ST}) than expected and appeared to have been shaped by selection acting differentially among populations. Three additional genes showing lower differentiation than expected may have been under stabilizing selection; two of these coded for dehydrin proteins known to be important to protecting cells from dehydration in a number of species. The colossal (21-page), richly detailed paper by Eveno et al. (2008) is exemplary in its application of molecular methods to the investigation of adaptive genetic differentiation in a well-studied tree species.

6.3.3 Climate Change

As one might predict, studies of the evolutionary ecology of plants in relation to climate have become increasingly relevant as a result of a growing interest in predicting the consequences of global climate change for microevolution and adaptation (Geber & Dawson 1993; Potvin & Tonsignant 1996; Ward et al. 2000; Etterson 2004a, b; Jump & Peñuelas 2005; Aitken et al. 2008; Briggs 2009; Kim & Donohue 2013). The agent responsible for the change is increasing atmospheric carbon dioxide (CO_2) levels resulting from anthropogenic activities worldwide. Although complex and variable geographically, the expected climate changes involve increased temperatures and altered patterns of precipitation, including an increased potential for drought in some regions (IPCC 2013).

Several reviews have explored the possible ecological and evolutionary responses of animal and plant populations to climate change (Bradshaw & McNeilly 1991; Geber & Dawson 1993; Pulido & Berthold 2004; Jump & Peñuelas 2005; Parmesan 2006; Aitken et al. 2008; Briggs 2009; Hoffmann & Sgrò 2011). One of the long-standing questions for plant ecologists has been whether species have the requisite genetic variation to enable microevolutionary responses to climate change, as emphasized in the early review by Bradshaw and McNeilly (1991). What is clear, as discussed in earlier parts of this chapter and in several reviews (Geber & Dawson 1993; Jump & Peñuelas 2005; Reusch & Wood 2007; Aitken et al. 2008; Shaw & Etterson 2012; Weinig et al. 2014), is that the populations of many plant species do show adaptive differentiation at both the phenotypic and molecular genetic levels in relation to climatic regimes, expressed as long-term patterns in temperature and precipitation at local sites. The focus here is on CO_2 specifically and whether plants exhibit genetically based variation in their responses to elevated CO_2 within populations.

6.3.3.1 Genotypic Responses to Elevated CO_2

A few early studies reported genotype-specific effects of elevated CO_2 levels on life history traits. Paternal families of the weedy annual *Raphanus raphanistrum* showed variable responses of flower and seed production per plant to elevated CO_s (Curtis et al. 1994). Although the means for both fitness traits were significantly greater at greater amounts of CO_2, there was considerable crossing of the norms of reaction for the different families (Fig. 6.9A), suggesting quantitative genetic variation for response to elevated CO_2 levels. Significant genotypic variation in fruit mass of the

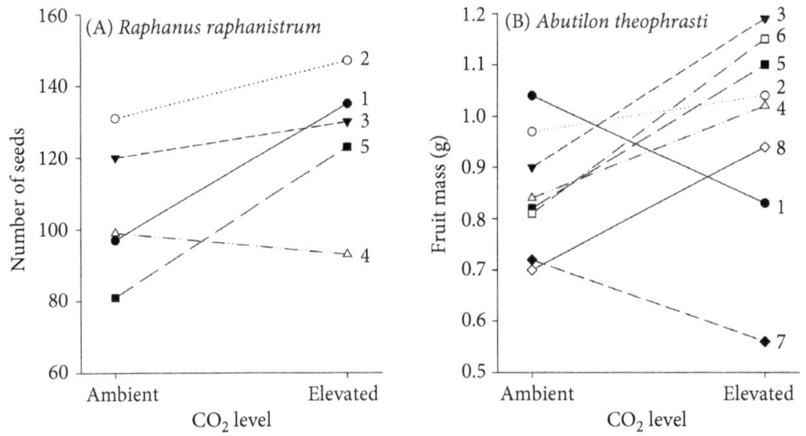

FIGURE 6.9

Norms of reaction for reproductive responses to elevated CO_2 in two weedy annual species. (A) Average number of seeds per plant for five paternal families of wild radish (n = 28–32 [Curtis et al. 1994]). (B) Average fruit mass of eight replicated genotypes of velvetleaf (n = 12 [Bazzaz et al. 1995]).

annual *Abutilon theophrasti* (Fig. 6.9B) and vegetative mass for seedlings of the tree *Betula alleghaniensis* in relation to elevated CO_2 was also found in another experiment (Bazzaz et al. 1995).

Research on *Arabidopsis thaliana* provided further evidence for genotypically based responses of fitness to changes in CO_2 levels (Ward & Strain 1997). Several significant genotype-by-CO_2 interactions for growth and allocation traits have also been found in the tree *Populus tremuloides* (Lindroth et al. 2001). A later review concluded that, collectively, the norm-of-reaction studies have supported the idea that changes in atmospheric CO_2 levels are likely to act as a significant selection agent for natural populations (Ward & Kelly 2004).

However, the picture appears to be more complicated. For some herbaceous species, genotype-by-CO_2 interactions have been found to be relatively weak or nonexistent (Volk & Körner 2001; Roumet et al. 2002; Wieneke et al. 2004). In a table compiled by Lau et al. (2007) that summarized results from 24 studies, there were only nine plant species in which significant genotype-by-CO_2 interactions were reported for a trait, and 23 species for which the interactions were insignificant. Note that four species were in both categories, including *Arabidopsis thaliana*. In addition, more recent selection-based analyses have not always supported the premise that CO_2 is a potent selection agent (Lau et al. 2007, 2010; Steinger et al. 2007).

6.3.3.2 Selection Analyses

Artificial selection for increased reproductive fitness over multiple generations under low or elevated levels of CO_2 has been attempted with a few annual herbs (Potvin & Tousignant 1996; Ward et al. 2000). The earlier study used rapid-cycling populations of *Brassica juncea* propagated over seven generations of selection; at the end, plants were grown reciprocally in control and elevated CO_2 environments (Potvin & Tousignant 1996). There was no evidence that selected populations performed best relative to control populations in either environment. Ward et al. (2000) selected for high fecundity in *Arabidopsis thaliana* grown for five generations at low or high CO_2 conditions. They then reciprocally planted populations from each selection regime into each of the two environments to test for genetic adaptation. They found that plants selected at high CO_2 did not show significantly greater fecundity when grown at high CO_2 levels than plants selected at low CO_2. However, plants selected at low CO_2 *did* show the greatest fecundity under low CO_2 levels (Ward et al. 2000).

A more complex, but perhaps more realistic, experiment involved growing *Brassica napans* (canola) for four generations at two CO_2 levels and two temperatures (Frenck et al. 2013). In this research, seeds were chosen randomly for each new generation, and thus artificial selection for greater fecundity was not used (this design has been called a "controlled natural selection" experiment [Reusch & Wood 2007, p. 3976]). Four accessions of *B. napans* were used to generate replicated lineages. After the four generations, lineages exposed to each selection regime were compared with a control lineage not subjected to environmental manipulations. The researchers reported a strong directional response for increased biomass and reproductive output under elevated CO_2 levels compared with ambient CO_2 (Frenck et al. 2013). This growth-chamber experiment is one of the first to show a strong differential response of a plant species to selection imposed solely by environments that differed in CO_2 levels.

Quantitative genetic parameters and measures of phenotypic selection have been provided in some studies in which plant populations were subjected to two levels of CO_2 (Bazzaz et al. 1995; Lau et al. 2007; Steinger et al. 2007; Lau et al. 2010; additional references in Shaw and Etterson [2012]). In general, results have not strongly supported the hypothesis that elevated CO_2 levels would result in different patterns of selective response (relative to ambient CO_2), as shown by the following example.

Lau et al. (2007) grew *Arabidopsis thaliana* outdoors in a free-air CO_2 enrichment facility at ambient and elevated CO_2 levels. This study involved an admirable sample size; in each treatment, 7 to 18 plants were grown per each of 164 recombinant inbred lineages. This supplied a broad range of genetic and phenotypic variation available for natural selection. Broad-sense heritabilities, selection differentials, and selection gradients were estimated for populations through a complete life cycle in each environment. Relative fitness was based on per-capita fruit production. As expected, elevated CO_2 levels increased plant biomass and fruit production significantly. However, heritabilities of 11 phenotypic traits did not differ between ambient and elevated CO_2 environments. This suggests that the rate of future population responses to selection would not change appreciably at elevated CO_2. In addition, although highly significant directional selection was apparent for most traits, the CO_2 environment has no effect on the patterns of selection (Table 6.3). There were strong genetic correlations across environments, indicating that the same genotypes were favored selectively under both ambient and elevated CO_2 levels. The authors concluded that future responses of *A. thaliana* to elevated CO_2 levels were more likely to be ecological in nature rather than evolutionary (Lau et al. 2007).

Phenotypic selection has also been evaluated for *Arabidopsis thaliana* under both intraspecific and interspecific competition in the free-air CO_2 enrichment facility in Minnesota (Lau et al. 2010). The direction of selection was consistent for a number of phenotypic traits, but their magnitude varied with competition treatment. However, selection differentials (S) and gradients (β) were broadly similar in ambient and elevated CO_2 conditions within a treatment. For example, for biomass in intraspecific

Table 6.3 Similarity in standardized selection differentials (S) and selection gradients (β) for three phenotypic traits of *Arabidopsis thaliana* grown outdoors at ambient or elevated levels of carbon dioxide.

Trait	*S value*		*β value*	
	Ambient	*Elevated*	*Ambient*	*Elevated*
Dry mass	0.25	0.25	0.30	0.24
Date of flowering	−0.21	−0.21	−0.11	−0.12
Rosette size	0.16	0.15	−0.12	−0.08

All values are significantly different from zero ($p < 0.05$) after Bonferroni correction.

Source: Lau, J.A., Shaw, R.G., Reich, P.B., Shaw, F.H. & Tiffin, P. 2007. Strong ecological but weak evolutionary effects of elevated CO_2 on a recombinant inbred population of *Arabidopsis thaliana*. *New Phytol.* 175: 351–362.

competition, β equaled 0.16 at ambient CO_2 and 0.17 at elevated CO_2 (the corresponding selection differentials were S = 0.24 and S = 0.18 for ambient and elevated CO_2, respectively). Although there was little doubt that elevated CO_2 tended to weaken the negative effects of competition on *A. thaliana* relative to that of ambient conditions (Lau et al. 2010), much remains to be learned about the general consequences of elevated CO_2 to future population microevolution in natural communities where plant–plant interactions are the norm (Brooker 2006).

6.3.3.3 Molecular Approaches

In an article titled "Molecular Ecology of Global Change," Reusch and Wood (2007) provided an overview of the molecular genetic basis of evolutionary changes that underlie organismal responses to a changing climate. They provided examples of selection experiments designed to simulate global change, mostly in insects and plants, and considered the utility of ecogenomics approaches to uncover the genetic basis of adaptation. Because molecular approaches to investigating adaptation to temperature and precipitation were described in previous sections (6.3.1 and 6.3.2), here the focus is on genomic studies of plants specifically in relation to elevated CO_2.

The timing of flowering is known to be an ecologically important trait that is readily altered by elevated CO_2 levels (Springer & Ward 2007). In an artificial selection experiment, for example, Burgess et al. (2007) showed how divergent selection could result in early- and late-flowering populations in only three generations in the herb *Campanulastrum americanum*. Using *Arabidopsis thaliana*, Ward et al. (2000) generated genotypes selected at low or high CO_2 that differed in the timing of flowering at elevated CO_2 levels; these were used in a later study to explore the expression of genes known to be important to flower initiation in *Arabidopsis* (Springer et al. 2008). The researchers found that delayed flowering at elevated CO_2 levels was likely a result of continuous expression of a floral repressor gene in the late-flowering genotype. They speculated that excess leaf sugars that accumulated in this genotype at elevated CO_2 levels may have influenced the expression of this gene, as reported by past research (Springer & Ward 2007). Changes in gene expression in response to

Table 6.4 The mean percent of the phenotypic variance in seven traits of a hybrid popular (*Populus*) explained by mapped QTL.

Trait	*Ambient CO_2*		*Elevated CO_2*	
	No. of QTLs	*Variance (%)*	*No. of QTLs*	*Variance (%)*
No. of branches	2	3.2	2	5.2
Primary root density	1	2.7	4	3.6
Root growth rate	2	5.2	1	3.4
Secondary root density	3	4.3	3	3.0
Stem diameter	5	3.1	4	4.5
Stem height	2	4.7	2	6.1
Stem volume	2	4.1	2	5.0
Grand mean	2.4	3.9	2.6	4.4

Plants were grown at ambient or elevated levels of carbon dioxide (CO_2). Adapted from Rae et al. (2007). QTL, quantitative trait loci.

climate-related factors have also been reported in other species (Taylor et al. 2005; Travers et al. 2010).

Molecular analysis of QTLs that could improve growth in response to elevated CO_2 levels, and thereby be selected for, have been studied in hybrids of poplar (*Populus*). Rae et al. (2006) identified dozens of QTLs that may be candidate genes important to the future adaptation of tree populations (Aitken et al. 2008). QTLs were also mapped for various growth-related traits in a hybrid poplar exposed to ambient or elevated CO_2 levels (Rae et al. 2007). The specific QTL detected for each trait often differed between plants in the two environments, and the proportion of the phenotypic variance explained by different QTLs was relatively low (Table 6.4). However, this study has clearly quantified molecular genetic variation in relation to elevated CO_2 levels and identified some of the loci of small effect that could evolve as selection favors specific variants in an atmosphere enriched with CO_2.

6.4 OTHER ABIOTIC FACTORS

There are other abiotic factors that can act as selection agents that are not easily categorized as climatic or edaphic in nature. Geographic variation in light regimes provides major differences in photoperiod, the daily cycle of light and dark periods, which are an aspect of regional climate, varying predictably along latitudinal gradients. In addition, at a local scale, individuals and populations can be exposed to variable light availability, depending on the extent to which they are shaded by other members of the plant community. These local light conditions are an aspect of fine-scale microclimate that can select for specific adaptive features of plant species (Givnish 1979, 1987; Dudley & Schmitt 1995; de Casas et al. 2011; Neufeld & Young 2003). Another agent of selection in some habitats that can affect growth and reproduction is salinity, whether in the soil solution or airborne as salt sprays in coastal ecosystems (Boyce 1954; Sykes & Wilson 1988; Cheplick & White 2002; Maun 2009). This chapter concludes with a brief consideration of how populations can evolve in relation to conditions of light or salinity.

6.4.1 Light

Morphological and physiological adaptation of plants in relation to sunny and shady environments has long been recognized (Björkman & Holmgren 1966; Givnish 1979, 1987). Plants adapted to low light availability typically exhibit a greater leaf area, thinner leaves of lower mass, and greater chlorophyll content than plants adapted to sunnier conditions. Populations in the herbaceous layer under the closed canopy of a forest clearly must have adaptive features that allow survival and growth in a highly variable light environment (Watling & Press 2000; Neufeld & Young 2003). For some herbaceous species, stem elongation in dense, competing populations can be an adaptive mechanism to avoid shading and increase light capture (Dudley & Schmitt 1996; Donohue et al. 2000a). Differentiation among populations from high and low light environments has been demonstrated in a number of species, mostly using common garden techniques (e.g., Björkman & Holmgren 1966; Leclerc & Blaise 1991; Dudley & Schmitt 1995; Cheplick 2005; de Casas et al. 2011; Cheplick 2014).

Phenotypic selection analysis (Lande & Arnold 1983) was applied to several leaf traits for two species of *Luzuriaga* that differ in their degree of shade tolerance (Gianoli & Saldaña 2013). The species are both climbing plants that grow in the evergreen temperate rainforest of southern Chile. In the highly shade-tolerant *Luzuriaga radicans*, there was positive selection on leaf area (selection gradient $\beta = 0.48 \pm 0.16$ [SE], $p < 0.01$) and negative selection on dark respiration rate ($\beta = -1.31 \pm 0.22$, $p < 0.001$), indicating the fitness benefits of maximizing light capture while minimizing respiratory costs in the low light of the forest understory.

Local adaptation of the selfing annual *Claytonia perfoliata* to sunny and shady environments in California has been demonstrated using reciprocal transplants (McIntyre & Strauss 2014). In an oak understory, plants with broader leaves (the form they normally take in shady conditions) showed greater fitness than the more linear leaves found in plants from open grassland. The reverse was true in the sunny grassland habitat. Phenotypic selection analyses showed a consistent pattern in selection gradients for leaf morphology in the two habitats across two years. Linear selection gradients were positive for leaf broadness under the oak canopy ($\beta = 0.34$ and 0.40 in first and second years, respectively), but were negative in the open grassland ($\beta = -0.45$ and -0.25 in the two years, respectively).

In the wild sea beet *Beta vulgaris* subsp. *maritima*, van Dijk and Hautekèete (2007) performed an artificial selection experiment for sensitivity to day length, starting with variable populations collected from southwest France. They were able to reduce the day length period needed for floral induction from more than 13 hours to less than 11 hours after nine generations of selection. Thus, in this outcrossing—perennial long-day species—it was demonstrated that it was possible to select for genetically based changes in flowering time in relation to photoperiod.

Another selection experiment used the outcrossing annual *Sinapsis arvensis* (wild turnip) grown under a neutral shade canopy or in full light for three generations (Stanton et al. 2004). After an intervening fourth generation under standard conditions, seedlings from both selection regimes were transplanted to a field site in California and subjected to partial shade versus full-sun (at two densities). A significant selection treatment-by-light condition interaction was detected for lifetime seed production ($F = 7.06$, $p = 0.0014$); plants from the high-light-selected population had reduced seed production under partial shade in the field compared with the shade-selected population (Stanton et al. 2004).

6.4.2 Salt

Airborne salt sprays in coastal habitats, and soil salinity in marshes, arid regions, and elsewhere, can have major effects on plant survival, growth, and reproduction, thereby acting as agents of natural selection (Boyce 1954; Aston & Bradshaw 1966; Cheplick & Demetri 1999; Lowry et al. 2009; Maun 2009). Evolutionarily significant quantitative genetic variation within and among plant populations for salt tolerance has been reported (Hannon & Bradshaw 1968; Ashraf et al. 1986; Hester et al. 1996; Lowry et al. 2008). For example, genotypes of the coastal grass *Spartina patens* were collected from 19 stands of marshes in Florida, Louisiana, and Texas, and were subjected experimentally to increasing levels of soil salinity. Highly significant differences in the lethal salinity level, as well as dry mass accumulation, were found between genotypes (Hester et al. 1996). This and other studies demonstrate that adaptation to salinity can occur in populations that are sufficiently variable (Rozema et al. 1985; Wang & Redmann 1996; Lowry et al. 2008; Maun 2009; Cordeiro et al. 2014).

The yellow monkeyflower (*Mimulus guttatus*) in the western United States has been used in an extensive set of investigations of adaptive tolerance to salt, involving common gardens, reciprocal transplants, salt spray trials, and molecular analysis of QTLs (Hall & Willis 2006; Lowry et al. 2008, 2009). Coastal and inland populations of *M. guttatus* appear to constitute reproductively isolated races distinguished by pronounced differences in life history, flowering phenology, and tolerance of salt sprays. In addition, there is strong molecular genetic divergence indicated by pairwise comparisons of populations of the two races (mean F_{ST} = 0.48 [Lowry et al. 2008]). Reciprocal transplants of seedling populations between coastal and inland habitats in California revealed that selection favored plants in their native habitats. Three QTLs appeared to be important to salt tolerance and had a significant positive effect on seed production in the coastal habitats (Lowry et al. 2009).

Table 6.5 The mean and total (summed) percent of the variance in several ion uptake measures and survival duration explained by several QTL markers in the salt-tolerant hybrid sunflower species *Helianthus paradoxus* transplanted into its usual saline habitat in New Mexico.

Trait	*No. of QTLs*	*Variance explained (%)*	
		Mean	*Total (Σ)*
Calcium uptake	3	26.0	78.0
Magnesium uptake	4	15.8	63.0
Potassium uptake	2	25.0	50.0
Sodium uptake	4	16.3	65.0
Sulfur uptake	1	21.0	21.0
Survival duration (days)	3	12.7	38.0

QTL, quantitative trait locus.

Source: Lexer, C., Welch, M.E., Durphy, J.L. & Rieseberg, L.H. 2003. Natural selection for salt tolerance quantitative trait loci (QTLs) in wild sunflower hybrids: implications for the origin of *Helianthus paradoxus*, a diploid hybrid species. *Mol. Ecol.* 12: 1225–1235.

Molecular analyses have revealed additional gene loci important to salt tolerance in other species (Lexer et al. 2003b; Puerma & Aguade 2013). In salt marsh habitats, several QTLs apparently influence ion uptake and thereby contribute to the success of the hybrid sunflower *Helianthus paradoxus* in saline soils (Lexer et al. 2003b, 2004). Some of the QTLs individually explained relatively high fractions of the variance in the uptake of calcium, potassium, and other ions (Table 6.5). Three QTLs important to survival duration (Table 6.5) were mapped to the same genomic regions as QTLs for the uptake of specific ions (Lexer et al. 2003b). Thus, these genomic regions may be critical to the adaptation of this species to the salt marsh habitat.

6.5 WRAP-UP

This chapter provided multiple examples of the diverse role of abiotic agents of selection in the evolutionary ecology of many plant species. These abiotic factors can be natural or anthropogenic in origin and vary greatly across scales—from local microsites to regional landscapes. The plethora of abiotic factors overlap greatly and interact in complex ways in terms of their effects on plant populations, rendering the categorization attempted in this chapter somewhat contrived and artificial. Nonetheless, there is abundant evidence that populations can differentiate in morphological, physiological, and molecular traits in response to climatic and edaphic factors, and show patterns consistent with the hypothesis of local adaptation brought about by habitat-mediated selection.

7 Biotic Interactions I: Competition and Facilitation

7.1 THE UBIQUITY OF BIOTIC INTERACTIONS

As sessile organisms, all plants interact with abiotic and biotic factors in their local environment. These factors act as agents of natural selection through their impact on the survival, growth, and reproduction of individuals in a population. Like the abiotic factors covered in Chapter 6, biotic factors show great spatial and temporal variation, and can interact in complex ways with variable ecological outcomes (Urban 2011; Chamberlain et al. 2014). Examples of biotic interactions involving plants include competition, facilitation, herbivory, pollination, and symbiosis. These were loosely classified in Table 2.1 based on whether the interactions were with macroscopic or microscopic organisms.

This chapter focuses specifically on plant–plant interactions, both negative (competition) and positive (facilitation). Subsequent chapters examine microbial symbionts (Chapter 8) and animal pollinators, herbivores, and seed predators (Chapter 9) as agents of natural selection with profound consequences for the evolutionary ecology of plant populations.

7.2 COMPETITION AND COMPETITIVE ABILITY

Competition is a pervasive interaction in plant communities and has been studied extensively for decades (Clements et al. 1929; Harper 1977; Antonovics & Levin 1980; Grace & Tilman 1990). It generally has a negative effect on the "big three" attributes of life history: survival, growth, and reproduction (Weiner 1988; Aarssen 1989; Aarssen & Keogh 2002; Aarssen 2005). Although many definitions of competition have been proposed, the one used by Keddy (2001, p. 5) is as good as any and encapsulates the major aspects of the concept: "[t]he negative effects that one organism has upon another by consuming, or controlling access to, a resource that is limited in availability." Plants require the same basic set of resources—light, water, and mineral nutrients—all of which are "consumed" but can be limited in any one environment. Also, plants cannot move to escape competitive stress. Thus, plants may be particularly susceptible to competition relative to other types of organisms.

Neighbors act as biotic agents of selection, and phenotypic traits that improve the production of offspring should evolve in populations under competitive stress (Aarssen 2005). That is, in a genetically variable population in which interactions occur with individuals of their own (intraspecific) and/or different (interspecific) species, genotypes with good competitive ability should be favored by natural selection. But, what is competitive ability in plants? This somewhat nebulous concept clearly is a composite of many possible physiological and morphological attributes that collectively improve Darwinian fitness under competitive conditions (Sakai 1961; Aarssen 1989; Goldberg 1996; Aarssen & Keogh 2002; Wang et al. 2010). Although the majority of studies have used growth traits (e.g., biomass) to assess competitive ability, survival and reproduction should also be included to encompass the "three fundamental components of competitive ability" (Aarssen & Keogh 2002, p. 535). In a thorough overview of competitive ability in plants, Aarssen (1989) identified 13 primary traits with direct relevance, along with 14 secondary traits and 15 tertiary traits that affect competitive ability indirectly (via their effects on the primary traits)! It is not difficult to imagine how many of these traits might improve the competitive ability of a genotype, such as having a relatively greater height, leaf area, or root density. Clearly, the distributions of phenotypic values for any traits that influence the three components of competitive ability in a population are likely to be shaped by natural selection.

To quantify the generally negative response of an individual to the presence of neighbors, one must standardize performance under competition to performance in the absence of competition (Goldberg 1996). Relative competition intensity (RCI) provides an easy way to gauge the extent to which performance of a target individual (or group) is reduced under a particular competitive regime (Grace 1995). It is expressed as

$$\mathrm{RCI} = \frac{P_{(-C)} - P_{(+C)}}{P_{(-C)}},$$

where P is performance, which can be any measured trait (such as biomass or fecundity) that shows different phenotypic values without ($-C$) versus with ($+C$) competition. This index normally ranges between zero and one, unless a plant performed better when interacting with neighbors (possibly a case of facilitation, not competition). It stands to reason that the RCI would be less for a competitively superior phenotype and more for a competitively inferior phenotype. Therefore, to turn this around and view it from the perspective of growth and reproduction under competition, one can subtract RCI from unity and consider this value to represent a measure of relative competitive performance (RCP):

$$\mathrm{RCP} = 1 - \mathrm{RCI}.$$

The RCP represents a measure of the extent to which a target plant is able to mitigate the negative effects of competition on its performance and is, in essence, an indication of competitive ability. In the next section, the RCP is used along with norms of reaction for fitness to illustrate genetic variation in competitive ability and genotype-by-competition interactions for several species.

7.3 GENETIC VARIATION IN COMPETITIVE PERFORMANCE

For a population to respond in an adaptive manner to the selection pressure imposed by competitors, there must be heritable variation in the traits that matter to competitive ability. Of course, there is abundant evidence that traits related to size and fecundity often show genetic variation in plant populations (Mazer & LeBuhn 1999), and significant effects of natural selection on many phenotypic traits are well documented (Chapter 2). The question here is: Are there significant differences among variable genotypes in their growth and reproductive responses to competitive stress? In other words, are there genotype (or family)-by-environment interactions for relative fitness when the environmental variation is strictly the result of differences in density?

Before reviewing some of the studies that have addressed these issues, I'd like to orient our thinking on the evolutionary effects of competition by presenting hypothetical populations of 10 variable genotypes growing at low or high density (Fig. 7.1). In terms of a practical experiment, each genotype replicated among the

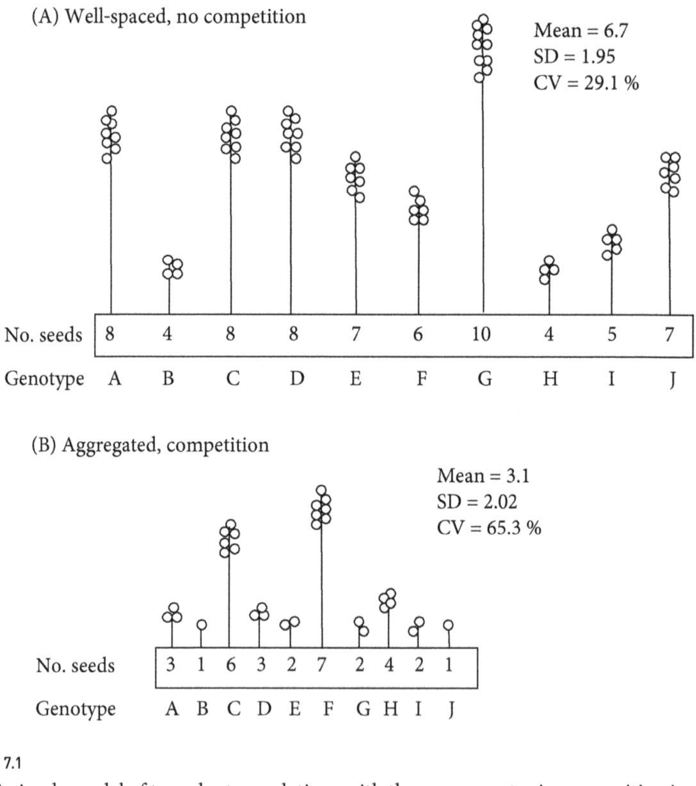

FIGURE 7.1
(A, B) A simple model of two plant populations with the same genotypic composition in a noncompetitive (A) or competitive (B) environment. The number of seeds produced is given for 10 genotypes (A–J), along with the mean, standard deviation (SD), and coefficient of variation (CV) for each population.

two densities may be genetically identical copies generated by separating ramets from the genet of a clonal plant (e.g., Goldberg 1988) or by cloning buds of an annual plant (e.g., Thomas & Bazzaz 1993). Alternatively, the genotypes may represent genetically cohesive sibling families collected as seeds from distinct maternal parents (e.g., Shaw 1986). In any event, the genotypes show variation in fecundity both in the well-spaced noncompetitive condition (Fig. 7.1A) and in the aggregated competitive condition (Fig. 7.1B). As expected, mean fecundity is greatly reduced in the aggregated population. The relative variation of fecundity in each population is easy to assess by calculating the coefficient of variation (CV), which is the SD divided by the mean. Typically, variation in size and fecundity (which depends on size) tends to increase with increasing density in competing monocultures (Weiner & Thomas 1986; Thomas & Bazzaz 1993). I have purposefully made this occur in these two imaginary populations; in the competitive environment, most genotypes have produced very few seeds, whereas two genotypes have produced most of the seeds (C and F in Fig. 7.1B produced 19% and 23%, respectively, of the population's seeds). Thus, the CV is 65.3% in the competitive population compared with 29.1% in the noncompetitive population. The evolutionary effect of this marked inequality in fecundity is that the alleles of these two genotypes are well represented in the gene pool of the next generation simply because the genotypes show high fecundity at high density (i.e., a high competitive ability). We presume that unique allelic variants contributed to this ability to maintain high fecundity under competitive conditions.

An example of an intraspecific competition experiment in which variation in fecundity increased with density is from research of a former graduate student (Jennifer Fox) on the annual *Microstegium vimineum* (Table 7.1). This invasive grass can form dense monocultures in the deciduous forest understory in the eastern United States (Cheplick 2005), and thus intraspecific interactions are common in nature. Seedlings were planted at four densities and were grown in a greenhouse until normal senescence in autumn (Cheplick & Fox 2011). Increasing density led to the expected decrease in the seed output of individuals and a marked increase in variation as quantified by CVs (Table 7.1).

A visual comparison of Figure 7.1A and Figure 7.1B reveals that the genotypes that were most fecund at high density were not necessarily the most fecund at low density. In other words, there were G × E interactions across the two densities.

Table 7.1 Mean, standard deviation (SD), and coefficient of variation (CV) for the number of seeds matured per plant in a greenhouse intraspecific competition experiment with the annual grass *Microstegium vimineum*.

Density per pot	*Density per m²*	*n*	*Mean*	*SD*	*CV (%)*
1	105	10	219.8	63.92	29.1
5	526	12	45.1	24.10	53.4
9	947	12	36.2	32.66	90.3
13	1368	12	28.7	36.79	127.9

Seedlings were planted at four densities in round pots with a diameter of 11 cm. Data are from Cheplick and Fox (2011).

This result can be depicted in norm-of-reaction diagrams (Mazer & Schick 1991a; Thomas & Bazzaz 1993; Cheplick & Gutierrez 2000; Donohue et al. 2000b; Cheplick 2003). This was done for the relative fitness of the 10 genotypes under low and high density. Relative fitness of a genotype was calculated as its fecundity divided by the mean fecundity of the 10 genotypes in each population. The diagram shows substantial crossing in the norms of reaction (Fig. 7.2) that would most likely stem from a statistically significant genotype-by-density interaction.

These sorts of genotype (or family)-by-density interactions for phenotypic traits have been demonstrated in some competition/density experiments (Shaw 1986; Mazer & Schick 1991a; Thomas & Bazzaz 1993; Vavrek 1998; Donohue et al. 2000b; Cheplick & Chui 2001; Cheplick 2003; Willis et al. 2010). In a greenhouse experiment with the perennial grass *Phleum pratense* (timothy), tillers of 11 genotypes were separated manually to be used as replicates and were planted alone or into pots (64 cm^2) containing a dense population of *Lolium perenne* (perennial ryegrass) equivalent to 1,875/m^2 (Cheplick 2003). There were five to seven replicates per genotype in each treatment (control vs. competition). Although growth measures were obtained at multiple times during the 16-week experiment, final shoot dry mass of the target genotypes is used here to estimate their relative fitness (= Mean shoot mass of a genotype ÷ Mean shoot mass of all genotypes in a treatment). In a repeated-measures analysis of variance, the genotype-by-treatment interaction was highly significant ($F = 2.94$, $p = 0.003$), indicating that genotype performance depended on the competitive environment. The substantial crossing of reaction norms for relative fitness (Fig. 7.3A) shows that the rankings of genotypes changed between noncompetitive and competitive conditions. Thus, selection would favor different genotypes in the two conditions. This is not an unusual finding, and similar results have been reported in several annuals: *Poa annua* (McNeilly 1984), *Raphanus sativus*

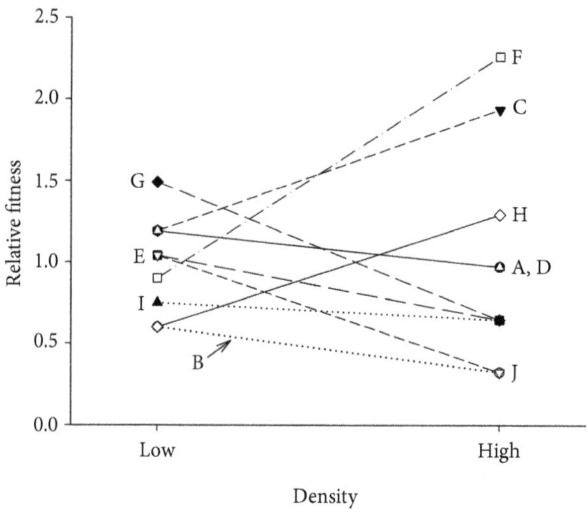

FIGURE 7.2
Norms of reaction for relative fitness based on seed production by the 10 genotypes (A–J) from Figure 7.1 growing at low and high densities.

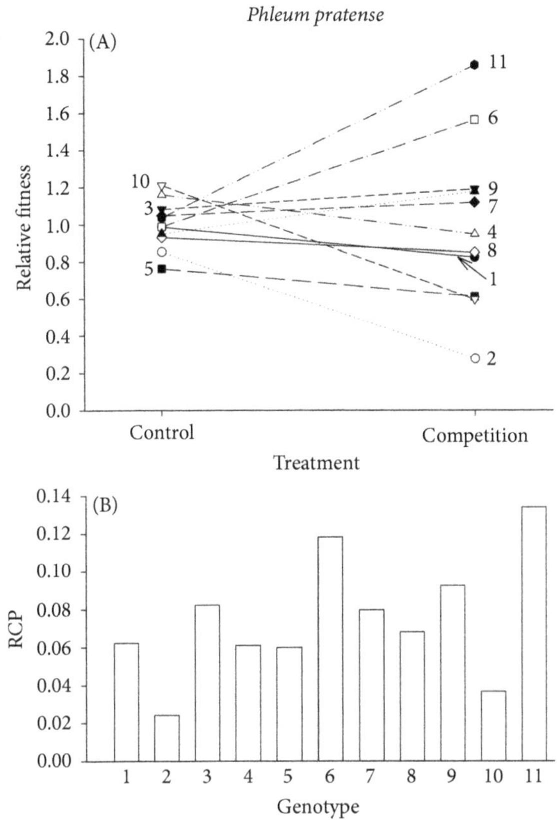

FIGURE 7.3

(A) Norms of reaction for relative fitness based on shoot mass by 11 genotypes of *Phleum pratense* growing without competition or within a dense population of *Lolium perenne* (Cheplick 2003). (B) Relative competitive performance (RCP) of the 11 genotypes based on shoot mass.

(Mazer & Schick 1991a), *Polygonum pensylvanicum* (Thomas & Bazzaz 1993), *Impatiens capensis* (Donohue et al. 2000b), and *Arabidopsis thaliana* (Willis et al. 2010). However, some studies have not detected significant variation among genotypes in their growth responses to competition in other species (Goldberg 1988; Cheplick & Gutierrez 2000; Bonser & Ladd 2011). In these cases, the performance of genotypes was relatively consistent across competitive treatments, suggesting that competitive ability was a stable property of a genotype.

Calculations of the RCP, as defined in the previous section, reveals the marked differences among genotypes of *Phleum pratense* in their ability to accumulate biomass in the presence of interspecific neighbors (Fig. 7.3B). Genotypes with the greatest relative fitness under competition (11 and 6) show the greatest competitive performance. Thus, the RCP provides another way of looking at the variation in fitness among genotypes under competitive, relative to noncompetitive conditions. Other indices have also been used to quantify competitive ability in a general way (e.g., Bonser & Ladd 2011).

More complex patterns in the norms of reaction can occur when the relative fitness of different genotypes is plotted against multiple densities. Kromer and Gross (1987)

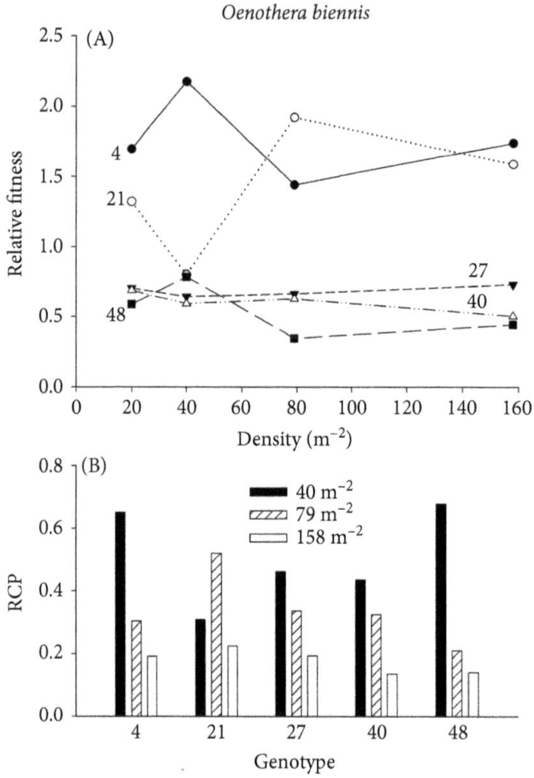

FIGURE 7.4
(A) Norms of reaction for relative fitness based on the number of seeds produced by five genotypes of *Oenothera biennis* grown at four densities in a greenhouse experiment (Kromer & Gross 1987).
(B) Relative competitive performance (RCP) of the five genotypes based on the number of seeds produced at three densities relative to the control density (1 plant per pot = 20/m^2).

performed a greenhouse experiment in which five genotypes of the biennial *Oenothera biennis* (evening primrose) were grown at five densities. Replicated genotypes were grown in pots (diameter, 25.4 cm) at densities of 1, 2, 4, 8, and 16 per pot, corresponding to densities of 20, 40, 79, 158, and 314/m^2, respectively. A variety of traits were recorded, including the number of seeds produced. I computed the relative fitness based on mean seed number for each genotype at each density, except the highest density (at which only two genotypes matured seeds). Two genotypes (4 and 21) consistently had a high fitness relative to the other three genotypes (Fig. 7.4A). The RCP was lowest at the highest density, where competitive stress was most limiting to seed production (Fig. 7.4B). The plot of RCP also shows the variation among genotypes in their ability to mitigate the negative effect of increasing density on their reproductive output.

Collectively, the studies just reviewed provide solid evidence that natural plant populations are likely to harbor significant genotypic variation for the ability to grow and reproduce under competitive conditions. This ability is heritable (Sakai 1961; Khan et al. 1976), and population differentiation in relation to environments that differ in competitive pressure (density) is a possible outcome (Linhart 1988; Mehrhoff & Turkington 1990; Miller & Fowler 1993; Donohue et al. 2000b).

7.4 DIFFERENTIATION, LOCAL ADAPTATION, AND COMPETITION

After reporting the results of interspecific competition trials using different populations of five perennial herb species, Mehrhoff and Turkington (1990) observed how experimental outcomes typically depended on the populations used for the competitive mixtures. They further generalized that competitive ability was not likely to be a species-specific feature, but rather could be expected to vary from population to population and from one environment to another. This sentiment echoes that of McNeilly (1981), who demonstrated population differentiation in *Poa annua* in response to different competition treatments. In general, one can expect population differentiation to be most likely when environmental factors vary spatially, but show relative stability in any given area over time (Turkington & Aarssen 1984). Thus, differences in the intensity of competition and the strength of density-dependent selection among habitats may result in the divergence of populations in relation to the traits that confer competitive ability.

7.4.1 Genetic Differentiation

I compiled a diverse set of studies that examined the differential performance of populations (or ecotypes or varieties) in competitive environments (Table 7.2). Although the variation in experimental techniques and treatments imposed is enormous, all the studies report at least some significant differences in the recorded traits among populations in intraspecific or interspecific competition (or both [Linhart 1988]). Common garden conditions in the field or greenhouse, as well as reciprocal transplants, have been used. Often, only two or three populations (or ecotypes) were examined, although for several species six or more populations were assessed (McNeilly 1981; Clay & Levin 1986; Miller & Fowler 1993; Mercer et al. 2006). Most studies compared performance with and without competition but did not vary density (e.g., Miller & Fowler 1993; Rice & Knapp 2008; Ariza & Tielbörger 2011). Note that one or more of the three components of life history (Aarssen & Keogh 2002) were typically used in all studies to assess plant performance: survival, growth (measured as biomass or some other size metric), and reproduction (measured as fecundity; Table 7.2).

Interpreting the patterns of differentiation as they relate to competitive pressures in different sites, and ascertaining the adaptive nature of the pattern are complex undertakings (Linhart 1988). Much of the interpretation is based on conjecture about the relative importance of competition in the habitats from which populations were sampled. Competitive outcomes, whether a result of intraspecific or interspecific interactions, are also highly contingent on environmental conditions (Clay & Levin 1986; Mehrhoff & Turkington 1990; Goldberg 1996; Donohue et al. 2001; Ariza & Tielbörger 2011). Populations may show significant differences in competitive ability as measured under a specific set of experimental conditions, but it can be a challenge to relate these differences back to the selection pressures operating in their natural habitats.

The use of ecotypes—groups of differentiated populations distinguished by differences in multiple traits (Lowry 2012)—has allowed results of a few competition experiments to be related to the habitat conditions to which the ecotypes are adapted (e.g., McNeilly [1981]). We saw in Section 4.4.4 that the long-lived arctic perennial *Dryas octopetala* consists of two well-defined, locally adapted ecotypes as determined by long-term results of reciprocal transplant experiments (Bennington et al. 2012). McGraw (1985) performed a 15-week competition experiment between the two ecotypes (fellfield and snowbed) in treatments that allowed root and shoot

Table 7.2 Significant differentiation in relation to performance in different competitive environments for some herbaceous plant species.

Species	*Life form*	*Unit of differentiation*	*No. of units*	*Traits recorded*	*Experimental environment*	*Reference*
Taraxacum officinale	Perennial herb	Ecotypes	2	Biomass, germination, fecundity	Greenhouse	Solbrig and Simpson (1974)
Ranunculus repens	Perennial herb	Populations	2	Biomass, number of leaves and ramets	Greenhouse	Lovett Doust (1981b)
Poa annua	Annual grass	Populations	6	Biomass	Greenhouse	McNeilly (1981)
Phlox drummondii	Annual herb	Varieties	5	Biomass, survival	Greenhouse	Heywood and Levin (1984)
		Populations	6	Biomass, number of flowers	Reciprocal transplants	Clay and Levin (1986)
Dryas octopetala	Perennial herb	Ecotypes	2	Biomass, number of ramets	Growth chambers	McGraw (1985)
Veronica peregrina	Annual herb	Populations	2	Biomass, height, fecundity	Greenhouse	Linhart (1988)
Trifolium repens	Perennial legume	Subpopulations	3	Biomass, survival	Reciprocal transplants	Turkington (1989)
Dactylis glomerata,	Perennial grass	Populations	3	Biomass, number of ramets	Gardens	Mehrhoff and Turkington (1990)
Holcus lanatus,	Perennial grass	Populations	3			
Lolium perenne,	Perennial ryegrass	Populations	3			
Trifolium repens	Perennial legume	Populations	3			
Anthoxanthum odoratum	Perennial bunchgrass	Populations	2	Fecundity. number of tillers, survival	Reciprocal transplants	Platenkamp and Foin (1990)
Bouteloua rigidiseta	Perennial bunchgrass	Populations	8	Fecundity, height, number of tillers	Gardens	Miller and Fowler (1993)

Impatiens capensis	Annual herb	Populations	2	Height, fecundity, size	Reciprocal transplants	Donohue et al. (2001)
Helianthus annuus	Annual sunflower	Populations	9	Fecundity, survival	Gardens	Mercer et al. (2006)
Elymus glaucus,	Perennial bunchgrass	Populations	2	Germination, number of flowering tillers, survival	Reciprocal transplants	Rice and Knapp (2008)
Nasella pulchra	Perennial bunchgrass	Populations	2			
Brachypodium distachyon,	Annual grass	Ecotypes	2	Biomass, number of inflorescences, survival	Reciprocal transplants	Liancourt and Tielbörger (2009)
Bromus fasciculatus	Annual grass	Ecotypes	2			
Biscutella didyma,	Annual crucifer	Populations	4	Biomass, fecundity, survival	Reciprocal transplants	Ariza and Tielbörger (2011)
Hymenocarpos circinnatus	Annual legume	Populations	4			

Reciprocal transplants and gardens are field environments in all cases. A species name followed by a comma indicates the next species below is part of the same study. Studies are arranged by date.

competition to be compared separately. Data were collected on a number of growth measures, including biomass. In general, the snowbed ecotype proved to be the better competitor, being able to increase the root-to-shoot ratio under root competition and to increase leaf length under shoot competition. Both of these responses may represent adaptations to the productive conditions (high nutrient availability and high standing biomass) of the snowbed habitat, where greater root allocation and larger leaves enhance competitive success (McGraw 1985).

In a classic study of the common dandelion (*Taraxacum officinale*), Solbrig and Simpson (1974) distinguished four genetic types based on allozyme profiles and subjected two of them to a set of competition experiments conducted under different environmental conditions. The two "biotypes" (denoted A and D) would likely be called ecotypes today, because they were abundant in sites characterized by specific conditions and they maintained their distinct morphology under cultivation (Solbrig & Simpson 1974). Ecotype A comprised 78% of the individuals in a population along a dry, highly disturbed pathway that received full sun. The area was mown weekly and vegetation consisted mostly of low-growing weeds. Ecotype D predominated at a relatively undisturbed floodplain where soils were moist and taller herbaceous plants were abundant. The researchers grew plants of each ecotype in monocultures or in mixtures (75%:25%, 50%:50%, and 25%:75%) in plastic trays at low and high densities, equivalent to 119 plants/m^2 and 476 plants/m^2, respectively. They also varied soil conditions and light levels (see lower part of Fig. 7.5). Based on survival and shoot dry mass, it was found that ecotype D was a better competitor under most conditions (Solbrig & Simpson 1974). Using their mean shoot mass data for the 50:50 mixture relative to the monocultures, I calculated the relative competitive performance of the two ecotypes; values were mostly greatest for ecotype D, especially under high density and high light (Fig. 7.5). However, ecotype A showed precocious flowering and

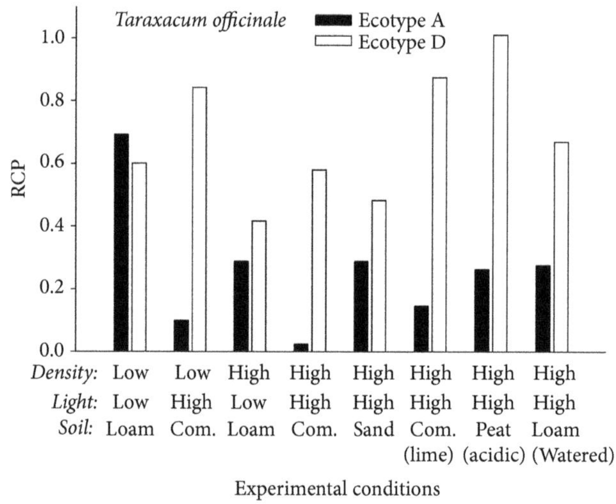

FIGURE 7.5

Relative competitive performance (RCP) of two ecotypes of the common dandelion based on shoot mass data presented by Solbrig and Simpson (1974). Plants of the ecotypes were grown in monoculture or 50:50 competitive mixtures at variable light and soil conditions.Com., commercially obtained growing medium.

produced more seeds. Solbrig and Simpson (1974) interpreted these results in terms of the classic concept of *r*- and *K*-selection (reviewed for plants in Gadgil and Solbrig [1972]), whereby under disturbed conditions, selection favors earlier and greater reproduction (ecotype A), whereas in undisturbed conditions, density-dependent selection favors larger size and greater performance under competition (ecotype D).

In a follow-up experiment, Solbrig and Simpson (1977) grew the two ecotypes separately or mixed together (50:50) in plots in an experimental garden in Cambridge, Massachusetts, for four years. Plots were subjected to three treatments: undisturbed, disturbed by removal of aerial plant parts (defoliation), and disturbed by soil tilling, which destroyed most plants. The disturbance treatments were applied at a single time in midsummer of two years. By the end of the experiment, ecotype D predominated in the undisturbed plots, comprising more than 84% of the plants and 91% of the biomass in plots with mixtures of the two ecotypes. Thus, ecotype D was confirmed as a more competitive genotype that was favored by natural selection over the four-year period. In disturbed plots, the opposite occurred, with ecotype A predominating in both numbers (>90%) and biomass (>93%). Thus, this early-flowering and more fecund ecotype was favored selectively when disturbances were severe.

These experiments and others with dandelions have shown strong evidence for genetically based differences in response to competitive stress (Solbrig & Simpson 1974, 1977; Vavrek 1998). It is apparent that neighbors can cause density-dependent selection that favors genotypes that grow and perhaps reproduce more effectively under competitive conditions. The expected outcome is local adaptation, as assessed indirectly by Solbrig and Simpson's (1977) common garden trial. However, as we know from Chapter 4, reciprocal transplants are a more powerful way to demonstrate local adaptation, and studies using this approach are considered next.

7.4.2 Reciprocal Transplants and Local Adaptation

To date, the evidence for local adaptation to competitive environments as revealed by reciprocal transplants has been relatively weak (compared with evidence of adaptation to abiotic factors). Part of the problem has been the difficulty in separating the effects of interactions among plants from the effects of abiotic differences between sites (Ariza & Tielbörger 2011). In addition, the history of the studied sites is often not known, so one is unsure of the length of time populations have had to evolve in response to competitive pressures. It is not usually known whether competition has been a consistent agent of selection at the sites chosen for study.

Platenkamp and Foin (1990) reciprocally transplanted cloned replicates of *Anthoxanthum odoratum* between xeric and mesic grassland sites in California. They included intraspecific and interspecific neighbor (*Holcus lanatus*) treatments. Neighborhood competition reduced reproductive output, but there were no population origin-by-planting site interactions, no differences among neighbor treatments, and no interaction of genotype with neighbor treatments. Thus there was "no evidence for specialization of genotypes with respect to immediate neighbors" (Platenkamp & Foin 1990, p. 207). In contrast, in the annual *Phlox drummondii*, reciprocal transplants (all maintained at a single intraspecific density) showed the greatest reproduction for two populations in their native site, but not for two other populations examined (Clay & Levin 1986). Thus, there was some evidence of home-site advantage at high intraspecific density, but the authors did not provide data on the usual densities that occur in the study populations. The assumption is that growth and reproduction of this species occur consistently under competitive conditions. In addition, because density was not varied, it was not possible to distinguish biotic (competitive) effects from abiotic effects resulting from environmental differences between sites.

A reciprocal transplant experiment using the clover *Trifolium repens* conducted by Turkington (1989) was described briefly earlier in Section 4.4. Clones were taken from sites within a permanent pasture in Wales and planted back into their native sites with or without the dominant grass species removed. Local (native) subpopulations of *T. repens* grew no better in their home site when their local grass neighbors were removed compared with when they were present. However, only when the competing grasses were present, was there a home-site advantage for *T. repens*. These results were interpreted as evidence for the coevolution of competitor species within the pasture (Turkington 1989), although later work showed that the specificity between *T. repens* and *Lolium perenne* genotypes may have been mediated primarily by *Rhizobium* bacteria in roots of the clover (summarized by Turkington [1996]). Nonetheless, the methods in Turkington (1989) underscore the usefulness of experimental manipulation of site conditions to interpret reciprocal transplant studies.

Intraspecific density at the planting sites was manipulated for reciprocal transplants of genetically differentiated populations of the annual *Impatiens capensis* between two sites less than 1 km apart in Rhode Island that differed in light levels (Donohue et al. 2000a, b, 2001). At each site, plots were cleared of vegetation, and *I. capensis* seedlings were planted at low density (53/m^2) or the "natural" high density typical for the site: 470/m^2 at the shady woodland site and 1,305/m^2 at the sunny open site (Donohue et al. 2000a). Based on seed production as a fitness measure, local adaptation was observed as a home-site advantage at both sites and densities (except at high density in the sunny site). Populations were differentiated with respect to their phenotypic plasticity in response to density, and population-by-density interactions were apparent for a number of phenotypic traits (Donohue et al. 2001). For some traits deemed to be important for adaptive responses to density, genotypic selection differentials and gradients were determined for each density and site (Donohue et al. 2000a). There was significant positive selection on length of the first internode, a measure of stem elongation (a shade avoidance mechanism), especially at high density in the sunny site (Table 7.3). Thus, plants with longer internodes showed greater fitness at high density in both sites, supporting an earlier study of density-dependent selection in the same species (Dudley & Schmitt 1996). At the shady site there was significant selection for early flowering at both densities, whereas at the sunny site selection favored later flowering (Table 7.3). The authors noted that density-dependent selection on the measured traits contributed to local adaptation; at both sites, this was attributed mostly to selection on traits important to shade avoidance at high density (Donohue et al. 2000a).

Interspecific competition appeared to enhance the expression of local adaptation of two perennial bunchgrasses to two sites in California (Rice & Knapp 2008). Seeds and seedlings were transplanted in separate reciprocal transplant experiments, and survival and reproduction were monitored for three years. Averaged across the species, survival of seedlings was greater for the local population most consistently when seedlings were in intact, undisturbed vegetation (compared with a vegetation removal treatment). A cumulative fitness index based on survival, and tiller and seed

Table 7.3 Genotypic selection gradient (β) estimates for first internode length (a measure of stem elongation) and date of flowering in the annual *Impatiens capensis*.

Trait	*Sunny site density*		*Shady site density*	
	High	*Low*	*High*	*Low*
Internode length	**0.61****	−0.07	**0.21****	0.10
Flowering date	0.22	**0.16****	**−0.46*****	**−0.49*****

Transplants were grown at a sunny or shady site in Rhode Island at low density (53/m²) or high density (1,305/m² at sunny site and 470/m² at shady site). Significant β values are in bold type. $^{*}p < 0.05$, $^{**}p < 0.01$, $^{***}p < 0.001$.

Source: Donohue, K., Messiqua, D., Pyle, E.H., Heschel, M.S. & Schmitt, J. 2000a. Evidence of adaptive divergence in plasticity: density- and site-dependent selection on shade-avoidance responses in *Impatiens capensis*. *Evolution* 54: 1956–1968.

production, showed evidence for home-site advantage in *Elymus glaucus*, a highly selfing species, especially under competitive conditions (Rice & Knapp 2008).

Attempts to separate the effects of abiotic conditions from those resulting from interactions with neighbors using modified reciprocal transplant experiments have been made with several annual species. In one study, the seeds of two annual grass species were reciprocally transplanted between two sites 12 km apart along a precipitation gradient in Jordan (Liancourt & Tielbörger 2009). The Mediterranean site had more precipitation (475 mm/year), a longer growing season (5.5 months), and supported greater plant biomass (300 g/m²), whereas the semiarid site had less precipitation (249 mm/year), s shorter growing season (2.5 months), and less plant biomass (169 g/m²). Thus, populations were exposed to a greater intensity of competition in the productive Mediterranean site. In half of the 48 plots (2 × 1.5 m) at each site, vegetation was removed to exclude neighbors. Plants from the two sites were also grown under standard common garden conditions in a greenhouse to ascertain potential genetic differentiation. Neighboring vegetation reduced survival, biomass, and inflorescence production significantly in the field. Regardless of whether neighbors were present, plants of both species from the Mediterranean site always produced more biomass than plants from the semiarid site. Populations were differentiated for growth as revealed by the greater biomass of plants of the Mediterranean site relative to that of the semiarid site when grown under common conditions. However, the populations did not differ in their response to competitive stress at the field sites (no significant population-by-neighbor treatment interaction). Home-site advantage was also not apparent (no significant population-by-site interaction for the measured variables). However, the authors speculated that the larger size of plants of both species from the Mediterranean site in the field (regardless of where they were planted), and maintained under greenhouse conditions, could represent selection for greater competitive ability in a productive environment ((Liancourt & Tielbörger 2009).

A second reciprocal transplant study that made use of a natural precipitation gradient was done in Israel using an annual crucifer (*Biscutella didyma*) and an annual legume (*Hymenocarpos circinnatus*) growing at four sites: arid, semiarid, Mediterranean, and mesic Mediterranean (Ariza & Tielbörger 2011). Seeds of the two species were reciprocally sown into control plots (diameter, 15 cm) with undisturbed, intact vegetation and into "no-neighbor" plots from which other plants had been removed. Data were recorded on survival, biomass, and reproductive output. Local adaptation to climate was assessed by examination of population-by-site interactions in the no-neighbor plots, where competition was absent. Adaptation to the overall local environment that included abiotic (climatic) *and* biotic (neighbors) factors was assessed in the analysis of control plots (details in Ariza and Tielbörger [2011]). Planting site had a strong effect on biomass and reproduction for *B. didyma* in plots with and without neighbors; however, there were no population-by-site interactions (expected under a scenario of local adaptation) for either species. Despite an extensive data set, the disappointing conclusion was that "there was neither evidence for local adaptation to neighbor presence, climate or overall (biotic + abiotic) environmental conditions" (Ariza & Tielbörger 2011, p. 937). Nonetheless, their methodological approach was informative and points the way toward future studies of adaptation to neighbors in sites with different sets of abiotic factors.

7.4.3 Fine-Scale Adaptation to Neighbors

In many environments with an adequate supply of light, water, and mineral nutrients, plants are members of species-rich communities with many individuals packed into every square meter. Interactions with neighbors of the same and different species for extended periods are likely when perennial plant communities persist. Selection may favor niche differentiation and a balance of competitive abilities among interacting species to reduce the negative fitness effects of competition for the same contested resources (Aarssen 1983; Aarssen & Turkington 1985). Over time, continuous competitive interactions can result in fine-scale differentiation within populations (Turkington & Aarssen 1984; Turkington & Mehrhoff 1990).

In a series of studies, Roy Turkington and colleagues have documented adaptive differentiation in white clover (*Trifolium repens*) in relation to competing grass species in permanent pastures (Turkington & Harper 1979; Aarssen & Turkington 1985; Turkington 1989; Mehrhoff & Turkington 1990, 1995; Turkington 1996). In general, genotypes of *T. repens* have performed best in competition when interacting with the grass neighbors with which they were associated in the field. As an example, I graphed some of the tabular data presented by Turkington and Harper (1979) on shoot mass of *T. repens* genets collected from sites dominated by different grass species and then transplanted back into the sites in a pasture in North Wales (Fig. 7.6). Transplants grew for a year in competition with the associated grasses or in plots cleared of vegetation with an herbicide. In cleared plots there were no differences between *T. repens* sampled from different sites, However, in the vegetated plots in which competitive stress was very high, *T. repens* transplants tended to grow best in their site of origin, although this was not true for the site dominated by *Holcus lanatus* (Fig. 7.6).

In a second experiment conducted in a greenhouse, *Trifolium repens* genets from the same sites were planted into dense monocultures of the four grass species that dominated the field sites. After one year, the biomass of *T. repens* was typically greatest when competing with the grass species dominant at the site from which the *T. repens* genets had been collected (Turkington & Harper 1979). Thus, the large population of *T. repens* in this pasture appeared to have undergone fine-scale genetic differentiation in relation to specific grass neighbors that continually

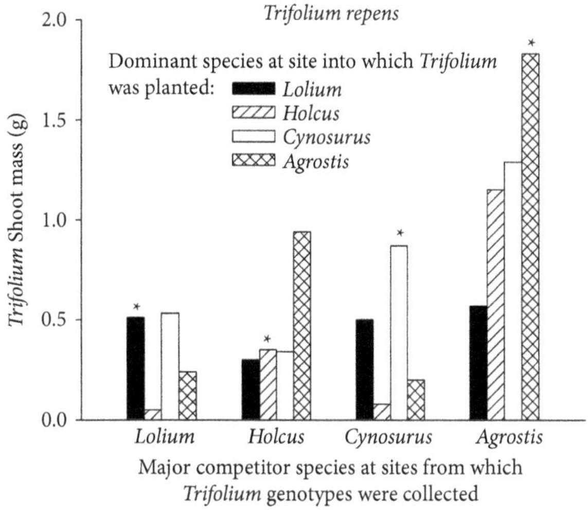

FIGURE 7.6
Shoot mass of *Trifolium repens* genotypes originally collected from four sites, each dominated by a different grass species (*Lolium perenne, Holcus lanatus, Cynosurus cristatus*, or *Agrostis capillaris*), and then transplanted back into the same vegetated sites to grow for one year. *Transplants growing in their home sites. Drawn from data in Turkington and Harper (1979).

imposed a competitive (selection) pressure on *T. repens* genotypes (Turkington & Aarssen 1984). However, such findings are not always detected in pasture communities. For example, no evidence could be found for "co-adaptation" between *Lolium perenne* genotypes sampled from pastures that were 10 and 40 years old, and their neighbors when in competition (McNeilly & Roose 1996). Nonetheless, Turkington and Mehrhoff (1990) have made a convincing argument for the importance of competitive interactions as an evolutionary force in structuring pasture communities, although the possible role of microorganisms in the apparently adaptive matching of competitor genotypes in earlier studies should not be discounted (Turkington 1996).

7.5 GENOTYPIC INTERACTIONS AND COMPETITIVE OUTCOMES

Within a population, individuals can compete with members of the same and/or different species. At a fine level of spatial resolution it is individual genotypes that are interacting. These individuals can range greatly in genetic similarity for any one species (Fig. 7.7). They may be clonal members of the same genet, siblings resulting from self-fertilization or inbreeding among relatives, full or half siblings resulting from outcrossing, or unrelated members of the same population. Greater genetic dissimilarity may be found among differentiated populations or ecotypes in which at least some members from previously isolated groups interact after dispersal. Of course, the greatest genetic dissimilarity during competition occurs among genotypes of separate species (Fig. 7.7).

The competitive outcomes of genotypic interactions are likely to depend on the genetic similarity (or dissimilarity) of the individuals involved and on environmental

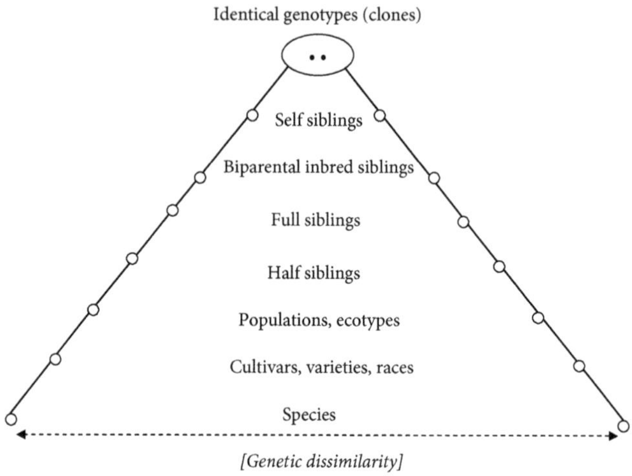

FIGURE 7.7
Genetically defined entities that may interact in nature or in competition experiments along an arbitrary scale of increasing genetic dissimilarity. "Self siblings" are the product of self-fertilization of a single plant whereas "biparental inbred siblings" are the product of mating between relatives.

conditions (Goldberg 1996; Turkington 1996), with implications for plant evolution and ecology (Harper 1977; Kelley & Clay 1987; Taylor & Aarssen 1990; Cheplick 1993a; Donohue 2003; Thorpe et al. 2011). Drawing from niche theory (Bazzaz 1996; Chase & Leibold 2003), it is maintained that in a situation when competition functions continually as a selection pressure, diversification in resource use (niche partitioning) between interacting groups will evolve, because reciprocal selection favors phenotypic trait values that maximize the fitness of individuals in both groups under competitive conditions (reviewed in Thorpe et al. [2011]). This idea has been referred to as *selection for ecological combining ability* and has been applied mainly to interspecific interactions in which relative yields are greater in species mixtures than in pure culture (Harper 1977; Aarssen 1983). For a demonstration of how competitive interactions between two plant species can lead to population differentiation, the experiments of Martin and Harding (1981) are informative. They competed co-occurring (interacting) and isolated (not interacting) populations of two annual *Erodium* species in a de Wit replacement series (includes monocultures and mixed-species cultures; see Gibson et al. [1999] for details on the design of competition experiments). Based on seed production, it was found that populations with a history of co-occurrence showed greater reproduction when competing relative to isolated populations with no prior history of contact. Results suggested "that evolution in response to competition between two species of plants occurs at the population level" (Martin & Harding 1981, p. 986).

Although Martin and Harding (1981) examined interspecific competition using closely related species, evolutionary changes had occurred within populations. Note that the genetic similarity between competing groups can be at any of the levels of resolution shown in Figure 7.7. The remainder of this section is devoted to interactions below the species level that affect the evolutionary ecology of plant populations.

7.5.1 Genetic Relatedness and Intraspecific Competition

One set of competition studies using barley and wheat varieties, and selfed siblings of barley, provides a classic, oft-cited early example of how selection may favor ecological combining ability below the species level (Allard & Adams 1969). The varieties of the two annual crop species were developed for high performance in pure culture and, although the more genetically diverse mixed cultures showed greater seed yield overall, differences between pure and mixed cultures were not great. However, Allard and Adams (1969) also examined the seed yield of eight sibling groups (families) of barley, each derived from selfing one maternal plant, in pure culture and in mixed stands of families. The maternal source plants were part of a heterogeneous population originally developed from intercrosses among 31 barley varieties and had been through 18 generations of natural selection in large, dense competitive stands under field conditions (populations had been propagated each generation as seeds collected in bulk and there was no imposed conscious selection). Compared with the yield of pure cultures of single families, mixtures of different families often showed significantly greater yield, suggesting that selection favored genotypes with "superior ecological combining ability" (Allard & Adams 1969, p. 630).

Since that time, other researchers have addressed experimentally the question of how genetic relatedness influences competitive outcomes in a number of plant species (Tonsor 1989; Cheplick & Salvadori 1991; Andalo et al. 2001; Cheplick & Kane 2004; Willis et al. 2010). Results from experiments comparing the outcome of intragenotypic and intergenotypic competition have often found genotypic variation in competitive ability and sometimes provided additional support for the idea of resource (niche) partitioning in mixtures of different genotypes. However, conflicting results can be found, even for similar experiments using the same species.

In the workhorse *Arabidopsis thaliana*, Bonser and Ladd (2011) performed an intraspecific competition experiment with nine inbred lines (considered to represent distinct homozygous genotypes) and reported no significant genotype-by-competition interactions for biomass or fruit number, but did not include an intragenotypic competition treatment. Willis et al. (2010) used 20 inbred lines (genotypes) of *A. thaliana* to reveal significant genotype-by-competition interactions for fruit number and other traits, illustrated in reaction norm diagrams. They also found greater performance of plants in intergenotypic versus intragenotypic competition, supporting the hypothesis of resource partitioning. Andalo et al. (2001) examined intra- and intergenotypic competition in five genotypes of *A. thaliana* under ambient and elevated CO_2 levels. In contrast to Willis et al. (2010), there was no support for resource partitioning among diverse genotypes; in ambient CO_2, fitness of a genotype was *greater* when surrounded by individuals of the same genotype! This was interpreted as possible evidence of kin selection (see next section).

A former graduate student and I investigated genotypic competition in the rhizomatous perennial *Amphibromus scabrivalvis*, a grass native to South America that is cloned readily by manual separation of ramets for each genet (Cheplick & Salvadori 1991). Four morphologically variable genets were chosen from a population originally collected in Louisiana and then maintained in greenhouse culture for six years. After that time, the genets were large enough to enable many ramets to be separated from each and relegated to three treatments (12 replicates each): alone, paired with a second ramet of the same genotype (intragenotypic competition), and paired with

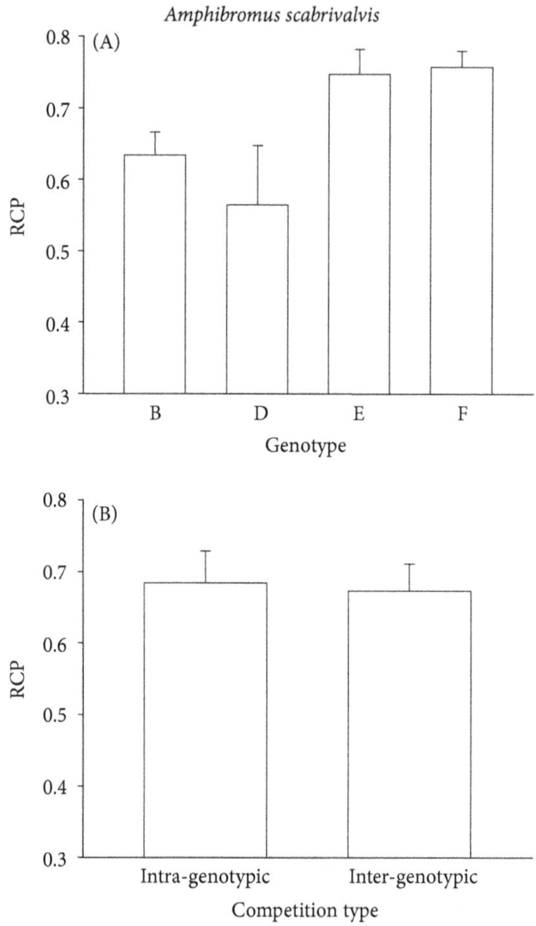

FIGURE 7.8
Mean relative competitive performance (RCP) ± standard error of four cloned genotypes of the perennial grass *Amphibromus scabrivalvis* calculated from total biomass data in Cheplick and Salvadori (1991). (A) RCP of each genotype averaged over intra- and intergenotypic competition treatments. (B) RCP of the intra- and intergenotypic competition treatments averaged over all genotypes.

a second ramet of a different genotype (intergenotypic competition). Ramets were grown in square pots (5 × 5 cm) in a greenhouse for four months. At harvest, plants were dried; separated into roots, shoots, rhizomes, corms, and seeds; and weighed.

Genotypes grown without competition showed highly significant ($p < 0.01$) differences in several biomass components (Cheplick & Salvadori 1991). Using total biomass, I calculated the RCP of the four genotypes based on their average growth in intra- and intergenotypic competition. Two genotypes (E and F) were superior competitors, whereas the two others (B and D) performed less well under competitive stress (Fig. 7.8A). I also calculated the average RCP for genotypes in intra- and intergenotypic competition and failed to find a significant difference between the two conditions ($t = 0.25$, $p > 0.50$; Fig. 7.8B). Thus, there was no support for the contention that competitive effects would be reduced in genotypic mixtures resulting from resource partitioning by the genets tested.

7.5.2 Sibling Competition and Kin Selection

Because the seeds produced by a maternal parent are at least related as half siblings, and dispersal can be limited in plants, after germination siblings may be in close proximity and compete (Cheplick 1992, 1993a, b). This is detrimental not only to the direct fitness of the competing siblings, but also to the parent(s) that produced them. Therefore, it is beneficial to both the parent and its offspring to have traits that minimize the negative fitness consequences of sibling competition. Sibling competition as a selection pressure has been invoked to explain the evolution of a variety of features, including seed packaging within fruits, sexual reproduction via outcrossing (vs. asexual reproduction or selfing), seed dispersal dimorphism, and germination patterns (Nakamura 1980; Cheplick 1993b, 1996; Hyatt & Evans 1998).

Sibling competition may favor the evolution of life history features that reduce its negative effects and increase the fitness of relatives that share 50% of their genes if they are full siblings. Any one individual has its own direct fitness and also an indirect fitness component resulting from the additional reproduction of genetic relatives. Selection that favors alleles that increase the indirect component of fitness is kin selection (Griffin & West 2002).

For an example of the methods typically used in studies of genetic relatedness and competition, we examine the experiment of Milla et al. (2009) in which the annual legume *Lupinus angustifolius* was grown within 15-cm-diameter pots in a greenhouse. Four treatments were established: (1) control (one plant per pot), (2) interpopulation competition (three plants per pot from different populations in western Spain), (3) interfamily competition (three plants per pot from different families in the same population), and (4) intrafamily competition (three siblings per pot from the same family). A number of phenotypic traits were scored when plants had matured; only seed production is detailed here (Fig. 7.9). The researchers analyzed traits as response ratios: the mean trait value in a competitive treatment divided by that of the control (Milla et al. 2009). Data analyses of seed number ratio revealed a highly significant effect of genetic relatedness ($F = 6.27$, $p = 0.002$). Individuals competing with nonrelatives produced more seeds than those competing with siblings (Fig. 7.9). This study, plus a later one by the same authors (Milla et al. 2012), could not find evidence that kin selection had been important in the evolution of this species.

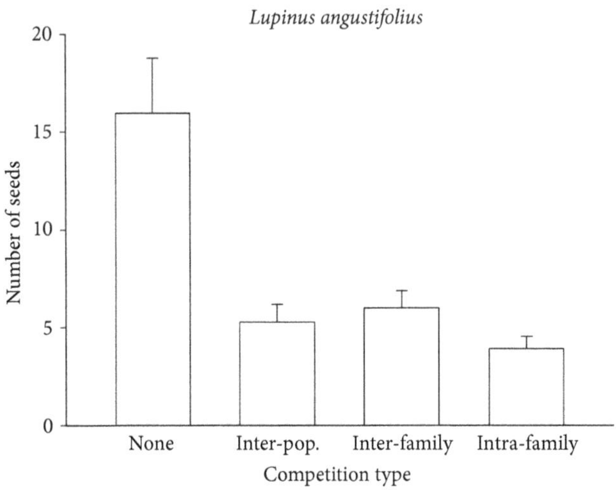

FIGURE 7.9
Mean number of seeds produced by the annual legume *Lupinus angustifolius* ± standard error when growing alone or when competing with plants from a different population, different family (within the same population), or same family (i.e., siblings). Redrawn from Milla et al. (2009).

So far, the evidence for kin selection as an evolutionary force in plant populations has been equivocal (File et al. 2012). In a review of sibling competition in 1992, I presented a table listing eight species in which outcomes had been compared for competing siblings versus competing nonsiblings (Cheplick 1992). For only two species did competing siblings outperform competing nonsiblings: *Phytolacca americana* (Willson et al. 1987) and *Plantago lanceolata* (Tonsor 1989). Since that time, there have been several similar studies with other species. Some have found no support for kin selection (Karron & Marshall 1993; Cheplick & Kane 2004; Koelewijn 2004; Milla et al. 2009; Masclaux et al. 2010; Milla et al. 2012), whereas others have reported that interacting kin can show evidence of cooperation rather than competition under some conditions (Kelly 1996; Andalo et al. 2001; Donohue 2003; Murphy & Dudley 2009; Biernaskie 2011).

The review by File et al. (2012) updates much of the available information and presents a table compiling the results of plant competition experiments that have used genetically related individuals. Note that their summary included not only sibling competition studies, but also studies of competition between cultivars (e.g., Allard & Adams 1969) and between identical genotypes (clones [e.g., Cheplick & Salvadori 1991]). In more than half of 41 studies, there was no difference in the performance of genetic relatives versus nonrelatives in competition (File et al. 2012). In 11 studies (27%), performance of unrelated plants exceeded that of genetic relatives (suggesting resource partitioning), and in nine studies (22%), performance of genetic relatives exceeded that of nonrelatives (suggesting kin selection). Clearly the issue of kin selection and its significance to plant microevolution is far from resolved.

7.6 SELECTION EXPERIMENTS

There have been a few experiments conducted to determine whether competitive abilities could be changed in populations after several generations of selection under controlled conditions. Aarssen (1989) grew the annual *Senecio vulgaris* for three generations in a low- or high-density monoculture, and in an equally proportioned mixture with a competing perennial grass (*Phleum pratense*). The seeds used to establish the experiment had been obtained from 21 locations worldwide and mixed to provide a highly variable gene pool available for selection. Competition with *P. pratense* apparently selected for early germination in *Senecio*. Also, by the third generation, *Senecio* plants had significantly greater biomass than *P. pratense* under interspecific competition, whereas in prior generations both species had similar biomass. This suggested that a variable *Senecio* population subjected to competition with a perennial species could evolve improved competitive ability in just three generations. However, growth responses to selection in low- versus high-density monocultures of *Senecio* were not detected (Aarssen 1989).

Note that the type of selection protocol used in Aarssen (1989) involves the experimenter choosing seeds randomly from one generation to use for the next generation. This is essentially a natural selection experiment conducted under artificial conditions and does not qualify as an artificial selection experiment (see Section 2.2.4). A more elaborate experiment using the annual *Brassica rapa* in intraspecific or in interspecific competition (with the annual *Raphanus sativus*) incorporated both random and artificial (divergent) selection across three generations (Miller 1995). The primary question addressed was: Will performance improve during selection under intra- and interspecific competition? The selection procedures were imposed for both types of competition. In controls, 10 of 100 plants were chosen randomly and intercrossed, whereas in artificially selected groups, the 10 individuals with the greatest number of flowers were intercrossed to provide seeds for each ensuing generation. There was no group in which plants were grown without competition. After data collection over the three generations, another generation was grown in a common environment to mitigate possible maternal effects, and the offspring were used to examine responses to prior selection by growing plants alone, and in intra- and interspecific competition.

The *Brassica rapa* that had been selected artificially for high flower production maintained greater flower production under intra- and interspecific competition in the second and third generations than the randomly selected control plants (Miller 1995). This shows that the selection procedure was effective in increasing fitness under competitive conditions. For all three conditions examined in the final generation (alone, intra- and interspecific competition), plants of the *B. rapa* population that had been selected artificially under intraspecific competition were taller and produced significantly more flowers. This research provided a solid demonstration that performance under intraspecific competition can evolve rapidly in an annual species, provided that a relatively strict selection process is applied.

The two final examples of density-dependent selection experiments to be discussed were conducted with the stoloniferous perennial *Ranunculus reptans* (van Kleunen & Fischer 2003; van Kleunen et al. 2005). Starting with a mixed group of 40 genets from 10 populations, ramets were propagated and allocated to low-density ($293/m^2$) or high-density ($1{,}466/m^2$) lines. Sixteen trays (31 × 44 cm) of each density were maintained in a growth room. In each generation, 40 ramets and 200 ramets from the low- and high-density lines, respectively, were chosen randomly to use for the next generation (van Kleunen & Fischer 2003). After four generations, plants were again grown at the same low or high densities. Plants from the lines selected at high density tended to have longer leaves under both densities and branched less frequently when grown at high density (relative to plants selected at low density). The authors concluded that density-dependent selection among genotypes had occurred.

In the second study of *Ranunculus reptans*, van Kleunen et al. (2005) again grew groups at low and high density, cross-pollinating flowering individuals to get seeds for each new generation. This was not deliberate selection per se, but genets with the greatest flower production were most likely to contribute offspring to the next generation, which likely simulates natural situations. After three generations, offspring were planted into a common environment (growth room) without competition. Although there were no significant evolutionary responses to density for size and reproductive allocation, plants from the high-density lines had longer leaves and fewer branches per ramet, suggesting a possible adaptive response to the shading that occurs in dense populations (van Kleunen et al. 2005).

Collectively, selection experiments performed over several generations in relation to competition or density treatments have added to the evidence that microevolutionary changes in fitness-related traits are possible over relatively short time frames. In many types of competitive plant communities, repetitive density-dependent selection across generations could result in subtle changes in phenotypic traits that provide better adaptation to a competitive environment.

7.7 OTHER GENETIC ASPECTS OF COMPETITION

We have seen how the outcome of intraspecific competition can depend on the genetic relatedness of the interacting individuals. The genetic relatedness is, of course, influenced by the parental breeding system. Efforts to investigate the fitness consequences of outcrossing versus selfing have sometimes involved growing progeny from both types of reproduction under competitive conditions (McCall et al. 1989; Gurevitch et al. 1996; Cheptou et al. 2001; Koelewijn 2004). Progeny produced via outcrossing is more variable genetically than those produced by selfing and, because of the lack of inbreeding depression and possible resource partitioning, the progeny resulting from outcrossing are expected to outperform those resulting from selfing in high-density situations.

McCall et al. (1989) did not find evidence that progeny from outcrossed chasmogamous flowers of *Impatiens capensis* were better able to avoid the negative effects of competition than those from selfed cleistogamous flowers in greenhouse and field experiments. However, in another annual (*Brassica rapa*), although the performance of progeny from outcrossing versus selfing was similar without competition, the former outperformed the latter in terms of biomass and flower production at intermediate (~200/m^2), but not high, densities in a greenhouse experiment (Gurevitch et al. 1996). In the annual composite *Crepis sancta*, survival and reproductive fitness were greatest for progeny from outcrossing when competing (1:1) with progeny from selfing at two test densities (Cheptou et al. 2001). For the short-lived perennial *Plantago coronopus*, Koelewijn (2004) found that the performance of seedlings derived from selfing versus outcrossing depended on the relatedness and frequency of neighbors. There was a frequency-dependent fitness advantage to plants from outcrossing; their shoot and reproductive mass increased with an increase in the relative proportion of competitor plants derived from selfing. Although generalizations are not easily made, these studies do provide a perspective on how the fitness outcomes of an outcrossing or selfing breeding system can depend on the competitive environment surrounding the resulting offspring.

Given that the evolutionary response to density-dependent selection relies on an available pool of genetic variation in phenotypic traits, quantitative genetic analyses have been applied to some plant competition studies (Mazer & Schick 1991a, b; Shaw & Platenkamp 1993; Thomas & Bazzaz 1993; Shaw et al. 1995; Donohue et al. 2000b). Collectively, these studies show that density clearly affects the components of genetic variance in competing groups, but not in any predictable or consistent way. As the heritability of quantitative traits differs across densities (Thomas & Bazzaz 1993; Mazer & Schick 1991b), quantitative genetic structure favors the maintenance of genetic variation within populations by variable selection among competitive groups (Shaw et al. 1995; Donohue 2004).

Molecular genetic approaches to investigating plant competition, such as identification of candidate gene loci important to competitive ability, are not common in the literature. Molecular markers (ISSR) were used by Matesanz et al. (2011) to examine the fine-scale spatial genetic structure of two closely related *Thymus* species that co-occur in the wild and potentially interact. Although not a competition experiment per se, spatial patterns of both species were dissociated; the genetic structure of the narrow endemic *Thymus loscosii* was related to abundance of the widespread *Thymus vulgaris*, and the authors suggested that interspecific competition was responsible.

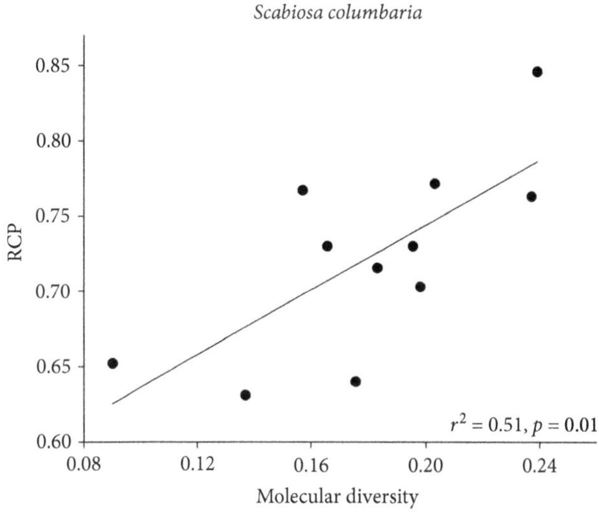

FIGURE 7.10
Relationship of relative competitive performance (RCP) for 11 populations of the perennial *Scabiosa columbaria* to its molecular diversity (expected heterozygosity). Redrawn from Pluess and Stöcklin (2004).

Molecular genetic diversity (based on RAPD) was assessed for 11 populations of the outcrossing perennial *Scabiosa columbaria* in Switzerland, and seeds from 89 maternal families were used in a greenhouse competition experiment (Pluess & Stöcklin 2004). Plants of *S. columbaria* were grown alone or in competition with the grass *Bromus erectus* for seven months. Populations did not differ significantly in their biomass response to competition; however, their relative competitive performance increased significantly with increasing molecular diversity, quantified as expected heterozygosity (Fig. 7.10). The authors suggested that genetic erosion in small populations of *S. columbaria* may lead to reduced competitive ability and a greater risk of local extinction (Pluess & Stöcklin 2004). This is an interesting result and future work should be aimed at characterizing the molecular genetic loci important to competitive success in different genotypes and under different environmental conditions.

7.8 ALLELOPATHY

Competitive interactions between plant species should select for traits that can be used to acquire key resources effectively, such as light or soil nutrients (Harper 1977; Tilman 1988; Bazzaz 1996). The resource reduction that occurs as individuals compete is an important reason for the negative effects of one species on the growth and reproduction of another. However, some plant species can also produce and secrete secondary metabolites or other chemicals that interfere directly with the survival, growth, and/or reproduction of neighboring plants of other species. This active process is known as *allelopathy*, and the compounds that mediate it are allelochemicals (Rice 1984; Inderjit & Weiner 2001; Chou 2006; Inderjit et al. 2011). Root exudates containing allelochemicals that act as phytotoxins are likely to play a major role in plant–plant interactions with allelopathy (Bertin et al. 2003; Bais et al. 2006).

The ecological mechanisms involved in allelopathy are complex and difficult to elucidate or distinguish from resource competition (Weidenhamer 2006). Although direct allelopathic interference resulting from the production of one or more allelochemicals can occur between adjacent plants, allelopathic interactions can be indirect and mediated by abiotic components of the soil or microorganisms (Inderjit & Weiner 2001; Bais et al. 2006).

Allelochemicals can play a role in the evolution of plant–plant interactions (Inderjit et al. 2011), and that is the focus here. For example, Callaway et al. (2005) investigated five native grass species at sites in western Montana that had been invaded or not invaded by the allelopathic species *Centaurea maculosa*. The roots of this invasive weed exude catechin, an allelochemical with known phytotoxic effects. Because catechin is present only in the soils of invaded sites, it was hypothesized that only populations of the grasses in invaded sites would show evidence of adaptation indicated by a greater resistance to *C. maculosa* and its root exudates. The results of interspecific competition and seed germination trials showed that the native grass populations from sites where invasion had occurred 20 to 30 years earlier were less affected by *C. maculosa* root exudates than conspecific grass populations from sites that had never been invaded (Callaway et al. 2005). Thus, in a relatively short time, these grass species had evolved some level of resistance to the allelochemical released by a problematic invasive species.

The terpenes produced by aromatic species of *Thymus* may have allelopathic functions and act as chemical selection agents on associated species in the community. Local adaptation to specific terpenes present within native sites was demonstrated in several plant species (Ehlers & Thompson 2004; Grøndahl & Ehlers 2008). In the annual *Brassica nigra*, production of greater levels of the allelochemical sinigrin correlated positively with the ability to compete with other species (Lankau 2008). Increased production of sinigrin by *B. nigra* was also shown to occur in response to competition with three other annual species (Lankau & Klienbenstein 2009). Results indicate that induction of allelochemicals by competition with other species is a likely adaptive response that improves the fitness of *B. nigra*.

Although these examples show that allelochemicals can act as selection agents and contribute to the success of some invasive species, much remains to be learned regarding the evolutionary significance of allelopathic interactions (Inderjit et al. 2011). Given the ability of plants to evolve quickly in response to edaphic factors such as heavy metals (Section 6.2), it is perhaps not unreasonable to expect evolutionary responses of populations to allelochemicals with potentially harmful effects in the soil. Allelopathic interactions may represent a coevolutionary arms race, or what Thompson (2005, p. 237) refers to as "coevolutionary alternation with escalation"—one species produces allelochemicals and improves its fitness; a second species evolves a way to mitigate its negative effects; the first species evolves a slightly different, more effective form of allelochemical; and so on.

7.9 FACILITATION

Not all plant–plant interactions have negative fitness effects on the individuals involved. There are many documented cases in which one species facilitates (i.e., enhances) the survival, growth, and/or reproduction of a second species (Hunter & Aarssen 1988; Callaway 2007). There has been much interest in facilitation as a critical feature in plant communities with ecological and evolutionary implications (Brooker et al. 2008; Kikvidze & Callaway 2009). Meta-analyses have shown that facilitative interactions are especially important in environmentally stressful habitats (He et al. 2013). There are many ways facilitation can occur, such as amelioration of the microenvironment by one species (e.g., reducing moisture loss by shading), rendering improved conditions for survival and growth of a second species. General mechanisms of facilitation have been detailed elsewhere (Hunter & Aarssen 1988; Callaway 2007).

It is worth noting that facilitation may be a result of the direct consequences of habitat modification by a facilitator species that is not affected by the interaction. If there is no fitness benefit to the facilitator species, then the relationship between the interacting species is a commensalism and there is no evolutionary explanation for why one species is a facilitator. Note that this does not preclude there being genotypic variation in a facilitator species in regard to its influence on recruitment and growth of associated species (Proffitt et al. 2005; Michalet et al. 2011). However, for the species being facilitated, selection should favor traits that improve the chances of a facilitative interaction because the plants show greater fitness as a result. For example, in an arid environment, wind-dispersed seeds with features (e.g., hooks or hairs) that make them more likely to be "caught" by a potential nurse plant species could be favored by selection if the nurse plant increases the chances for germination and seedling establishment.

If two species are each facilitated by the presence of the other, then the interaction is a mutualism and coevolution is expected (Thompson 2005; Bronstein 2009). For example, selection should favor synchronized flowering in both species, if this improves pollinator visitation and the chances of successful fruit and seed set (Rathcke 1983). Evolutionary aspects of facilitation and mutualism are explored further by Bronstein (2009). Another scenario that can occur is when the facilitated species do so well in the presence of the facilitator species that they grow large and become effective competitors, thereby reducing the fitness of the facilitator (Holzapfel & Mahall 1999; Michalet et al. 2011). Callaway (2007) indicated that this situation may be the most common—facilitation by one species with a reciprocal competitive effect of the other species on the first (see Holzapfel and Mahall [1999] for an example with a desert shrub and an annual plant community in the Mojave Desert, California). In these cases, selection should favor traits in the facilitator species that *reduce* the likelihood of facilitation! Most ecologists now recognize that competition and facilitation often co-occur in natural plant communities (Callaway & Walker 1997) and vary geographically across diverse habitats (e.g., Pennings et al. 2003; He et al. 2013).

Although it is challenging to envision how a facilitative interaction might arise via natural selection, evolutionary consequences should follow when they do, provided there exists genetic variation in the ability of a species to facilitate and in how organisms respond to facilitation. Liancourt and Tielbörger (2011) investigated two differentiated populations of the annual grass *Brachypodium distachyon* transplanted into a stressful arid site in Jordan to quantify its facilitative interaction with a nurse shrub species (*Gymnocarpos decander*). One population was from a mesic site with a Mediterranean climate whereas the other was from a desert site with arid conditions (the same as the transplant site). The expectation was that the Mediterranean population, which is not adapted to arid conditions, would benefit more from facilitation by shrubs than the arid population (note that these populations were considered distinct ecotypes by Liancourt and Tielbörger [2011] based on prior studies cited therein). Seeds of both populations were transplanted to locations underneath or away from nurse shrubs at the environmentally stressful arid site. Seedling emergence, survival to reproduction, and final aboveground biomass were assessed for populations in the two locations.

As predicted, facilitation by shrubs enabled individuals from both populations of *Brachypodium distachyon* to survive and grow better at the arid site, but the fitness benefits were greater for plants from the Mediterranean population that were not adapted to the stressful conditions at locations away from the shrubs (Liancourt & Tielbörger 2011). In fact, no Mediterranean plants survived at all unless they were beneath nurse shrubs, although plants from the arid population showed more than 20% survival without nurse plants (Fig. 7.11A). Both shrub presence ($\chi^2 = 12.45$, $p < 0.001$) and population origin ($\chi^2 = 7.26$, $p = 0.007$) had highly significant effects on survival. Aboveground biomass was significantly different under shrubs versus away ($F = 78.65$, $p < 0.001$), and there was a significant interaction of plant location

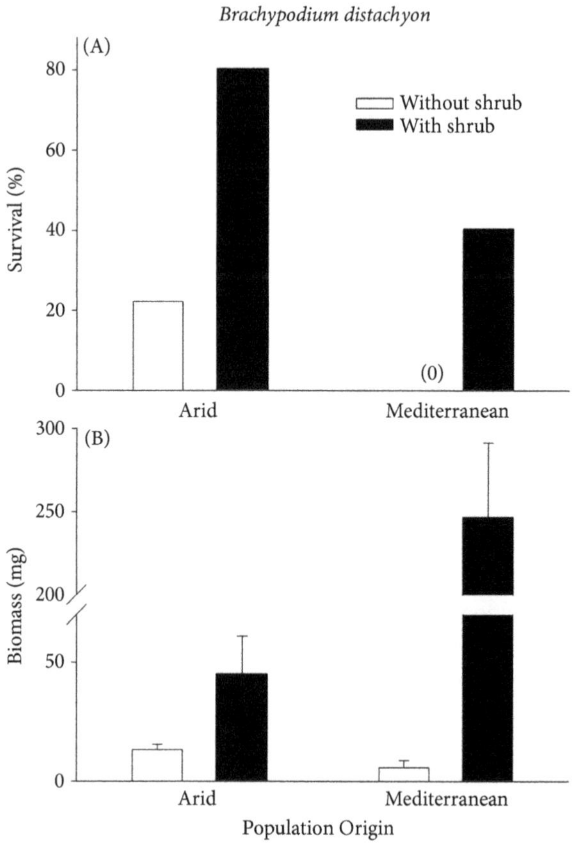

FIGURE 7.11
Population differentiation in response to facilitation in the annual grass *Brachypodium distachyon*. Populations were from arid or mesic (Mediterranean) ends of an ecological gradient in Jordan, and seeds were transplanted underneath or away from nurse shrubs (*Gymnocarpos decander*) at an environmentally stressful arid site. (A) Survival to reproduction. (B) Mean biomass at maturity ± standard error (note the break in the *y*-axis). Redrawn from Liancourt and Tielbörger (2011).

and population origin ($F = 10.02$, $p = 0.004$). The latter is the result of the huge biomass benefit to Mediterranean plants (relative to the arid population) when they were growing beneath the facilitator shrub species (Fig. 7.11B). The authors concluded that local adaptation within a species can result in differential responses to facilitation under field conditions (Liancourt & Tielbörger 2011). This finding agrees with that of an earlier study of serpentine and nonserpentine populations of the annual *Plantago erecta* reciprocally transplanted into serpentine and on-serpentine soil (Espeland & Rice 2007). Plants from the population *not* adapted to stressful serpentine soils produced more biomass at greater densities when grown on serpentine soil, suggesting a net facilitative effect among plants growing together in a stressful environment.

Interactions falling under the banner of "facilitation" are broad and include mutualistic, commensal, and partially antagonistic relationships (Brooker & Callaway 2009). Interactions are often, of course, reciprocal, and a net benefit to one or more of the interacting species identifies facilitation. Compared with the plethora of examples of facilitation in many ecosystems worldwide (Callaway 2007), aspects of the evolutionary ecology of the interaction are only beginning to be explored (Bronstein 2009). The fitness costs and benefits of facilitation for the facilitator species have been only rarely addressed (Michalet et al. 2011), and the genetics of the relevant traits expected to evolve as a consequence of the interaction are not resolved (Bronstein 2009).

7.10 WRAP-UP

There is little doubt that plant–plant interactions play an important role in the evolutionary processes that affect all communities (Harper 1977; Bazzaz 1996; Callaway 2007; Thorpe et al. 2011). However, the hypothesis that communities of species that have coexisted for a long time will behave differently than species without a history of interaction has rarely been adequately tested. There is clearly a genetic component to competitive performance within natural populations, and neighbors—whether competitive or facilitative—can be expected to act as agents of natural selection (Table 2.1) that affect adaptive evolution in many ecosystems. Unfortunately, the quantitative trait loci involved have yet to be characterized adequately. Furthermore, other factors—both abiotic and biotic—can affect the outcomes of competitive and facilitative interactions, adding an additional level of complexity to interpretation. Future research focused on the evolutionary implications of plant–plant interactions and using diverse approaches, including quantitative and molecular genetic analysis, should provide additional insight.

8 Biotic Interactions II: Microbial Symbiosis

8.1 THE UBIQUITY OF PLANT–MICROBE INTERACTIONS

Microbes are ubiquitous associates of all plant species, growing on and in leaves, stems, and roots (Monier 2006; Müller et al. 2006; Arnold 2007; Barton & Northup 2011). Diverse groups of many bacteria, fungi, and viruses are known to infect plants. For example, Schmit and Mueller (2007) estimated that more than 82% of the 700,000+ species of fungi worldwide are microfungi associated with terrestrial plants! The intimate interactions between microbes and plants provide the opportunity for coevolutionary changes in natural populations (Thompson 2005; Thrall et al. 2007; Burdon & Thrall 2009; Thompson 2009). The focus in this chapter is on the effects of microbial symbionts on host plant populations in an effort to understand their role as agents of natural selection that cause microevolutionary change.

Because they are microscopic, for a long time the potential significance of microbes to plant population dynamics and evolution was often overlooked by ecologists (Harper 1977; Burdon 1987; Gilbert 2002). Given the widespread prevalence of microbial symbionts, one may wonder how much of the unexplained phenotypic variation among individuals subjected to experimental treatments by plant ecologists is the result of unknown variation in the diversity, densities, and general effects of the bacteria and fungi living on and in them (Friesen et al. 2011; Gaiero et al. 2013). However, there is little doubt that symbionts can affect attributes of plant life history—including reproductive fitness—in many ways. In addition, it is well established that these diverse effects are contingent on environmental conditions and the interacting genotypes (host and symbiont).

Using a broad definition of symbiosis, one or both of the interacting species may be harmed by, or benefit from, the association (Goff 1982; Paracer &

Ahmadjian 2000; Hirsch 2004; Thrall et al. 2007). However, the microbial symbiont is mostly presumed to benefit through its use of carbohydrates (or other carbon sources) produced by the host plant. Three general types of host–symbiont relationship are possible, each identified by its fitness consequences to the host (Table 8.1). Parasitic symbionts reduce host fitness and are specified as pathogens when they cause disease. The adjective "antagonistic" is often applied to this type of symbiosis (e.g., Saikkonen et al. 1998; Thompson 2005). Gilbert (2002) has reviewed aspects of the evolutionary ecology of plant diseases, providing an update of Burdon's (1987) book.

When an infected host suffers no loss or gain of fitness relative to uninfected individuals, the relationship is commensalistic, and evolutionary effects on host populations are not expected (Table 8.1). Hosts that have evolved complete tolerance for a (formerly) parasitic microbe may indicate a commensalism (Miller et al. 2006). When host fitness is improved by symbiont infection, both host and symbiont benefit, and the relationship is a mutualism (Table 8.1). There seems to be increasing fascination with this symbiosis by evolutionary ecologists inclined to review the topic, and my personal "mutualism" file bulges with dozens of review articles spanning decades. For a few general overviews, see Bronstein (2001, 2012), Gomulkiewicz et al. (2003), and Thompson's (2005) coevolution text (with three chapters on mutualism). Some of the most widely studied plant–microbe interactions are predominantly mutualistic in nature (e.g., mycorrhizal fungi or rhizobial bacteria in roots).

As with any categorization scheme for biological phenomena, the symbiotic relationships outlined in Table 8.1 describe a continuum that lacks well-defined boundaries. This continuum from parasitism to mutualism has been recognized for decades (Ewald 1987; Saikkonen et al. 1998; Hirsch 2004). The "blurring of boundaries" along

Table 8.1 Overview of plant-microbe symbiotic relationships that range from parasitic (antagonistic) to mutualistic as defined by their consequences to host fitness, and their possible evolutionary effects on host populations.

Symbiotic relationship	*Effects on microbe, host*	*Proximate consequences*	*Evolutionary effects*	*Microbial examples*
Parasitic	+, —	Reduced fitness of infected host	Selection favors resistant or tolerant hosts	Bacterial and fungal pathogens
Commensal	+, 0	Fitness of infected and uninfected hosts is equivalent	None	Some asymptomatic fungal endophytes
Mutualistic	+, +	Enhanced fitness of infected host	Selection favors hosts that establish/maintain symbiosis, vertical transmission (via seeds)	Rhizobial bacteria, mycorrhizal fungi, some asymptomatic fungal endophytes

this continuum (Saikkonen et al. 1998, p. 320) is the result of the interaction of environmental and genetic factors that cause widespread variation in all symbiotic associations.

This chapter examines four major groups of microbes involved in plant–microbe interactions from the perspective of the evolutionary ecology of the host: parasites/pathogens, rhizobial bacteria, mycorrhizal fungi, and systemic fungal endophytes.

8.2 PARASITES/PATHOGENS

Pathogens are microbial bacteria and fungi (or viruses) with typically negative effects on the big three of plant life history: survival, growth, and reproduction (Gilbert 2002). As such, they function as "potent selective forces" (Burdon 1991, p. 423) capable of changing population genetic structure and causing microevolution. Much of what is known about plant–pathogen interactions has come from agricultural species and their economically important pests; thus, there is a long history of research recorded in plant pathology journals (e.g., *Phytopathology* has been published continuously since 1910, whereas *Molecular Plant–Microbe Interactions* launched in 1988). However, Harper (1977) recognized their probable importance to natural populations and included a chapter on pathogens in his monumental treatise *Population Biology of Plants*. Ten years later, Burdon (1987) published his highly influential text *Diseases and Plant Population Biology*, with a predominant focus on nonagricultural systems. It included chapters on the genetic basis of disease resistance, the effect of pathogens on host population genetic structure, and environmental modification of host–pathogen interactions. Since that time, the field of disease ecology has emerged as a major discipline (Real 1996; Collinge & Ray 2006; Ostfeld et al. 2008), with much research focused specifically on the coevolutionary relations of plants and their pathogens, and the role of pathogens as selection agents (Burdon 1991; Schmid 1994; Clay & Kover 1996; Burdon & Thrall 2001, 2009).

8.2.1 Genetic Variation in Host Resistance

One key to establishing the potential for host populations to evolve when pathogens occur is genetic variation in the ability to resist pathogen attack, or to at least minimize the negative fitness consequences of infection. Such genetically based differences have been demonstrated both within and between host populations attacked by a variety of infectious agents. Approaches that entail traditional ecological genetics as well as those that use the tools of molecular genetics have provided insight into the complexity of host resistance in natural plant populations.

Early studies by Parker (1985) on the selfing annual legume *Amphicarpaea bracteata* suggested population differentiation for resistance to the fungal pathogen *Synchytrium decipiens* after the transplanting of plants from three host populations in Illinois into a heavily infested site. Plants that grew normally at that site became heavily infected by *S. decipiens*, whereas two populations from elsewhere (1 km or 100 km away) were mostly immune to infection, indicating genetic specialization at the population level. However, later analyses of this system showed that the pathogens were most likely adapted to specific host lineages that might be regarded as cryptic species within *A. bracteata* (Parker 1996; Spoerke et al. 1996).

Significant within-population variation (i.e., among families) was also found for growth measures of *Amphicarpaea bracteata* hosts made on infected plants in a greenhouse environment (Parker 1986). In addition, directional selection differentials for infection intensity were calculated for two infected populations at field sites near Chicago, Illinois (following Lande and Arnold [1983]), using total seed mass as a fitness measure. Standardized differentials were −0.34 and −0.78 for the populations, indicating strong selection for increased resistance to fungal attack at both sites (Parker 1986). Later analysis of F_2 plants from a cross between resistant and susceptible populations suggested quantitative genetic inheritance (seven to eight loci) for disease resistance in *A. bracteata* (Parker 1991).

Significant genetic variation for disease resistance has been well documented in the model plant *Arabidopsis thaliana*. Kover and Schaal (2002) examined both resistance and tolerance (i.e., the ability of the host to reduce the negative effect of infection on fitness) to the infectious bacterium *Pseudomonas syringae* in a worldwide collection of 19 *A. thaliana* accessions. Pathogen growth, disease symptoms, and host fitness (seed number) were measured as components of resistance in a growth-chamber experiment. The accessions differed greatly in the expression of disease symptoms ($F = 3.39$, $p < 0.01$) and the size of bacterial populations present in their leaves ($F = 13.43$, $p < 0.0001$). The negative fitness consequences of infection also varied among accessions (accession-by-infection interaction, $F = 2.66$, $p < 0.01$). Interestingly, there was no correlation between disease symptoms or bacterial growth and host fitness, implying that tolerance plays an important role in the response of *A. thaliana* to this pathogen (Kover & Schaal 2002). Although it has been somewhat understudied relative to resistance, tolerance to disease may be a particularly important evolutionary strategy for long-lived perennials exposed continually to pathogens over long time periods (Roy et al. 2000).

In another study (Salvaudon et al. 2005), selfed *Arabidopsis thaliana* lines derived from five European populations were used in a cross-inoculation experiment with two strains of a downy mildew (*Hyaloperonospora parasitica*, Oomycota). Parasitic virulence was estimated as the difference between the mean number of seeds made by uninfected control plants and that made by inoculated plants. Virulence differed among combinations of parasite strains and host lines, and some host lines actually showed *increased* seed production when infected! There was also a significant strain-by-host line interaction for the number of sporulating leaves ($F = 27.13$, $p < 0.001$)—evidence of strong genotype-specific interactions between parasite and host (Salvaudon et al. 2005).

The genetic diversity of host populations can be critical to determining the extent to which pathogens prevail. For example, the diversity of resistant phenotypes in populations of *Linum marginale* correlated negatively with the prevalence of disease caused by a rust fungus (Burdon & Thrall 2001). In the goldenrod *Solidago altissima*, more genetically diverse groups supported lower levels of infection by a species of mildew, although there was pronounced variation among families in their levels of infection (Schmid 1994).

Genetic variation in resistance to pathogens is undoubtedly widespread within and between plant populations. Although fungal pathogens have been most studied, bacterial (e.g., Kover & Schaal 2002; Goss & Bergelson 2006) and viral (e.g., van Mölken & Stuefer 2011) pathogens have also revealed genetically based host resistance. In *Trifolium repens*, the effects of the white clover mosaic virus on ramet production and biomass showed great variation among host genotypes (identified by AFLPs), as shown by reaction norm diagrams in van Mölken and Stuefer (2011). In general, host genotype-by-infection interactions provide the setting for pathogen-mediated effects on the microevolution of plant populations.

For some plant–pathogen associations, molecular approaches have been used to characterize the gene loci involved in host resistance (Karasov et al. 2014). This is especially true for long-studied diseases of agricultural species and model plants such as *Arabidopsis thaliana* (Bergelson et al. 2001; Rausher 2001; de Meaux & Mitchell-Olds 2003). The so-called plant resistance (R) genes include loci that segregate for many allelic variants, and this polymorphism is a major part of the variation in pathogen resistance (Bergelson et al. 2001; de Meaux et al. 2003; Karasov et al. 2014). Some of the molecular genetic interactions between host and pathogen follow the basic gene-for-gene model proposed long ago by Flor (1951), which maintains that, for any one gene that determines host resistance, there is a corresponding gene in the pathogen with which it interacts (Burdon 1987; Simms 1996). For example, the highly variable *Rps2* gene in *A. thaliana* confers resistance to *Pseudomonas syringae* bacteria having a corresponding avirulence gene *avrRpt2* (Mauricio et al. 2003). In a worldwide survey of 27 accessions of *A. thaliana*, *Rps2* alleles could be grouped into those conferring mild, intermediate, or strong resistance to *P. syringae*; evidence was found for selection maintaining different groups of alleles in disease-resistant versus susceptible accessions (Mauricio et al. 2003). Polymorphism in a candidate gene family important to resistance to a fungal pathogen that causes anthracnose was found for 15 populations of wild beans (*Phaseolus vulgaris*) examined in Argentina (de Meaux et al. 2003). Furthermore, populations were differentiated with respect to molecular markers for disease resistance.

It is worth noting that not all plant–pathogen relationships are determined by interactions between single gene loci (and their molecular products), although they may be the easiest to track via evolutionary models (Barrett 1985; Clay & Kover 1996; Burdon 1997; de Meaux & Mitchell-Olds 2003). Quantitative genetic inheritance of disease resistance can also be expected in natural populations (Parker 1991; Simms 1996). Clearly, much remains to be learned about the subtleties of the complex molecular genetics of host resistance to pathogen attack and how that information would improve our understanding of evolutionary ecology. As Burdon and Thrall (2009) pointed out, it is necessary to integrate knowledge about the molecular basis of host resistance and pathogen virulence with an analysis of how variation in resistance and virulence genes affect disease and host dynamics in nature to advance understanding of host–pathogen coevolution.

8.2.2 Local Adaptation

Given the variation in pathogen virulence and host resistance found in natural populations (e.g., Thrall et al. 2002; Goss & Bergelson 2006) and the potential for reciprocal selection pressures acting on pathogen and host, it might be expected that local adaptation of both interacting partners would arise (Kaltz & Shykoff 1998). Although most investigators focus on one side of a cross-species interaction, there are now many studies of species that show evidence of local adaptation to other species and "the list continues to grow" (Thompson 2005, p. 56). Greischar and Koskella (2007) provide an overview of the issues involved in showing parasite/pathogen adaptation to many kinds of hosts, including plants.

In this section, space precludes presenting more than just a few studies on local adaptation in pathogen–plant associations. Often, the adaptive traits examined are those of the pathogen species that allow it to infect and grow on host plants. Cross-inoculation procedures (essentially reciprocal transplants of pathogen strains) may be used to determine the performance of pathogen strains on sympatric versus allopatric host plants (Thrall et al. 2002; Capelle & Neema 2005; Sicard et al. 2007; Tack et al. 2014). Superior performance of a pathogen strain, however assessed, with the sympatric host population it infects in nature (relative to other [allopatric] populations) should indicate local adaptation of the pathogen to its host. Some pathogen–plant systems have shown remarkably fine-scale pathogen adaptation at the level of the individual host plant (e.g., Capelle & Neema 2005).

Thrall et al. (2002) investigated local adaptation of the rust *Melampsora lini* to its perennial host *Linum marginale* at several spatial scales in six populations in Australia. Using an extensive set of cross-inoculation trials, strong adaptation of *M. lini* was detected, especially at regional scales of host distribution. Typically, the average virulence of a rust strain was greatest on the population of *L. marginale* with which it had coevolved (Thrall et al. 2002).

Three strains of the fungus (*Colletotrichum lindemuthianum*), which causes anthracnose disease, were isolated form three wild bean (*Phaseolus vulgaris*) populations in Mexico and were used for a cross-inoculation experiment (Sicard et al. 2007). A detached leaf assay, developed for this system, was used to determine the proportion of leaves that became infected after inoculation and the extent of leaf surface damage. Evidence for local adaptation of the pathogen to its source host population was found for both of these "fitness traits" (Sicard et al. 2007). For example, after experimental cross-inoculation, each of the pathogen strains isolated from the three host populations infected the greatest proportion of the leaves of the host population with which it had coevolved (Fig. 8.1). Thus, strains were more infectious on their "home" host populations (Sicard et al. 2007, p. 33). Note that statistical comparisons

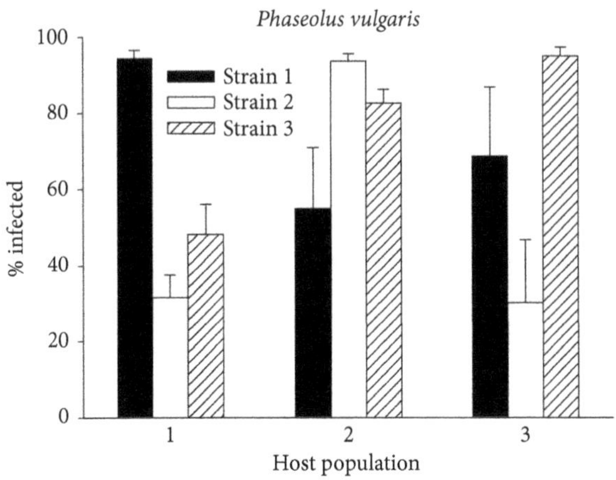

FIGURE 8.1
Percentage of leaflets of plants from three populations of wild bean (*Phaseolus vulgaris*) that were infected after inoculation with strains of the fungal pathogen *Colletotrichum lindemuthianum*. The pathogen strains were originally isolated from the three bean populations in Mexico. Redrawn from Sicard et al. (2007).

of infection proportions (Fig. 8.1) were all significant, whether between home versus away or resident versus nonresident. This type of local differentiation in pathogen–host compatibility among host populations may be especially likely in self-pollinating species such as wild beans (Capelle & Neema 2005; Sicard et al. 2007), annual legumes (Parker 1985), and some perennial herbs (Thrall et al. 2002).

Geographic variation in coevolutionary selection pressures is one of the hallmarks of the mosaic theory of coevolution and is expected to lead to mosaics of local adaptation across the landscape (Thompson 2005). A geographic cline in resistance to the rust fungus *Melampsora lini* was found in the annual dwarf flax *Hesperolinon californicum*, a species endemic to California (Springer 2007). A cross-inoculation experiment was used to characterize resistance of 16 host populations that spanned its latitudinal range. Northern populations that field surveys showed had the highest levels of rust infections had the lowest levels of resistance. Co-occurring clines in infection levels in the field and genetically based disease resistance suggested "strong links between pathogen-mediated selection and host evolution" (Springer 2007, p. 1820). Across the entire experiment, significantly greater infection levels occurred in pairings of sympatric pathogen–host combinations (compared with allopatric combinations), indicating local adaptation of the rust to its host.

The relative scale at which populations are sampled clearly plays a role in the ability to detect local adaptation in plant–pathogen associations (Davelos et al. 1996). For example, Laine (2005) found local adaptation of a powdery mildew to its perennial host *Plantago lanceolata* at scales of tens of kilometers, but not at smaller scales, perhaps because of the strong dispersal capacity (gene flow) of the pathogen. However, in areas within some host populations where encounter with the disease was highly probable, resistance to the pathogen was greatest, indicating relatively small-scale adaptive responses to pathogen-mediated selection were possible (Laine 2006). A later study of the same pathogen–host system in Finland corroborated the findings of scale-dependent patterns of local adaptation (Tack et al. 2014).

This section provided a small sampling of the studies that have provided evidence for local adaptation of pathogens to their host populations. Genetically based variation in resistance (of hosts) and virulence (of pathogens) is widespread geographically in most systems, creating the opportunity for ongoing reciprocal selection to fine-tune coevolutionary interactions. Pathogen-mediated selection, when strong and consistent over time, may mold the evolution of specific host traits important to life history (beyond simple resistance traits).

8.2.3 Host Sexual Reproduction

Through their impact on the survival and reproduction of individuals, pathogens have great potential to affect the ecological genetics of plant populations and their future evolution. Foliar diseases, by reducing the photosynthetic capacity of leaves, can limit the carbohydrates needed for flower, fruit, and seed production, thereby reducing fecundity (e.g., Parker 1987). In addition, by attacking floral organs or developing seeds directly, host fecundity can be reduced or prevented completely by pathogen infection (Burdon 1987; Alexander 1989; Clay 1991).

Classic examples familiar to many ecologists include choke disease caused by the fungus *Epichloë typhina*, with stromata that cause complete suppression of the inflorescences of infected grasses (Fig. 8.2) and anther–smut disease caused by *Ustilago*

FIGURE 8.2
Inflorescences of barnyard grass *Dactylis glomerata* covered by the stromata of the choke disease fungus *Epichloë typhina* (arrows). These inflorescences are completely sterile, never setting seeds. Photo courtesy of Adrian Leuchtmann.

violacea, which results in sterility of both male and female flowers of the short-lived perennial *Silene alba* (Alexander 1989).

Choke disease caused by *Epichloë typhina* was already referred to as a "common and well-known disease" in 1959, when A.D. Bradshaw published a seminal article on its incidence and ecological significance to the grass *Agrostis capillaris* (Bradshaw 1959, p. 310). Although infected hosts were sterilized completely by the pathogen, they made more tillers relative to uninfected plants, suggesting that infection could increase vegetative growth (a finding that also appears in some species of asymptomatic fungal endophytes that inhabit grass leaves; see Section 8.5; Cheplick & Faeth 2009). From the standpoint of the pathogen, this makes evolutionary sense if greater vegetative vigor of the host provides more opportunities for fungal reproduction via spores and their subsequent dispersal to new host plants.

In addition to the complete host sterility enacted by diseases like choke and anther–smut disease (studied in detail by Helen Alexander and colleagues in a series of articles too numerous to enumerate here [e.g., Alexander 1989; Alexander et al. 1996]), other types of parasitic castration of plants by fungi have been demonstrated (Clay 1991). This circumvention of host reproduction by parasites/pathogens may have selective value for the parasite because offspring produced by the host via sexual reproduction may be more likely to include rare, resistant genotypes to which the parasite population is not adapted (the "Red Queen Hypothesis" [Clay & Kover 1996; Thompson 2005]). Also, production of spores and their dispersal may be enhanced on inflorescences that are typically elevated above the foliage in many plants (Fig. 8.2). The coevolutionary arms race that ensues involves selection for increased virulence in the parasite/pathogen (to reduce host outcrossing and increase spore output) and selection for increased resistance or tolerance in the host (to maximize sexual reproduction). However, outcrossing can also lead to greater susceptibility to pathogen attack resulting from the breakup of resistant genotypes previously adapted to local pathogens (Koslow & Clay 2007). In fact, regarding the Red Queen Hypothesis, Bell (2008, p. 429) stated that "the flaw in this idea is that recombination is just as likely to create inferior [genetic] combinations, so that there is no net advantage of sex." Nevertheless, pathogen-mediated selection has long been thought to be important to the evolutionary maintenance of sexual reproduction in plants (Levin 1975; Kelley 1994; Clay & Kover 1996).

The association between outcrossing rate of different host species and the number of fungal pathogen species that attack them has been examined (Busch et al. 2004). Data sets on outcrossing rates and infecting pathogen species were obtained for 182 plant host species. Meta-analysis revealed a significant positive correlation between outcrossing rate and the number of fungal pathogen species attacking each plant species (Spearman's $\rho = 0.23$, $p < 0.001$, $n = 182$ [Busch et al. 2004]). Although this result accords with the hypothesis that genetic exchange via outcrossing during sexual reproduction by plants could be an adaptive response to selection imposed by pathogens, empirical studies of specific pathogen–host associations may provide more compelling evidence. For example, field experiments with the perennial grass *Anthoxanthum odoratum* have shown that sexually produced genotypes from particular parents were less likely to get infected by a particular virus transmitted by aphids (and suffer the documented negative fitness consequences) than asexual clones of the same parents (Kelley 1994). Experiments like these provide strong support for the idea that there is a selective advantage for rare genotypes produced by sexual populations under pathogen pressure, and therefore lend support to the Red Queen Hypothesis (Clay & Kover 1996).

8.3 RHIZOBIAL BACTERIA

Although the interaction of pathogens with plants is, by definition, an antagonistic one, that of rhizobial bacteria and roots is one of the textbook examples of mutualism (e.g., Paracer & Ahmadjian 2000; Smith & Smith 2012). Rhizobial bacteria in soil can form unique structures (nodules) on roots, mostly in plants of the legume family (Fabaceae). The bacteria in these nodules fix atmospheric nitrogen into ammonia that can be used by plants to produce amino acids and proteins; in return, the bacteria obtain carbohydrate products of host photosynthesis for their own energy metabolism. As with plant–pathogen systems, much of the extensive work on legume–rhizobia interactions has been done with agricultural species, and much is known about the genetically controlled molecular signals from both partners needed to establish the symbiosis. Some of the key literature is summarized by Sachs et al. (2013). Here, the emphasis is on the population genetic variation in microbe–host compatibility found in a few natural legume species.

Intraspecific variation among host plants in their responses to beneficial microbes has been documented in a diversity of plant–microbe interactions (Smith & Goodman 1999). Genotype-by-genotype interactions between partners that determine plant fitness provide evidence for operation of the reciprocal selection pressures expected during coevolution. Heath and Tiffin (2007) examined 10 populations of *Medicago truncatula* across its geographic range and two strains (denoted A and W) of the symbiotic bacterium *Sinorhizobium medicae* commonly used to examine signaling in legume–rhizobia associations. In this greenhouse inoculation experiment, host populations were found to vary significantly ($p < 0.0001$) for the number of leaves and fruits produced; more important, there was a significant ($p < 0.05$) interaction between host population and bacterial strain for both fitness measures. This is evident in the crossing of population-level reaction norms for host fitness (fruit production) that depended on the bacterial strain (Fig. 8.3). Data on the number of nodules produced and their size provided estimates of bacterial fitness (Heath & Tiffin 2007). Statistical analysis showed that the fitness benefits the rhizobia gained from the interaction depended greatly on the plant population it infected. Thus, genetic variation in both partners of the symbiosis can affect coevolutionary dynamics in this plant–rhizobia system. A later study of the same system used gene expression microarrays to show that genetic variation in the transcriptome (i.e., messenger RNA transcripts of active gene loci) also occurred, and host genotype had a large influence on the expression of bacterial genes used in nodule formation (Heath et al. 2012).

Working with the annual legume *Amphicarpaea bracteata*, which has a nitrogen-fixing rhizobial symbiont (*Bradyrhizobium* sp.), Parker (1995) uncovered differential compatibility of host and symbiont genotypes in a common garden experiment at a field site in New York. Plants from eight populations were inoculated with *Bradyrhizobium* isolates from sympatric (native) or allopatric sites. The design entailed each population from a site being paired with a population from a different site within the same geographic region and reciprocally inoculated with bacterial isolates from each population site. Total seed biomass was the estimate of host fitness. Most hosts had greater fitness when growing with bacteria from their native sites, suggesting local adaptation of hosts with their co-occurring symbiont populations. In a related study, cross-inoculation trials were conducted using 11 host populations from Illinois, Indiana, New York, and Virginia (Wilkinson et al. 1996). Again, host and bacterial genotypes from the same sites grown in a greenhouse resulted in significantly greater

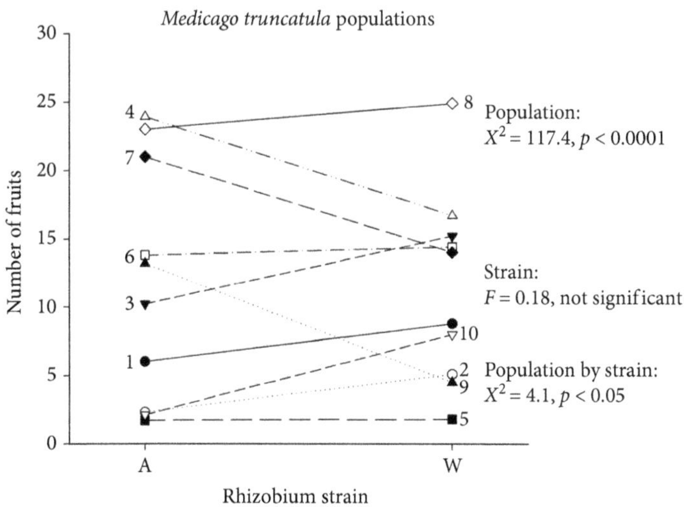

FIGURE 8.3

Mean number of fruits produced by 10 populations of *Medicago truncatula* infected experimentally by two strains of the rhizobial bacteria *Sinorhizobium medicae*: ABS7 (W) and WSM540 (W). Results from mixed-model analysis of variance are presented (population and population by strain are random effects).

Source: Heath, K.D. & Tiffin, P. 2007. Context dependence in the coevolution of plant and rhizobial mutualists. *Proc. R. Soc. B* 274: 1905–1912.

biomass relative to hosts inoculated with nonnative bacterial genotypes. Also, two genetically distinct host lineages (assessed by multilocus enzyme electrophoresis) from different sites showed better growth with symbiotic bacteria isolated from the same lineage. These results suggest intraspecific specialization in this plant–rhizobium mutualism and strong geographic structure in coevolutionary adaptation. In addition, more recent molecular analysis of 14 legume species associated with *Bradyrhizobium* indicate that specific distinct groups of bacterial gene loci (important to nodulation and nitrogen fixation) are associated consistently with particular legume species (Parker 2012). Thus, the specificity of genetically driven interactions in coevolved plant–rhizobium symbioses may occur between, as well as within, legume species.

8.4 MYCORRHIZAE

The evolutionarily ancient symbiosis between the roots of most terrestrial plants and mycorrhizal fungi is another plant–microbe interaction that is viewed mostly as mutualistic (Allen 1991; Smith & Read 1997; Brundrett 2002). The fungi benefit from the carbohydrates provided by the plant that can be used as an energy source, whereas the host plant benefits from improved uptake of soil minerals (and possibly water [Augé 2001]). As with any plant–microbe interaction, hosts can vary in terms of the relative benefits received (Gange & Ayres 1999; Smith & Goodman 1999; Klironomos 2003; Hegason & Fitter 2009), and it is well documented that the benefits versus costs of mycorrhizal infection can change with environmental conditions (Fitter 1991; Johnson 1993; Johnson et al. 1997; Neuhauser & Fargione 2004; Piculell et al. 2008; Hoeksema et al. 2010). It is the benefits (improved fitness) relative to the costs (carbohydrates used by the fungi) of mycorrhizal infection that will ultimately determine where the symbiosis falls along the parasitism–mutualism continuum (Johnson et al. 1997; Morgan et al. 2005; Cheplick 2009). An overview of the key literature on mycorrhizal ecology is provided by Ji and Bever (2012).

There are several distinct groups of mycorrhizal fungi that can influence plant communities and ecosystems (Johnson & Gehring 2007). The arbuscular mycorrhizal (AM) fungi (Phylum Glomeromycota) include several hundred species that are very widespread and abundant; in terms of their effects on plants, they have been the focus of much ecological research. AM fungi are characterized by forming highly branched hyphal structures (arbuscules) within penetrated cells of the root cortex (Smith & Read 1997). Most of the examples in this section are from AMs because most angiosperm species (at least 80%) are infected by them (Brundrett 2002). The ectomycorrhizal fungi, a diverse group of thousands of species (mostly Phylum Basidiomycota, but also Ascomycota), are characterized by a fungal mantle that envelops the root, forming a net of densely intertwined hyphae. These fungi occur in less than 10% of plant species, predominantly in woody angiosperms and gymnosperms, especially in temperate and boreal forests (Johnson & Gehring 2007).

To demonstrate ongoing, reciprocal (coevolutionary) selection in mycorrhizal symbioses, one needs to show significant genetic variation within plant and fungal species for their effects on, or responses to, one another (Hoeksema 2010). In other words, one expects a host genotype-by-fungal genotype interaction to determine the fitness of both participants. Not only are these genotype-by-genotype interactions important to the ecological and evolutionary outcome of the mycorrhizal symbiosis, they also are expected to depend on environmental conditions and to show geographic (between-population) variation (Thompson 2005; Hoeksema & Thompson 2007; Hoeksema et al. 2010, 2012). General background information on coevolutionary selection and the empirical evidence for its occurrence in mycorrhizal interactions has been cogently reviewed by Hoeksema (2010).

There is now good evidence that, despite their asexual mode of reproduction, genetic variation in AM fungi occurs both between and within species (Klironomos 2003; Koch et al. 2006; Hegason & Fitter 2009), with highly variable effects on host populations. Although AM fungi have a broad range of host species they can infect, there is evidence for genetic specificity in some plant–AM fungal associations (Sanders 2002). Variation in host plant responses to infection by different genotypes or taxa of ectomycorrhizal fungi has also been reported for a few species of pine (Piculell et al. 2008; Hoeksema et al. 2009, 2012).

Klironomos (2003) reciprocally inoculated 10 plant species from a field in Canada with 10 species of AM fungi isolated from the same field (home) or from other locations (foreign) in southern Canada. While examining the proportional change in host biomass after 16 weeks in a greenhouse for mycorrhizal versus nonmycorrhizal treatments, Klironomos (2003) found extensive variation in the direction and magnitude of fungal-mediated effects on host growth. No specific plant or AM fungal species was linked consistently to either positive or negative responses of host growth to mycorrhizal infection. Thus, host–fungal interactions were not consistently mutualistic, as commonly supposed (see also Koch et al.]2006]). Furthermore, positive host responses tended to be more likely when inoculated with AM fungi from their home site, suggesting that some plant species and their fungal associates may be locally adapted.

In another insightful experiment, reciprocal cross-inoculations were performed using the dominant prairie grass *Andropogon gerardii* and its AM fungal community across three sites: Konza Prairie in Kansas, Cedar Creek in Minnesota, and a restored prairie at Fermi National Laboratory in Illinois (Johnson et al. 2010). One objective was to determine whether coadaptation of host populations and AM fungi had evolved. If so, then the fitness of both host and fungi should be greatest for the plant–fungi combinations that normally co-occur at any one site. Rather than examine specific AM fungal species as Klironomos (2003) had done, Johnson et al. (2010) used a whole-soil inoculum that contained many species of AM fungi. Samples of sterilized soil from the three prairies were reciprocally inoculated with AM fungi from the same prairies; then, plants from each *A. gerardii* population were planted into all combinations of soil and AM fungi origin (as well as in uninoculated soil). The originally sterile soil was also inoculated with the natural soil microbes it normally contained by sieving out the larger AM fungal spores before establishing the experiment. The levels of nitrogen and phosphorus in plant tissues were determined at harvest (14 weeks), as well as the dry mass of roots, shoots, and inflorescences. The proportion of root length colonized by fungal arbuscules and density of hyphae were quantified as indicators of the success in establishment of host–fungal symbiosis.

For soils from Konza and Fermi, there were highly significant ($p < 0.0001$) positive relationships between both arbuscule colonization or hyphal density and shoot mass (Johnson et al. 2010). At both these sites, available soil phosphorus is very low, so the benefits of mycorrhizal infection were expected to be high in soils from them. Indeed, tissue phosphorus levels also showed a significant ($p < 0.0001$) positive correlation with arbuscule colonization and hyphal density in the greenhouse experiment for soils from Konza and Fermi (but not Cedar Creek), indicating that AM fungi improved phosphorus uptake from soils where phosphorus was limited. In Cedar Creek soils, phosphorus levels are naturally very high and there was no detectable host growth benefit to increasing levels of infection; indeed, shoot mass showed a significant ($p < 0.0001$) *negative* correlation with arbuscule colonization and hyphal density in soils from Cedar Creek. Thus, mycorrhizal infection was detrimental to host plants at a site where soil phosphorus was readily available, supporting earlier research into the benefits and costs of mycorrhizal symbiosis for other plant species (Smith & Read 1997). One might speculate that host traits that favor mycorrhizae formation would be selected in the Konza and Fermi prairie populations of *Andropogon gerardii*, but not in the Cedar Creek prairie. This variable pattern of selection across isolated populations of symbiotic pairs of species (i.e., a geographic mosaic of selection) is expected because coevolutionary interactions vary among environments (Thompson 2005; Piculell et al. 2008; Hoeksema 2010).

The study by Johnson et al. (2010) also provided good evidence that populations of *Andropogon gerardii* and their associated mycorrhizal fungi show local coadaptation. The reciprocal cross-inoculation experiment revealed that the proportion of root length colonized by AM fungal arbuscules was consistently greatest for plants growing in their local (home) soil containing the AM fungal community that normally inhabits that soil (Fig. 8.4). For example, plants from the Fermi prairie showed the greatest root colonization by mycorrhizal fungi (and consistently had the greatest reproductive biomass) when growing in soil from Fermi inoculated with the AM fungal community from the same site (Fig. 8.4A). Although this home-site effect occurred at Cedar Creek, too (Fig. 8.4C), recall that mycorrhizal fungi did not benefit host growth in the high-phosphorus soils from this site. Thus, AM fungi caused a reduction in reproductive biomass (relative to uninoculated controls), although the reduction was least when Cedar Creek plants grew with their local AM fungal community (Johnson et al. 2010).

The research on *Andropogon gerardii*, an important tall-grass prairie species, and its AM fungal community (Johnson et al. 2010) was considered here in depth because it is one of the only studies demonstrating a genotype-by-genotype-by-environment interaction in a mycorrhizal symbiosis (Hoeksema 2010). This type of study provides evidence of the possible role of coevolutionary selection in causing adaptive differentiation among plant and fungal populations by demonstrating local coadaptation of plants and their AM fungal associates (Johnson et al. 2010; Ji et al. 2013). However, it should be noted that not all studies of mycorrhizal fungi and their plant hosts have found evidence of host–fungi specificity or local coadaptation (Hoeksema et al. 2012; Schechter & Bruns 2013). For example, coadapted host–symbiont specificity was not detected in a common garden study of populations of the native annual *Collinsia sparsiflora* and its AM fungal community from serpentine and nonserpentine sites in northern California (Schechter & Bruns 2013). However, this may be because most annuals are not strongly mycotrophic and hence may not benefit greatly from association with AM fungi (N.C. Johnson, pers. comm., May 7, 2014).

Because mycorrhizal fungi spend some of their life cycle in the soil rather than as plant symbionts, selection could favor fungal traits that improve survival and fitness in the soil community independent of their symbiotic interactions with plants (Hegason & Fitter 2009). This may be especially true for ectomycorrhizal fungi that function as free-living saprotrophs, although certainly less so for AM fungi that, as obligate biotrophs (Smith & Read 1997), cannot complete their life cycle without

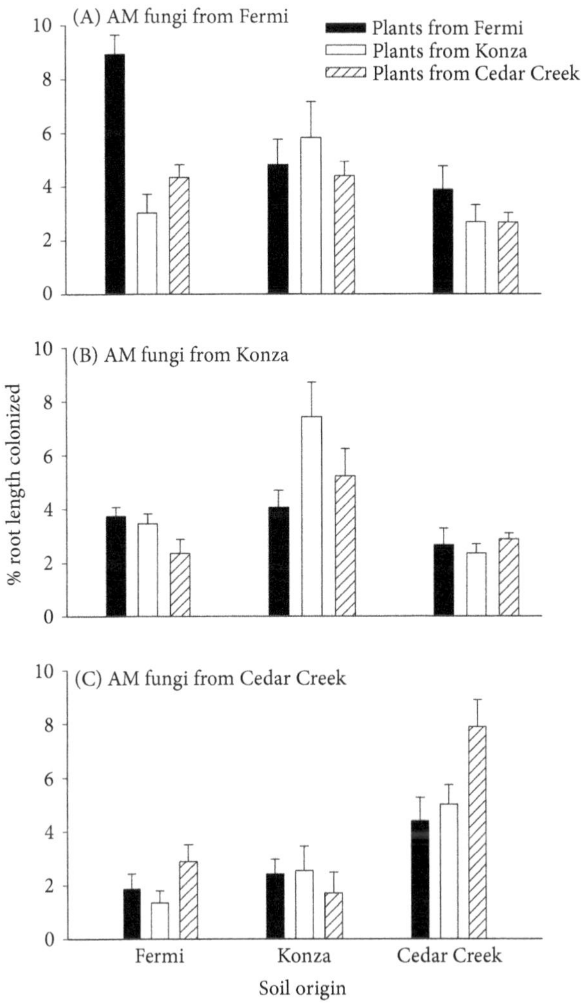

FIGURE 8.4

(A–C) Percent of the root length of the perennial grass *Andropogon gerardii* containing arbuscules formed by arbuscular mycorrhizal (AM) fungi (Glomeromycota) from three prairie sites in the midwestern United States. In this reciprocal cross-inoculation experiment, plants from each site were inoculated with AM fungi from all sites and grown in soil from all sites. Redrawn from Johnson et al. (2010).

infecting a plant host. In any event, the complete soil microbial community (which includes potential mycorrhizal species) can function as a selection agent on natural plant populations (Lau & Lennon 2011; Sherrard & Maherali 2012).

Thus, the list of edaphic factors that affect the evolutionary ecology of plants (Chapter 6) must include a biotic component—that is, a community rich in microbes (Rout & Southworth 2013)—and also invertebrate animals. Some members of the microbial community, including the rhizobial bacteria and mycorrhizal fungi, can form symbiotic associations with plants and alter the survival, growth, or reproduction directly of individuals in a genetically variable gene pool. Long unnoticed by ecologists, these microbial symbionts could be indirectly responsible for many habitat-mediated selection effects in plant populations and clearly deserve much additional study.

8.5 SYSTEMIC LEAF ENDOPHYTES

Like AM fungi, the fungal endophytes considered in this section are biotrophs that live symbiotically within tissues (Isaac 1992) and also have effects on hosts that range from antagonistic to beneficial (Saikkonen et al. 1998; Schulz & Boyle 2005; Cheplick & Faeth 2009). However, they differ in that they are predominantly ascomycetes (Phylum Ascomycota) and grow between the cells of leaves and stems (Arnold 2007; Christensen & Voisey 2007; Bacon et al. 2009). In grass leaves, endophytic hyphae have a characteristic appearance and are easily stained in situ to be viewed under a light microscope (Fig. 8.5). Unlike mycorrhizal fungi, leaf endophytes are unable to exist in soil independent of their hosts (Isaac 1992) and are therefore engaged in a symbiotic relationship that is obligatory for the endophyte, but facultative for the plant.

Leaf endophytes are extraordinarily common in plants in ecosystems worldwide and only a small subset of plant–endophyte symbioses have been investigated (Arnold 2007; Rodriguez et al. 2009a). Grass–endophyte interactions are the most thoroughly characterized, although even then much of the emphasis has been on a small number of widely distributed forage grass species and their asexual fungal symbionts (Saikkonen et al. 2006; Cheplick & Faeth 2009; Cheplick 2015). This section considers briefly how clandestine, asymptomatic fungal endophytes might function as selection agents in populations, providing the opportunity for coadaptation of host and symbiont. The emphasis is on research done since my book on grass–endophyte symbioses, coauthored with S.H. Faeth, was completed (Cheplick & Faeth 2009). More updated information can be found in 11 articles collected in a special issue of *Fungal Ecology* (2012; 5: 287–378) devoted to "The Secret World of Endophytes" and in an online book edited by Young et al. (2012) with 21 chapters (available in pdf form at www.noble.org/global/research/isfeg/isfeg7.pdf).

8.5.1 Genetic Variation in Host Responses

Endophyte infection can influence host growth, metabolism, and physiology, as well as vegetative or reproductive fitness (Belesky et al. 1987; Rice et al. 1990; West 2007; Faeth 2009; Saari et al. 2010; Gundel et al. 2012; Torres et al. 2012). As pointed out earlier in Section 2.1.6, by their effects on plant phenotypes, endosymbionts can act as hidden, internal agents of natural selection with effects that are highly contingent on environmental conditions and host (and endophyte) genotype. The significant genotype-by-endophyte interactions for a variety of quantitative traits in a single

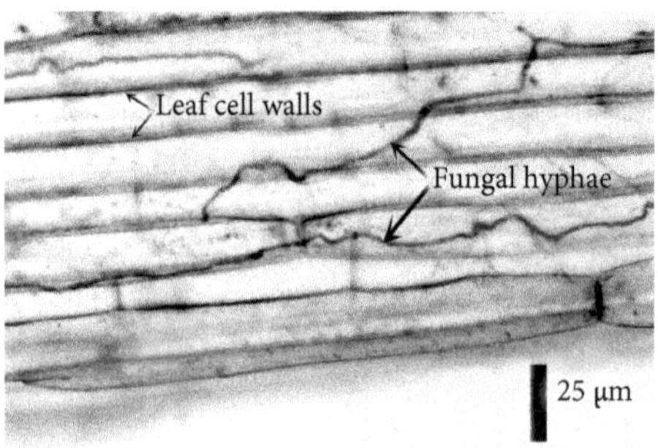

FIGURE 8.5
Hyphae of the endophytic fungus *Epichloë clarkia* within a leaf of the grass *Holcus lanatus*. Photo courtesy of Adrian Leuchtmann.

grass species (tall fescue) were summarized in Table 2.2 and testify to variation in how phenotypic traits of host genotypes are moderated by endophyte presence. The three-way interactions of host genotype-by-endophyte presence-by-environmental conditions (often varied in experimental treatments) documented for some grass–endophyte symbioses (e.g., Cheplick 1997; Cheplick et al. 2000; Morse et al. 2007) set the stage for endophyte infection to function as a selection agent within and between plant populations. In other words, endophyte infection can favor different host genotypes in different environments, paving the way for local adaptation of host populations to specific habitats (see next section).

Complicating the scenario further is variation in the haploid genotype (haplotype) of the endophyte itself, which can alter host growth and physiology (Morse et al. 2007), and affect the genetic compatibility of specific host–fungal combinations (Saikkonen et al. 2010). As with mycorrhizal symbioses (Section 8.4), host genotype-by-fungal genotype interactions can affect the ecological and evolutionary outcome of plant–endophyte interactions (Cheplick & Faeth 2009).

As an example of how endophytes can affect phenotypic responses of host genotypes to environmental conditions, consider the number of tillers produced by genotypes of perennial ryegrass (*Lolium perenne*) with (E+) or without (E–) fungal endophyte (*Epichloë festucae* var. *lolii*; formerly *Neotyphodium lolii*) infection when grown under competitive and noncompetitive (control) conditions. This genotypic competition experiment, conducted by former graduate student Amelia Harrichandra, used six morphologically variable genotypes of *L. perenne* distributed between two cultivars (Palmer and Repell) as target plants (Cheplick et al. 2014). Genetically identical individuals of this bunchgrass are replicated readily by manual separation of ramets (tillers) from a genotype. Note that genotypic variation is extensive in cultivars of *L. perenne*, which is a highly variable species because of its outcrossing breeding system. However, molecular genetic analyses of the endophyte have shown that most collections of *L. perenne* harbor a "single endophyte multilocus genotype" (van Zijll de Jong et al. 2008, p. 1492), so it is assumed that endophyte genotypes were homogeneous within each host cultivar.

An E + or E– ramet of each *Lolium perenne* genotype was paired with an uninfected ramet of a competitor *L. perenne* genotype. All competitor genotypes were from another accession originally collected from Turkey, part of the native range of this species. Note that E– ramets were obtained by systemic fungicide treatment following standard procedures (Cheplick & Faeth 2009). After three months in a greenhouse, the number of tillers, and dry shoot and root mass were recorded (Cheplick et al. 2014). Only tiller production is considered here.

A complete analysis of variance indicated that the number of tillers made by *Lolium perenne* genotypes was strongly reduced by competition ($F = 56.5, p < 0.0001$), and affected by cultivar ($F = 4.3$, $p = 0.04$) and genotype nested within cultivar ($F = 3.7, p = 0.01$). There was no main effect of endophyte (present vs. absent) on tiller production ($F = 1.0, p = 0.32$). However, there was a highly significant interaction of host genotype-by-infection status ($F = 6.4$, $p < 0.0001$). In addition, the genotype-by-infection-by-competition term was significant ($F = 4.0$, $p = 0.004$), indicating that the effect of endophytes on tiller production in host genotypes depended on whether plants were competing. This is readily observed in a plot of the symbiotic interaction norms (*sensu* Cheplick & Faeth 2009) for the six genotypes in control and competition treatments (Fig. 8.6). One genotype (Repell 27) showed much greater

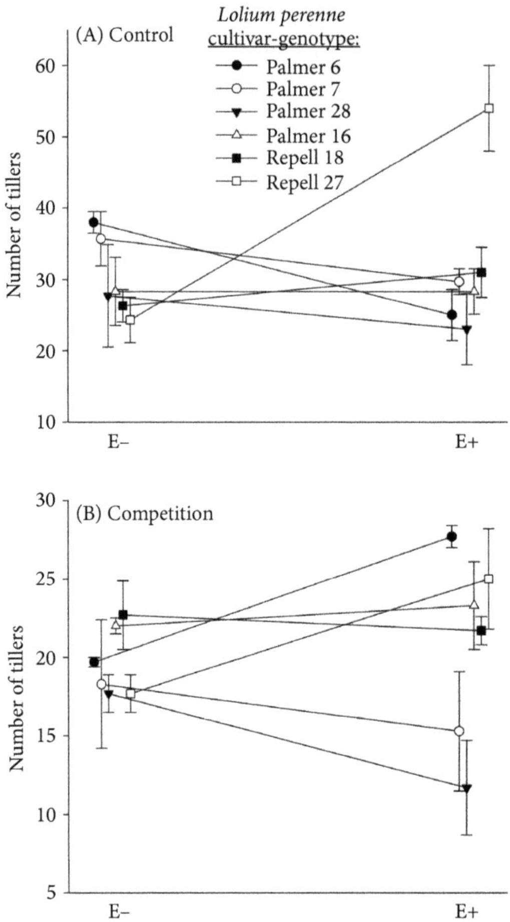

FIGURE 8.6
(A–B) Symbiotic interaction norms for the mean number of tillers ±standard error produced by six genotypes of *Lolium perenne* (cultivars Palmer and Repell) when uninfected (E–) or infected (E+) with the fungal endophyte *Epichloë festucae* var. *lolii* (formerly *Neotyphodium lolii*). In this greenhouse experiment, target plants of each genotype were grown alone (A) or in intraspecific competition (B) with a second, uninfected genotype of a native accession (Cheplick et al. 2014). Note that points are displaced to avoid overlapping standard errors.

tiller production when infected, both with and without competition. Endophyte also improved the growth of another genotype in a different cultivar (Palmer 6), but only when a competitor was present (Fig. 8.6B). Other genotypes did not show any growth improvement when infected.

The results from the perennial ryegrass genotypic competition experiment (Cheplick et al. 2014) underscore the remarkable variability that occurs in most plant–endophyte studies when host genotypes are retained as a separate variable and grown under diverse conditions (Faeth & Saikkonen 2007). These endophyte-mediated effects on plant quantitative traits, as in other types of microbially mediated traits (Friesen et al. 2011), are also likely to be conditioned by variation in endophyte genotypes, at least in some systems (Morse et al. 2007; Saari & Faeth 2012). The opportunity for ongoing coevolution exists and, theoretically, can lead to coadaptation of plants and their endophytes to particular habitats.

8.5.2 Local Adaptation

What is the evidence that endophytic fungi play a role in the adaptation of wild plant populations to specific habitats? The short answer is that there is very little direct experimental evidence, in marked contrast to plant–pathogen interactions

(Section 8.2.2). Despite a plethora of studies showing how fungal endophytes can deter herbivores (Popay 2009) and sometimes confer tolerance to abiotic stresses (West 2007; Rodriguez et al. 2009b), there have been surprisingly few investigations of local adaptation in hosts containing the specific, beneficial endophyte genotypes with which they presumably coevolved. From cross-inoculation and molecular genetics studies, however, we do know that populations/races of endophytes can show adaptation to specific populations or species of wild host plants (references in Cheplick & Faeth 2009; Saikkonen et al. 2010; Cheplick 2015). Nonetheless, the question of whether endophyte infection aids in the adaptation of host genotypes to local habitat conditions has rarely been addressed experimentally.

Many aspects of plant–endophyte symbiosis have been explored by Stanley H. Faeth and students or colleagues using the native American rangeland grass *Festuca arizonica* and its asexual, vertically transmitted *Epichloë* (formerly *Neotyphodium*) endophytes (e.g., Morse et al. 2007; Faeth 2009; Saari & Faeth 2012; and many others). Although these endophytes are asexual, hybridization by hyphal fusion is relatively common in this group of fungi and might provide new opportunities for coevolution with grasses because novel genetic variants are formed with possible selective value to their hosts (Moon et al. 2004). Morse et al. (2007) showed that two distinct endophyte genotypes interacted with host genotype and environmental conditions to influence several growth and physiological traits in *F. arizoni*ca. The possibility of local adaptation of *F. arizonica* populations when infected by genetically distinct endophytes was examined subsequently for two populations in Arizona (Sullivan & Faeth 2008). Specific host population–endophyte combinations, referred to as "symbiota" by the authors, were reciprocally transplanted between two distinct field sites (separated by ~90 km); Clint's Well is heavily forested by Ponderosa pine and has soils low in nitrogen and moisture, whereas Flagstaff is an open grassland and has soils high in nitrogen and moisture. Identified with molecular markers, both hybrid and nonhybrid haplotypes were distinguishable. The Clint's Well plant population has a mixture of both haplotypes (81% hybrid, 19% nonhybrid), whereas the Flagstaff population contains a single, nonhybrid haplotype. Transplanted hosts (Clint's Well plants with hybrid endophyte and Flagstaff plants with nonhybrid endophyte) grew at both field sites for three years. Plant volume was measured to estimate size nondestructively, and flowering stems with panicles were counted.

A significant interaction between site and the type of symbiotum ($F = 6.2, p = 0.01$) was found for the size of plants relative to their dry mass (Sullivan & Faeth 2008). At Flagstaff, there was no difference in plant size for populations with hybrid versus nonhybrid endophytes, whereas at Clint's Well, local plants (i.e., from Clint's Well) with the hybrid endophyte had greater size and panicle production than foreign plants (from Flagstaff) with the nonhybrid endophyte. Thus, the native symbiota, which included the hybrid endophyte, appeared to be locally adapted to the suboptimal habitat at Clint's Well, where light, soil nitrogen, and soil moisture were potentially limiting to growth. These positive effects of hybrid endophytes within the plants at Clint's Well (and the lack of such effects at Flagstaff) may explain in part the high frequency (81%) of hybrid haplotypes in Clint's Well. However, the complete symbiotum (host–endophyte) had to be considered a single treatment in the statistical analyses (Sullivan & Faeth 2008), and therefore effects of host and endophyte genotypes could not be assessed independently (i.e., there were no transplanted plants from Clint's Well with nonhybrid endophyte or plants from Flagstaff with hybrid endophyte).

Another investigation of potential adaptation in a grass–endophyte symbiosis involved the well-studied perennial *Lolium perenne* and its fungal endophyte in populations along a climatic gradient in southern France (Gibert et al. 2012). For 22 populations, the frequency of endophyte infection correlated negatively with a soil moisture index ($r^2 = 0.73$, $p < 0.001$), suggesting a benefit for endophyte-infected (E+) *L. perenne* in dry environments. From four climatically distinct sites, 18 E+ and 18 E– plants were collected, cloned into ramets, and subjected to experimental drought in a greenhouse experiment. A variety of morphological and physiological traits were recorded, 17 variables in all. Neutral DNA markers (simple sequence repeats) were also examined to characterize population genetic structure.

Principal components analyses of the phenotypic traits revealed differentiation among *Lolium perenne* populations and between E+ and E– plants (Gibert et al. 2012). Although molecular analysis showed evidence for differences among populations (pairwise F_{ST} range, 0.02–0.14), no neutral genetic differentiation was found between E+ and E– plants (F_{ST} not significantly different from zero). The greenhouse drought experiment also indicated significant population differentiation for most phenotypic traits that was unrelated to endophyte infection when water availability was not limited (although specific leaf area was greater and leaf dry matter was lower overall in E+ plants). However, under drought stress, two populations had significantly greater root biomass production when infected, whereas another population showed *lower* root production when infected. In addition, one population showed greater shoot biomass when infected. For the most drought-tolerant populations, survival rate after the most severe stress imposed (63 days without supplemental water) was significantly greater for E+ plants (10.5%) relative to E– plants (3.4%). Through the positive effects on survival and biomass production under drought, the authors concluded that “the symbiosis probably reinforces local adaptation of the grass populations” (Gibert et al. 2012, p. 569). Unfortunately, experimental tests of this hypothesis using reciprocal transplants of E + and E– plants among the different populations and field sites were not conducted, rendering this conclusion tenuous at best.

Additional experiments testing the hypothesis of endophyte-mediated local adaptation of host populations are needed. The critical test of reciprocal transplantation of hosts containing the usual (native) endophyte *and* containing endophytes isolated from other (foreign) populations has yet to be performed. Such an experiment would entail isolation of fungi from distinct populations and then cross-inoculating the isolates into plants from the different populations in a reciprocal fashion. The genetic distinctiveness of the isolates would need to be established by molecular analyses, and inoculated plants from each population should be distributed among the sites chosen for study. Measures of plant fitness under field conditions would provide the data needed to address potential coadaptation of host and endophyte to their local habitat.

8.6 WRAP-UP

Microbial interactions with plants are pervasive in nature. No doubt much of the “web of life” is shaped by coevolutionary processes resulting from many types of interspecific interaction (Thompson 2009). Bacteria and fungi can be found on the surfaces and within the tissues of all plant species. By altering plant physiological processes, microbes can influence the phenotypic values of a diverse set of functional traits (Friesen et al. 2011). They can also modify plant responses to abiotic and biotic conditions in complex ways. Whether their effects on host fitness are positive (mutualistic) or negative (parasitic), microbial species associated intimately with plants have tremendous potential to function as microscopic agents of natural selection that drive microevolutionary change and adaptation in plant populations.

9 Biotic Interactions III: Animals

9.1 ANIMALS AS AGENTS OF NATURAL SELECTION

Among the macroscopic biotic agents than can shape the microevolution of plants are the many thousands of animal species, from insects and other invertebrates to birds and mammals. A collection of papers edited by Herrera and Pellmyr (2002), titled *Plant–Animal Interactions: An Evolutionary Approach*, provides a solid introduction to the diversity of plant–animal interactions and their evolutionary implications. These include herbivory on roots, leaves, flowers, fruits, and seeds; pollination of flowers; and dispersal of fruits and seeds. Throughout the years there have been a good number of edited volumes or special journal issues containing articles on various aspects of different plant–animal interactions (Table 9.1). These sources provide an excellent introduction to the different approaches used to investigate these topics and to the most current specific plant–animal case studies.

By "choosing" which plant species, individuals, or specific tissues on which to feed, herbivores (including seed predators) can act as major agents of selection that favor the evolution of traits that impart herbivore resistance or tolerance (Strauss & Agrawal 1999; Fornoni et al. 2003a; Núñez-Farfán et al. 2007; Carmona et al. 2011; Agrawal et al. 2012). In an analogous way, by choosing which particular flowers on which individuals to visit, pollinators exert selection pressure that may favor specific types of floral traits (Pellmyr 2002; Elle 2004; Conner 2006; Harder & Johnson 2009; Schiestl & Johnson 2013). Last, by choosing which fruits (and seeds) to consume, animals may favor the evolution of adaptive features that improve the likelihood of successful seed dispersal and seedling establishment (Herrera 2002; Cousens et al. 2008). The fur or feathers of vertebrates may also select passively for features of the fruit and/or seed that promote dispersal by adhesion (Sorensen 1986).

Table 9.1 A sampling of edited collections of papers on plant–animal interactions.

Topic	*Editors*	*Title*
General	Gilbert and Raven (1975)	*Coevolution of Animals and Plants*
	Futuyma and Slatkin (1983)	*Coevolution*
	Price et al. (1991)	*Plant–Animal Interactions: Evolutionary Ecology in Tropical and Temperate Regions*
	Herrera and Pellmyr (2002)	*Plant–Animal Interactions: An Evolutionary Approach*
	Seckbach and Dubinsky (2011)	*All Flesh Is Grass: Plant–Animal Interrelationships*
Herbivory	Palo and Robbins (1991)	*Plant Defenses against Mammalian Herbivory*
	Fritz and Simms (1992)	*Plant Resistance to Herbivores and Pathogens: Ecology, Evolution and Genetics*
	Rosenthal and Berenbaum (1992)	*Herbivores: Their Interactions with Secondary Plant Metabolites. Vol. II: Ecological and Evolutionary Processes*
	Johnson (2011)	"Evolutionary ecology of plant defences against herbivores"
Pollination	Lloyd and Barrett (1996)	*Floral Biology: Studies on Floral Evolution in Animal-Pollinated Plants*
	Harder and Barrett (2006)	*Ecology and Evolution of Flowers*
	Waser and Ollerton (2006)	*Plant–Pollinator Interactions: From Specialization to Generalization*
	Ayasse and Arroyo (2011)	"Pollination and plant reproductive ecology"
	Patiny (2012)	*Evolution of Plant–Pollinator Relationships*
Dispersal	Estrada and Fleming (1986)	*Frugivores and Seed Dispersal*
	Fleming and Estrada (1993)	*Frugivory and Seed Dispersal: Ecological and Evolutionary Aspects*
	Clobert et al. (2001)	*Dispersal*
	Levey et al. (2002)	*Dispersal and Frugivory: Ecology, Evolution and Conservation*
	Dennis et al. (2007)	*Seed Dispersal: Theory and Its Application in a Changing World*
	Clobert et al. (2012)	*Dispersal Ecology and Evolution*

Within a topic, references are listed in order of publication date. Complete citations are in the list of references.

This chapter examines some of the many studies designed to address questions pertinent to the evolutionary significance of animal-mediated selection on plant populations. Three predominant types of plant–animal interaction are included: herbivory (including flower and seed consumers), pollination, and fruit and seed dispersal.

9.2 HERBIVORY

The herbivore–plant relationship is predominantly an antagonistic one (+, –), similar to a parasite–host association (Strauss & Zangerl 2002); the consumer benefits from the food nutrients received whereas the plant usually shows a reduction in fitness because of the loss of photosynthetic and/or reproductive tissues. Indeed, some of the same terminology used with parasite or pathogen–host systems is also used to describe the defensive strategies plants have evolved to cope with herbivore attack (Fritz & Simms 1992; Rausher 2001; Walters 2011). A trait conferring *resistance* reduces herbivore attack and/or damage, whereas one conferring *tolerance* allows plants to buffer the negative fitness effects of damage, often through adjustments in photosynthetic rate, allocation patterns, or branching architecture (Strauss & Agrawal 1999; Stowe et al. 2000; Núñez-Farfán et al. 2007). It is worth noting that the effects of herbivores on plants are highly variable and not always antagonistic, ranging from intensely detrimental to barely discernible to beneficial, in some instances. An overview of topics important to understanding the evolutionary ecology of plant defenses can be found in the collection of papers edited by Marc Johnson (2011) in a special issue of *Functional Ecology*.

Many types of plant traits have been considered to have evolved in response to herbivore pressure. These include structural features such as dense hairs, spines, or thickened cuticles on leaves and a wide array of secondary compounds that deter feeding or act as toxins to herbivorous insects and mammalian grazers (Palo & Robbins 1991; Rosenthal & Berenbaum 1992; Strauss & Zangerl 2002). However, casual observations that a specific trait reduces herbivore attack or population size does not necessarily imply that herbivore-mediated selection was responsible for its evolution. For example, many secondary compounds in plants have other functions besides herbivore deterrence, and their evolutionary origins may be unrelated to herbivore pressure (Carmona et al. 2011). Even when herbivores reduce plant fitness significantly, there must be genetic variation for any potentially defensive trait if it is to evolve in response to herbivore-mediated selection and lead to local adaptation.

9.2.1 Quantitative Genetic Variation and Selection for Resistance and Tolerance

What evidence is there that plant populations show genetic variation in their ability to resist or tolerate herbivore damage? This is important to know because a lack of such variation may limit the ability of a population to respond to selection imposed by herbivores, as reported for a population of the annual weed *Datura stramonium* in central Mexico (Núñez-Farfán & Dirzo 1994). The traits that could be important to herbivore resistance (Carmona et al. 2011) have long been considered to have a mostly quantitative genetic basis (Simms & Rausher 1992,) as shown by QTL mapping in *Arabidopsis thaliana* (Weinig et al. 2003b) and hybrid sunflowers (Dechaine et al. 2009). The evidence from QTL mapping, summarizing data from more than 50 studies, shows that one to nine genes are typically involved in host plant resistance (Núñez-Farfán et al. 2007). Thus, many studies analyzing host variation have used the standard techniques of quantitative genetics. However, it should be noted that sometimes traits that affect herbivory have a simple Mendelian inheritance, involving only one or two loci (Dirzo & Harper 1982; Kivimäki et al. 2007; Züst et al. 2012).

The classic case of cyanogenic forms of white clover (*Trifolium repens*) capable of releasing hydrogen cyanide after attack by invertebrate herbivores involves only two gene loci (Dirzo & Harper 1982); in addition, populations vary greatly in the relative proportions of cyanogenic and acyanogenic types (Richards & Fletcher 2002). Variation among populations for quantitative traits used in herbivore defense and tolerance has also been demonstrated in several plant species (Juenger et al. 2000; Fornoni et al. 2003b; Karley et al. 2008).

To show genetic variation for resistance and/or tolerance, generally cloned genotypes or sibling families generated from controlled crosses are used to obtain genetically definable groups of a host plant species. These are then exposed to the herbivore species that commonly attacks it in a greenhouse or common garden(s). A specific quantity of herbivores may be applied to each plant (Service 1984; Strauss 1990; Horner & Abrahamson 1992) or plants may be exposed to natural levels of herbivores in the field (Berenbaum et al. 1986; Maddox & Root 1987; Fritz 1990; Núñez-Farfán & Dirzo 1994; Wise & Rausher 2013). Note that, sometimes, herbivory is simulated by artificial defoliation treatments (e.g. Fornoni & Núñez-Farfán 2000). Depending on the objectives, a pesticide may be used to generate a control plant group lacking herbivory. For example, insecticides have been used to eliminate herbivorous insects in some experiments (Simms & Rausher 1989; Mauricio & Rausher 1997; Shonle & Bergelson 2000; Carmona & Fornoni 2013).

Host resistance to invertebrate herbivores is often quantified by subtracting the proportion of the damaged leaf area from unity, thereby expressing the relative amount of the available tissues that was *not* consumed. Alternatively, resistance has sometimes been based simply on the number of herbivores attacking a plant or their survival and growth when feeding on a plant. A highly resistant host genotype would support fewer herbivores and/or tend to reduce their survival or growth more than a genotype with lower resistance. In all these measures, resistance is not a plant trait per se, but rather some unknown composite of features that enables a genotype to reduce herbivore attack and damage or survival and growth. It should be recognized that the level of damage caused depends on traits of the herbivore as well as the host, so resistance is really a composite feature of the herbivore–plant interaction (J. Fornoni, pers. comm., June 5, 2014).

A number of early investigations indicated how host genotypes could show significant variation in their ability to resist attack by herbivorous insects (Service 1984; Maddox & Root 1987; Fritz 1990; Horner & Abrahamson 1992). For example, Maddox and Root (1987) grew 18 cloned genotypes of goldenrod (*Solidago altissima*) in a common garden in New York state and recorded the densities of 16 insect herbivore species attacking them for four years. For 15 of the insect species, significant differences were found among goldenrod genotypes in terms of insect abundance on them, implying quantitative genetic variation in resistance. Parent–offspring regression was used in another experiment using half-sibling groups collected from cross-pollinated parents to show that resistance to 10 insect species had significant heritability, ranging from 0.50 to ~1.0 (Maddox & Root 1987). In another study of *S. altissima*, the number of plants punctured by a gall-making herbivore and the number of its surviving larvae were significantly different among host genotypes (Horner & Abrahamson 1992).

Other studies using the proportion of undamaged tissues as a metric for resistance have also found quantitative genetic variation within plant populations

(Berenbaum et al. 1986; Simms & Rausher 1989; Wise 2007; Carmona & Fornoni 2013). For example, Wise (2007) documented significant genetic variation for resistance to 11 herbivore species, including leaf, flower, and fruit consumers, using 40 genets of horsenettle (*Solanum carolinense*) planted into an agricultural field in Virginia. Phenotypic selection analyses (Lande & Arnold 1983) of this same system showed significant ($p < 0.05$) negative selection gradients (β) for resistance to 6 of the 11 herbivore species (Wise & Rausher 2013). These gradients (±SE) ranged from $\beta = -0.142 \pm 0.036$ for a flower consumer to $\beta = -0.456 \pm 0.040$ for a fruit predator. The negative β values indicate that damage by these herbivore species significantly decreased *S. carolinense* fitness (based on seed number), and thus directional selection favored increased resistance to each. One selection gradient on a stem-boring insect species was significantly positive ($\beta = 0.087 \pm 0.032$, $p = 0.006$), indicating selection for *decreased* resistance, but this was apparently a result of indirect selection for increased leaf and flower production, which correlate positively with the chances of attack by this pest (Wise & Rausher 2013). Similar positive selection gradients for resistance have sometimes been found in other plant–herbivore systems as well (Rausher & Simms 1989; Núñez-Farfán & Dirzo 1994).

Wise and Rausher (2013) also found that the magnitude of selection on *Solanum carolinense* differed among herbivore feeding guilds, grouping species by whether they consumed leaves, flowers, or fruits (Fig. 9.1A). Fruit predators (three species) imposed the strongest selection on resistance, "presumably because their damage directly destroys developing seeds" (Wise & Rausher 2013, p. 1777), and plants may be able to compensate more readily for damage to other tissues such as leaves. Hence, selection for resistance to leaf feeders was lowest and only marginally different from zero ($p = 0.10$; Fig. 9.1A). This study illustrates how diffuse evolution (involving multiple interacting herbivore species acting collectively as selective agents) may be a more realistic portrayal of plant–herbivore relationships than simple pairwise comparison of a single herbivore species attacking one host species (Strauss & Irwin 2004).

The ability to tolerate herbivore damage by reducing or buffering fitness losses has also been shown to be under quantitative genetic control and to show variation within and between plant populations (Juenger et al. 2000; Fornoni et al. 2003a; Núñez-Farfán et al. 2007; Manzaneda et al. 2010). For example, genetic variation for both resistance and tolerance was found within and between two populations of *Datura stramonium* (jimsonweed) in central Mexico (Fornoni et al. 2003b). A later study of the same species (Carmona & Fornoni 2013) found that selection favored tolerance to a leaf-feeding beetle (*Lema daturaphila*)—a species that can do extensive damage to this plant (Fig. 9.2) and thereby lower its reproductive fitness. The significant positive selection gradient (Fig. 9.1b) indicates that tolerance of this beetle species is expressed as improved fitness of *D. stramonium*. Resistance to this herbivore was also favored by selection, although this was not statistically significant. This study (Carmona & Fornoni 2013), which used 64 full-sibling maternal families grown in a common garden, revealed highly significant ($p < 0.001$) quantitative genetic variation for both resistance and tolerance. In another study, microevolutionary changes in tolerance to the leaf-feeding beetles during a 20-year period was documented by comparing reaction norms of fitness in undamaged versus damaged conditions for plants reared from seeds collected from the same population in 1987 and 2007 (Bustos-Segura et al. 2014). Levels of insect damage, leaf trichome density, and levels of three alkaloids did not differ between the two samples, suggesting no

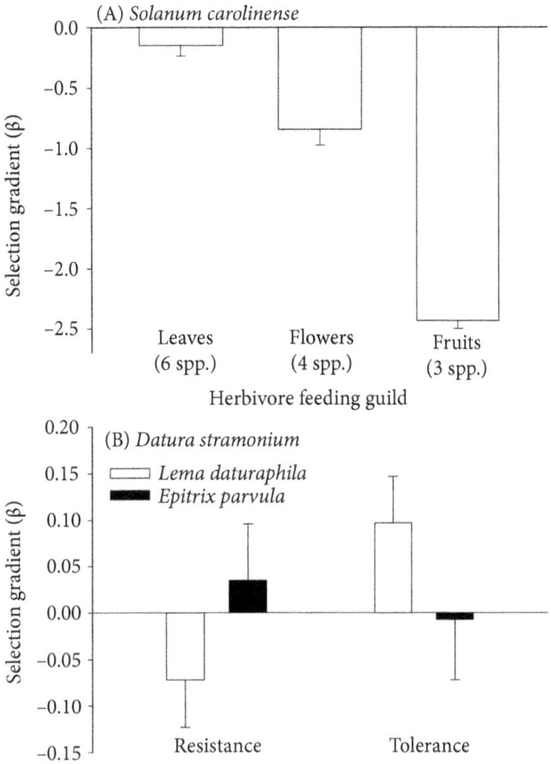

FIGURE 9.1
(A, B) Standardized directional selection gradients (β) ± standard error for resistance to species guilds of leaf ($p = 0.10$), flower ($p < 0.0001$), or fruit-feeding herbivores ($p < 0.0001$) (A) in horsenettle (Wise & Rausher 2013), and resistance and tolerance to two leaf-feeding beetle species in jimsonweed (B) (Carmona & Fornoni 2013). In (B), only β for tolerance to *Lema daturaphila* is marginally different from zero ($p = 0.057$).

changes in resistance had evolved. However, when herbivores were present, plants from 2007 showed a significantly ($p = 0.025$) greater fitness than those from 1987, suggesting that the population had evolved greater tolerance during the 20-year period.

Note that tolerance was assessed by a statistically significant interaction between leaf area damage and plant family in the *Datura stramonium* studies (Carmona & Fornoni 2013). This means that the norms of reaction for tolerance will show considerable crossing in a plot of plant fitness against herbivore damage for the different families (or genotypes [Fornoni et al. 2003a]). The same pattern occurs when comparing fitness of undamaged control plants (without herbivores) with plants damaged by herbivores if there is genetic variation in tolerance (Juenger & Bergelson 2000). A hypothetical set of reaction norms are plotted for 10 plant genotypes in Figure 9.3. The crossed reaction norms indicate that genotypes vary in their ability to tolerate herbivory. Most slopes are expected to be negative—that is, most genotypes suffer a loss of fitness when subjected to herbivory (undercompensation). However, the extent of the fitness reduction can differ between genotypes (compare A and B in Fig. 9.3). A horizontal reaction norm (slope, ≈0) indicates that a genotype

FIGURE 9.2
Severely damaged leaf of *Datura stramonium* being consumed by larvae of the herbivorous beetle *Lema daturaphila*. Leaf width is (was?) about 1 cm. Photo taken by Sergio Ramos-Castro and permission provided by Juan Fornoni.

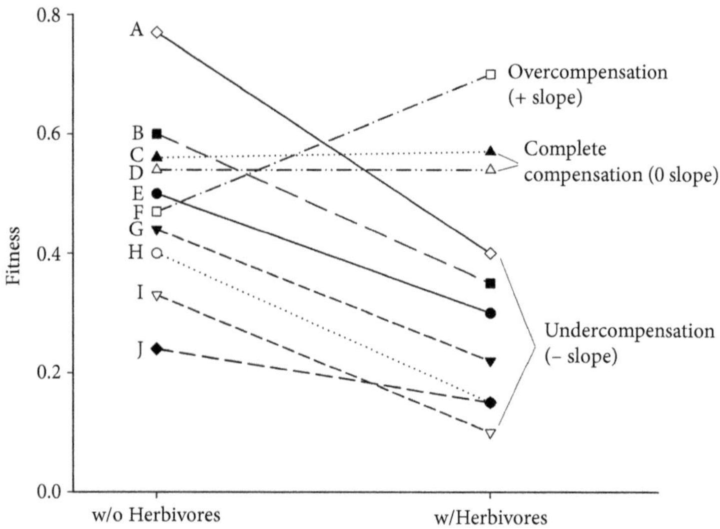

FIGURE 9.3
Norms of reaction for fitness in a hypothetical population of 10 plant genotypes with and without herbivores. Most genotypes have negative slopes, indicating relatively low tolerance to herbivory (undercompensation). Two other genotypes show complete compensation (zero slope), whereas one genotype shows overcompensation (positive slope). Modified from Juenger and Bergelson (2000) and Fornoni et al. (2003a).

can compensate completely in its response to herbivory, and therefore not suffer a fitness reduction (genotypes C and D in Fig. 9.3). A positive slope can also occur and indicates that a genotype shows greater fitness when herbivores are present (genotype F in Fig. 9.3). This phenomenon, known as overcompensation, has been reported in some species, especially those in which damage to the primary apical meristem triggers the growth of lateral, flowering branches (e.g., Paige 1992; Weinig et al. 2003a). Examples of all three types of compensatory response were found in populations of the biennial *Gentianella campestris* grown in a common garden in Sweden and subjected to an artificial clipping treatment to mimic browsing damage (Juenger et al. 2000).

That insect herbivores can drive microevolutionary change in just a few generations was considered for the native species *Oenothera biennis* (evening primrose) in field plots with and without insects (Agrawal et al. 2012). Eighteen host genotypes varied greatly in their resistance to moth species that act as major seed predators, causing substantial loss of fruits. Suppression of the insects (using an insecticide) resulted in major changes in the genotypic composition of the *O. biennis* populations in four generations. The researchers also found reduced resistance resulting from earlier flowering and lower levels of defensive tannins in fruits compared with populations with ambient insect levels (Agrawal et al. 2012). Elimination of the seed predators effectively caused a relaxation of the selection pressures that would normally favor flowering later and producing greater levels of defensive compounds in fruits.

9.2.2 Selection on Quantitative Candidate Traits

As noted earlier, there are many possible quantitative candidate traits that could be important to resistance to herbivory (Stowe et al. 2000; Carmona et al. 2011; Loranger et al. 2012). These include morphological traits such as size, branching, and trichomes; the timing of flowering; and a plethora of secondary chemicals long thought to be critical to herbivore deterrence (see the early, much-cited papers by Fraenkel [1959] and Ehrlich and Raven [1964]).

Significant selection for increased levels of defensive compounds has been reported in a few species. In a now-classic study (Berenbaum et al. 1986), the parsnip webworm, which feeds on the flowers and seeds of *Pastinaca sativa* (wild parsnip), was shown to select for greater production of several defensive compounds (furanocoumarins). In *Arabidopsis thaliana*, natural enemies (herbivores and pathogens) affected selection on the concentration of glucosinolates and trichome density, but directional selection gradients were negative and weak regardless of whether the natural enemies were present (Mauricio & Rausher 1997). Of two alkaloids in *Datura stramonium* examined by Shonle and Bergelson (2000), only one showed a positive selection gradient ($\beta = 0.155$), suggesting it could act in resistance to herbivorous insects; the other alkaloid showed a significant *negative* β value (−0.110), suggesting it may act as a feeding stimulant for some insects! These types of equivocal results are not uncommon in plant–herbivore selection studies and the evidence supporting "the conventional wisdom that secondary metabolites are the most important anti-herbivore defense" (Carmona et al. 2011, p. 358) is weak at best. Although it is possible to select artificially for higher or lower levels of defensive compounds in genetically variable species (Stowe 1998; Marak et al. 2000; Lankau & Klienbenstein 2009), indicating the potential for biochemical evolution, meta-analysis has not supported a relationship between secondary compounds and herbivory; rather, morphological and life history traits such as flowering time are better predictors of herbivore resistance (Carmona et al. 2011).

In species with strong apical dominance regulating lateral branch production, herbivory can favor greater branching (to yield more flowers, fruits, and seeds) and later flowering (Juenger & Bergelson 2000; Jeunger et al. 2000; Weinig et al. 2003a). In *Ipomopsis aggregata* (scarlet gilia) grown in a common garden in Colorado, damage to the apical bud (whether by the experimenters or mammalian browsers) strongly selected for a greater number of branches relative to undamaged control plants (Fig. 9.4 [Juenger & Bergelson 2000]). Furthermore, in both treatments, selection gradients were significantly negative for flowering date, indicating greater fitness (estimated by fruit production) for plants that flowered earlier, especially for the damaged group (Fig. 9.4). Juenger and Bergelson (2000) concluded that both early flowering and regrowth via branching were likely to function as candidate tolerance traits that had evolved in response to selection mediated by mammalian herbivores that remove apical buds as they feed.

As specialized herbivores, seed predators can also act as selective agents and mediate the evolution of characteristics such as seed size and the extent of protective tissues or chemicals (Janzen 1969; Smith 1975; Hare 1980; Hulme & Benkman 2002). One of the striking examples of coevolution between plants and seed predators is that of the lodgepole pine (*Pinus contorta*) and its vertebrate seed consumers. This complex system has been studied extensively by C.W. Benkman and colleagues in the western United States (e.g., Benkman 1999; Talluto & Benkman 2013). Seed predators include red squirrels, a moth species, and two species of bird: red crossbills (*Loxia curvirostra*) and hairy woodpeckers (*Picoides villosus*). Most of the quantitative traits measured on pinecones show significant heritability, and red squirrels are likely to be important selection agents wherever they occur (Benkman 1999). In places lacking the squirrels, crossbills were found to be more potent agents of selection than woodpeckers or moths (Benkman et al. 2013). Estimates of standardized selection differentials (Section 2.2.3) showed that, by preferring smaller cones, crossbills favored the evolution of larger, heavier cones ($S = 0.021 \pm 0.006, p < 0.001$) with

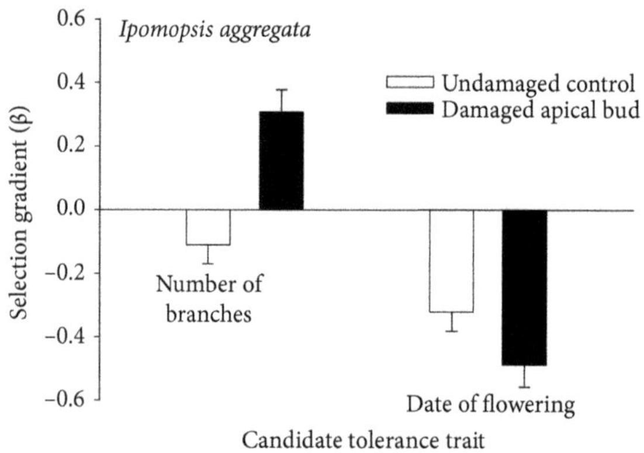

FIGURE 9.4

Standardized directional selection gradients (β) ± standard error for two candidate herbivore tolerance traits in the scarlet gilia for undamaged control plants and for plants with their apical bud damaged (Juenger & Bergelson 2000).

heavier seeds ($S = 0.016 \pm 0.067$, $p = 0.04$). Directional selection by woodpeckers was for smaller cones with a high ratio of seed mass to cone mass ($S = -0.007 \pm 0.002$, $p = 0.004$), and the abundance of woodpeckers among seven sites increased linearly with increases in the ratio (Benkman et al. 2013). Thus, the evolution of quantitative traits of lodgepole pinecones has been shaped by a number of seed predator species acting as agents of natural selection.

9.2.3 Plant Adaptation

Clearly, plant populations can be differentiated for resistance or tolerance to herbivores, and some studies have explored the possibility of local adaptation to herbivory by using methods such as common gardens or reciprocal transplants. In discussing the evolutionary ecology of plant defenses, Agrawal (2011) maintained that reciprocal transplant experiments with differentiated populations in which herbivory was manipulated experimentally could shed light on microevolution in relation to herbivore-mediated selection. If particular herbivores such as leaf feeders or seed predators act continuously as selection agents on populations for long enough, then one expects adaptive responses such as the evolution of traits involved in resistance or tolerance (Sork et al. 1993; Crémieux et al. 2008; Garrido et al. 2012; Benkman et al. 2013).

9.2.3.1 Common Gardens

The possibility that herbivores over time might select for specific plant traits has long been recognized, especially in species exposed repeatedly to grazing mammals (Gregor & Sansome 1927; Kemp 1937). These early experiments were common garden trials in which grasses or clovers were removed from pastures with variable histories of grazing pressure and planted into the garden. Erect and prostrate morphological types were distinguished in several plant species, with the shorter, prostrate form being from areas of heavy grazing pressure and maintaining its genetically based morphology in the common garden. Differences in growth form were attributed to genetically based, adaptive responses of the populations to natural selection imposed by the grazing animals. Many similar studies have since been conducted with a diversity of herbaceous species, revealing similar patterns of morphological differentiation in relation to both grazing and mowing (e.g., Warwick & Briggs 1978, 1980; McNaughton 1984; McKinney & Fowler 1991).

An experiment by Warwick and Briggs (1980) with *Plantago major* is included here to illustrate one approach used by plant ecologists who study differentiation of populations in relation to mammalian herbivory (e.g., McNaughton 1984) or analogous selection imposed by human activities such as repeated mowing or clipping of herbaceous communities. The approach of Warwick and Briggs (1980) entailed a single common garden with one manipulated variable (see Section 3.4). Seedlings of *P. major* were placed into (1) plots that were never cut, with grasses (six species) averaging 23 cm height, or (2) plots that were mown to 2 to 3 cm weekly during spring and summer. Four source populations provided the *P. major* seeds used to obtain seedlings planted into the garden. Two were from regularly mown university lawns (about 80–100 years old), whereas the other two were from unmown roadsides (note that prior work had shown the populations from lawns vs. roadsides to be differentiated with respect to prostrate and erect habit). Throughout the next 13 months, a number of variables were recorded, but only dry mass of reproductive tissues at harvest is examined here.

In the dense, unmown community of grasses, *Plantago major* plants of the roadside populations clearly had much greater reproductive success than those from lawn populations (Fig. 9.5). The selection coefficient (Section 4.5) against lawn plants was 0.77 (Warwick & Briggs 1980). In contrast, in the regularly mown, short-stature community, plants with prostrate growth from the lawn populations produced more reproductive mass (Fig. 9.5). The selection coefficient against roadside plants was 0.66. The pattern of population differentiation of *P. major* was similar to that found for *Poa annua* (Warwick & Briggs 1978), suggesting that regular mowing of the vegetation in lawns had selected for prostrate genotypes better able to withstand defoliation relative to the more upright genotypes selected in unmown communities.

Use of the common garden approach with experimental manipulation has also provided insight into local coevolutionary adaptation of plants and insect herbivores. Garrido et al. (2012) investigated local adaptation of the annual *Datura stramonium* and its major insect herbivore *Lema daturaphila* in central Mexico (Fig. 9.2). Four populations of *D. stramonium* known to be differentiated genetically (mean $F_{ST} = 0.39$) and separated by 52–245 km were grown in a common garden. All plants were within mesh cages, some of which were supplied with herbivore larvae; others were not. Plants from each of the four populations were exposed to each of the four associated herbivore populations. Resistance (based on leaf damage) and tolerance (based on the difference in seed production between damaged and undamaged plants) were determined along with herbivore performance and population growth. All variables for both plant and herbivore showed significant ($p < 0.01$) population differentiation. Two herbivore populations had the best performance on the host population they normally attacked in nature, suggesting local adaptation. For *D. stramonium*,

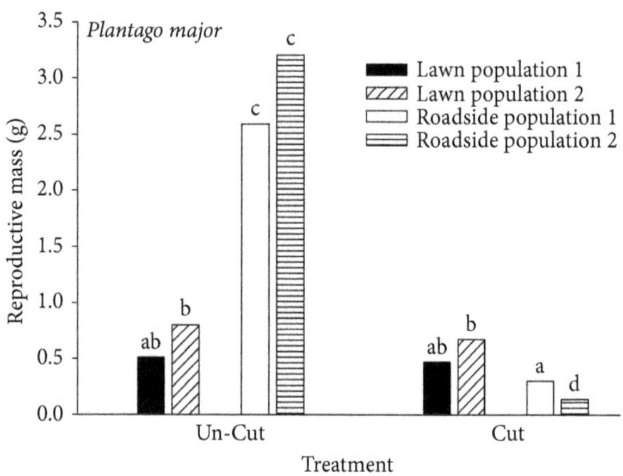

FIGURE 9.5
Reproductive mass of four populations of common plantain grown for 13 months in an outdoor plot in which plants were either left undisturbed (uncut plants grew to an average height of 23 cm) or mowed weekly in spring and summer (plants cut to a height of 2–3 cm). Two populations were from regularly mown lawns whereas two others were from roadsides that were never mown. Bars that share letters indicate means that were not significantly different ($p > 0.05$). Data from Warwick and Briggs (1980).

there was a significant ($p < 0.0001$) interaction between herbivore and host population for resistance; one host population showed significantly *lower* resistance when consumed by its native herbivore population (compared with the other populations). This suggests that this herbivore population had adapted to the particular defenses of its native host population. In short, there was more evidence for herbivore adaptation to its host (rather than the reverse), and Garrido et al. (2012) speculated that the abiotic environment may have imposed greater selection on the plant populations than the insect herbivore species. This difficulty in separating adaptation to the abiotic versus biotic environment is a common problem in coevolutionary studies of plants and their enemies, including those that use reciprocal transplants (Abdala-Roberts & Marquis 2007; Biere & Verhoeven 2008).

9.2.3.2 Reciprocal Transplants

Some investigators have performed reciprocal transplants of plant populations to examine local adaptation to insect herbivory (Table 9.2). Most studies have used sites separated by many kilometers; however, one of the earliest studies of this kind used three adjacent subpopulations of red oak (*Quercus rubra*) occupying a 4-ha plot and demonstrated that seedlings in their native (home) site had the lowest leaf herbivory (Sork et al. 1993). Although this was a novel and interesting study with clean results, subsequent work in other systems using well-isolated populations has often failed to produce convincing evidence of local plant adaptation to insect herbivory (Table 9.2). However, one study (Crémieux et al. 2008) conducted across three European countries did find that home populations of *Plantago lanceolata* were significantly ($p = 0.005$) less damaged by a specialist beetle herbivore than foreign populations, although this pattern did not occur in populations from contrasting habitats at a smaller, regional scale (7–100 km) within each country. In addition, Bischoff and Hurault (2013) found lower aphid infestation on *Brassica nigra* plants in their home sites in France, suggesting herbivory could sometimes be a driver of local adaptation. Kalske et al. (2012) also found two of four populations of *Vincetoxicum hirundinaria* to show evidence of local adaptation to the leaf-feeding larvae of a moth species.

There are a number of reasons why it may be especially difficult to detect local adaptation of plant populations to their local herbivores. As noted in the prior section, abiotic environmental selection pressures (e.g., soil type) may override the selective effect of herbivores (Ortegón-Campos et al. 2012), which can vary greatly in space and time, and not always provide continual pressure on plants. Repeated exposure to a particular herbivore may have not been severe enough or to have occurred over a long enough time for local adaptation to arise, and the spatial scale at which reciprocal selection between plant and herbivore occurs is not usually known. Phenotypically plastic plants may be able to compensate mostly for fitness losses resulting from herbivory, reducing greatly the potential of herbivores as selection agents. Also, some plant populations may not have the requisite genetic variation necessary for evolutionary responses to herbivore pressure (Núñez-Farfán & Dirzo 1994).

The results obtained from a reciprocal transplant experiment depend, in general, on the stage of the coevolutionary arms race of the plant and herbivore. If the local herbivore population has adapted well to its host and has overcome its resistance, then the host population will likely show more damage when attacked by local herbivores than when attacked by herbivores from other allopatric populations (Garrido

Table 9.2 Reciprocal transplant experiments to detect potential local adaptation of plant populations to their insect herbivores.

Species	*Life form*	*No. of populations*	*Distances between populations (km)*	*Location*	*Local adaptation?*	*Reference*
Brassica nigra	Annual	4	10–110	France	Yes	Bischoff and Hurault (2013)
Chamaecrista fasciculata	Annual	3	≥35	USA	No	Abdala-Roberts and Marquis (2007)
Datura stramonium	Annual	2	30	Mexico	No	Fornoni et al. (2003b)
Plantago lanceolata	Perennial herb	3	600–1100	CH, CZ, UK	Yes	Crémieux et al. (2008)
Quercus rubra	Tree	3	Adjacent	USA	Yes	Sork et al. (1993)
Ruellia nudiflora	Perennial herb	3	50–200	Mexico	No	Ortegón-Campos et al. (2009)
Ruellia nudiflora	Perennial herb	2	55	Mexico	No	Ortegón-Campos et al. (2012)
Vincetoxicum hirundinaria	Perennial herb	4	14–85	Finland	Yes	Kalske et al. (2012)

Studies listed alphabetically by genus. CH, Switzerland, CZ, Czech Republic.

et al. 2012). Alternatively, in cases when *less* damage is inflicted on native versus foreign plant populations at a particular site, we can consider the host population has evolved some level of resistance to its local herbivore population. That is, the plant is currently winning the arms race and is better adapted to the local herbivores than populations from elsewhere that have not been exposed to this particular herbivore population.

9.2.4 Molecular Genetic Approaches

Plant–herbivore interactions are described in an enormous, largely ecological and entomological literature into which molecular techniques have only recently been interjected. (Kessler & Baldwin 2002, p. 302)

Since that review of the "emerging molecular analysis" of plant responses to insect herbivory was written, there have been more examples of researchers adopting molecular genetic approaches to characterize the candidate gene loci important to the evolution of herbivore defenses in plants. An overview of the various ecogenomics techniques used is now available (Anderson & Mitchell-Olds 2011). Here, several studies are summarized to illustrate the kind of information on plant–herbivore interactions that can be generated by molecular genetic analyses.

As one might predict, much of the molecular analysis has focused on the model plant *Arabidopsis thaliana* and its allies (Weinig et al. 2003b). The glucosinolates of *A. thaliana* are secondary compounds important to insect resistance that have variable biochemical structures controlled by a variable group of gene loci (Kroymann et al. 2003). One gene family (*GS-Elong*) consists of different alleles that code for enzymes needed for glucosinolate synthesis. It was shown that one particular form of an allele caused increased concentration of glucosinolate and greater resistance to a generalist insert herbivore (*Spodoptera exigua*) compared with an alternative allele (Kroymann et al. 2003). Thus, genetic variation in *GS-Elong* results in biochemical phenotypes with differential resistance to herbivory.

Later work revealed that marked geographical patterns in *Arabidopsis thaliana* throughout Europe at this polymorphic locus (*GS-Elong*) correlated strongly with the relative abundance of two aphid species (Züst et al. 2012). Furthermore, a natural selection experiment in which many *A. thaliana* populations were exposed to the aphid species for five generations resulted in a progressive reduction in the negative effects of aphids on plant biomass, suggesting adaptive responses of the host populations to herbivore-mediated selection had occurred. Interestingly, molecular analysis showed that different aphid species had selected for specific biochemical genotypes in the host populations (Züst et al. 2012). In another study on the related plant *Boechera stricta* at field sites in Colorado and Montana, particular genotypes at another locus (*BCMA*) important to glucosinolate production showed reduced loss of leaf area and fitness resulting from insect herbivory relative to other genotypes, suggesting that geographic variation in herbivory contributes to the maintenance of polymorphism in this gene (Prasad et al. 2012).

Molecular genetic variation in loci important to quantitative morphological traits involved in herbivore resistance has also been examined in *Arabidopsis thaliana* and its allies (Weinig et al. 2003b; Kivimäki et al. 2007). Leaf hairs (trichomes) have long been thought to be a significant defense against certain herbivores (Strauss

& Zangerl 2002). Indeed, in the selection experiment (Züst et al. 2012) mentioned in the preceding paragraph, the density of trichomes remained high in plants exposed to the two aphid species for five generations, but declined to significantly lower levels in plants of the control group not exposed to aphids. This suggests relaxed selection for trichomes in the absence of herbivore pressure. In the outcrossing perennial herb *Arabidopsis lyrata* (a close relative of *A. thaliana*), a particular regulatory gene (*GLABROUS1*) is required for trichome production, and specific mutations at this locus result in hairless (glabrous) leaves (Kivimäki et al. 2007). Because natural populations of *A. lyrata* with trichomes show reduced leaf damage by insect herbivores, selection would not favor glabrous mutants where herbivore pressure is high.

Molecular genetics approaches have been used to characterize population differentiation in genes associated with defensive responses in the European aspen (*Populus tremula*), an outcrossing tree species (Bernhardsson & Ingvarsson 2012). Trees were sampled from 12 sites throughout Sweden, and SNPs from seven candidate genes known to be upregulated after herbivore attack were scored and compared with control SNPs from genes of unknown function. Ten of the putative defense SNPs were identified as outliers (see Chapter 5), and SNPs from the defense genes showed greater levels of population differentiation than the control SNPs (Bernhardsson & Ingvarsson 2012). Three of the defense genes showing differentiation code for protease inhibitors that are well known as an induced response to herbivory and that interfere with protein digestion by insect herbivores (Strauss & Zangerl 2002). Another gene codes a polyphenol oxidase that causes biochemical conversions into substances that affect herbivores negatively. Additional study of the same plant–herbivore system (Bernhardsson et al. 2013) was able to link the population differentiation found for the defense genes in *P. tremula* to geographic variation in the structure of a diverse herbivore community.

Molecular genetics clearly has much to offer regarding questions about the evolutionary ecology of plant defense against herbivores (Kessler & Baldwin 2002; Anderson & Mitchell-Olds 2011). When combined with traditional approaches to evaluating natural selection and adaptation, techniques such as genome scanning (e.g., Herrera & Bazaga 2011), mapping of QTLs, transcription profiling, and the identification of candidate genes with adaptive functions should be valuable tools in future studies of defensive trait evolution in plants.

9.3 POLLINATION

The two remaining plant–animal interactions discussed in this chapter are predominantly mutualistic (+, +). Both animal pollination and seed dispersal are familiar examples of mutualisms often illustrated in introductory ecology texts (e.g., Smith & Smith 2012).

Although there are thousands of wind-pollinated flowering plants such as grasses and most temperate trees, most angiosperm species produce flowers adapted for the transport of pollen by diverse types of animal vectors, the most important being insects (Fig. 9.6), birds, and bats (Baker 1963; Pellmyr 2002; Fenster et al. 2004; Ollerton et al. 2011). In this mutualistic relationship, the plant benefits from improved chances of cross-fertilization and seed set after successful pollen transport whereas the animal pollinator typically receives some type of high-quality reward such as sugar-rich nectar or the pollen itself.

FIGURE 9.6
A yellow swallowtail butterfly (*Papilio glaucus*) performing pollinator-mediated selection. Photo taken in Orange, New York, courtesy of S. Mitra.

The topic of plant–pollinator interactions is a vast one with a lengthy history of ecological and evolutionary investigation, going back to Sprengel's famous work in the late 1700s and Darwin's own classic studies on orchids (Darwin 1888; Harder & Johnson 2009; Yam et al. 2009). Several edited volumes provide a rich sampling of the topics and questions addressed by researchers (Table 9.1), and Rafferty (2013) provides many key references. Here the emphasis is on pollinators as agents of biotic selection on phenotypically variable populations, affecting the reproductive success of individuals differentially.

9.3.1 Genetic Variation in Floral Traits

Like all traits measured by ecologists, floral traits can be discrete with phenotypic forms being controlled by one or a few gene loci (e.g., color dimorphism) or, more likely, the traits are continuously variable and quantitative, being controlled by multiple loci (Conner 2006). There are many examples of quantitative floral traits that can be important to pollination biology: the area and length of petals or tubular corollas, the lengths of stamens and carpels, the number of pollen grains in an anther, and the quantity of available nectar (and its biochemical constituents). Note also that there are additional traits that can be subject to pollinator-mediated selection such as the timing and duration of flowering, size of the floral display at any one time, and height of the inflorescence.

Using heritability (h^2) estimation to describe the genetic component to the phenotypic variation in floral traits, quantitative floral traits (and associated features such as phenology) have in general been found to be genetically variable within many plant species and populations (Mazer & LeBuhn 1999; Ashman & Majetic 2006; Conner 2006; Sahli et al. 2008; Harder & Johnson 2009). In their survey of 41 species, Ashman and Majetic (2006) reported a mean h^2 of 0.39 for all floral traits, including significant genetic variation for the number of flowers and their size, and traits important to male and female function. Plants used for estimates of heritability are often derived from experimental crosses and the rearing of offspring in standard, controlled environments. For example, Andersson (1996) performed a crossing experiment with the perennial herb *Saxifraga granulata* and grew the offspring in a garden in Sweden. He then estimated h^2 after partitioning the phenotypic variance for all traits into within- and between-population components. For two years of data he found statistically significant additive genetic variation for most floral traits including petal length ($h^2 = 0.26$, $p < 0.01$) and width ($h^2 = 0.34$, $p < 0.001$), stamen length ($h^2 = 0.54$, $p < 0.001$), style length ($h^2 = 0.98$, $p < 0.001$), and flowering date ($h^2 = 0.41$, $p < 0.001$).

Of course, heritability should not be treated as if it was a property of a species, and estimates of h^2 for floral traits can vary greatly among populations (e.g., Sahli et al. 2008) and with environmental conditions (e.g., Mazer & Schick 1991b). For example, among eight populations of *Raphanus raphanistrum* (wild radish) from Australia, Europe, and the United States, h^2 for floral size ranged from 0.11 (not significant) to 0.38 ($p < 0.001$), whereas h^2 for the number of ovules ranged from 0.17 ($p < 0.05$) to 0.49 ($p < 0.001$ [Sahli et al. 2008]). Thus, there is considerable potential for populations of this widespread weed to show microevolutionary responses to selection mediated by its insect pollinators.

Besides flower size, color, and sexual traits, related traits that are likely to be subject to pollinator-mediated selection, such as the quantity and concentration of nectar, or the size of nectar guides, are also genetically variable and thus capable of response to selection (Medel et al. 2003; Mitchell 2004). The studies summarized by Mitchell (2004) reported h^2 values from 0.24 to 0.64 for nectar volume and h^2 values from 0.37 to 0.62 for total sugar concentration for herbaceous plants grown under controlled conditions. It is recognized that, in a more variable field environment, h^2 estimates are likely to be substantially less.

Perhaps the most convincing evidence that floral traits have the requisite genetic variation for selection-driven evolution comes from artificial selection experiments (Section 2.2.4) designed to change the phenotypic trait distribution of one or more floral traits (Conner 2003, 2006). Artificial selection over several generations with a number of herbaceous species has revealed the ability of populations to evolve with respect to flower size metrics, the number of floral parts, and flowering time (Mazer et al. 1999; Conner 2006; Holeski & Kelly 2006; Delph & Herlihy 2012; Galloway & Burgess 2012). Some of these studies were described briefly in Section 2.2.4. Here, only one additional example is given.

Selection for high and low numbers of anthers and ovules per flower was performed for two generations using the selfing annual *Spergularia marina*, starting with a wild population from California (Mazer et al. 1999). The species shows much phenotypic variation in all floral traits that have been examined, including the numbers of both anthers ($n = 0–8$) and ovules ($n = 46–182$).

The first generation of seeds made by 30 parents grown in a greenhouse was used to establish a base population (F_1) of 1,200 individuals. Five selection lines were established: high versus low anther number, high versus low ovule number, and a randomly selected control. Seeds were collected from 40 plants at the high or low extremes to produce the second generation (F_2), and then seeds from 12 plants of the F_2 generation at each extreme were selected to produce the final generation (F_3). The heritabilities of anther and ovule number were moderate and the two traits were clearly responsive to artificial selection, with the exception of low ovule number (Fig. 9.7). The researchers also found evidence for a negative genetic correlation between anther and ovule production, suggesting a possible evolutionary constraint on the total number of gametes made per flower (Mazer et al. 1999). Note, however, that

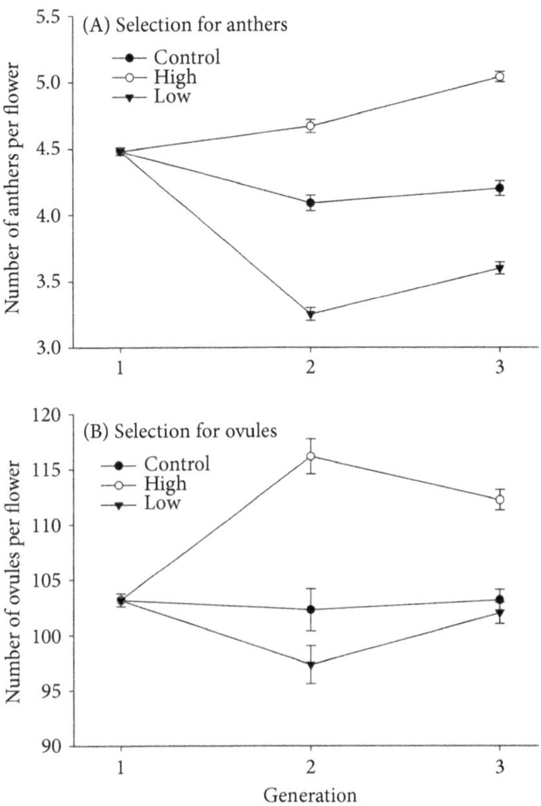

FIGURE 9.7
(A, B) An artificial selection experiment to increase or decrease the number of anthers (A) and ovules (B) in flowers of the annual *Spergularia marina*. The starting population is generation 1. Each point is a mean ± standard error of more than 200 individuals. Drawn from data in Mazer et al. (1999).

in most other studies, *positive* genetic correlations have been documented between pairs of floral traits (Ashman & Majetic 2006), implying that the flower might evolve as an integrated unit in response to selection mediated by pollinators (or other biotic agents). However, phenotypic correlations among floral traits is generally low (relative to those among vegetative traits), and there is limited evidence for overall floral integration to date (Conner et al. 2014).

9.3.2 Pollinator-Mediated Selection

Although genetic variation in floral traits is undoubtedly available and provides opportunity for population-level responses to selection, this does not mean that pollinating animals are necessarily the key selection agents. This is, in part, because other nonpollinating animals (e.g., floral herbivores, nectar robbers) can also act as agents of selection on floral traits (Strauss & Whittall 2006; Parachnowitsch & Caruso 2008; McCall et al. 2013). Furthermore, abiotic factors (e.g., edaphic conditions) have been shown to influence the evolution of some floral traits (Elle 2004; Meindl et al. 2013; Pope et al. 2013). Despite these complications, there have been numerous studies that have used the classic method of phenotypic selection analysis (Section 2.2.3) to show significant directional selection mediated by pollinators on traits such as the size, color, and number of flowers produced (reviewed in Harder and Johnson [2009]). Only a few of the many examples available can be detailed here to illustrate this approach.

Significant phenotypic selection on floral traits can sometimes be attributed to pollinator behavior, such as the discrimination against albino flowers of *Delphinium nelsonii* by hummingbirds and bumblebees, thereby favoring the usual blue-flowered genotypes (Waser & Price 1981). Selection on floral traits has also been documented in populations of some plant species in their natural habitats with exposure to their usual pollinators (e.g., Murren et al. 2009; Temeles et al. 2013); however, experiments in which pollination conditions are manipulated provide more powerful evidence for the specific role of pollinator-mediated selection on floral traits (Galen 1989; Fishman & Willis 2008; Parachnowitsch & Kessler 2010; Sletvold & Ågren 2010; Sletvold et al. 2010; Bartkowska & Johnston 2012). Typically these studies entail comparison of directional selection gradients (β) between plants that were open-pollinated by their usual pollinators to plants that received supplemental hand-pollination. The expectation is that when pollinators are the agents of selection for any one floral trait, then β for the open-pollinated group will be significantly greater than that of the hand-pollinated group (in which pollination is ensured regardless of floral phenotype). In fact, Sletvold and Ågren (2010) quantify pollinator-mediated selection ($\Delta\beta_{poll}$) by subtracting the β value for the hand-pollinated group from the β value of the open-pollinated group. Note that $\Delta\beta_{poll}$ is essentially a measure of how much more selection on a specific trait is occurring when pollinators are determining seed set versus when seed set is not limited by pollinator service.

Examples of selection gradients on three floral traits estimated in several pollination studies are shown in Figure 9.8. With the exception of the *Mimulus guttatus* population, which is annual (Fishman & Willis 2008), all species in the graph are perennial herbs. The two orchids *Dactylorhiza lapponica* (bee-pollinated) and *Gymnadenia conopsea* (butterfly- and hawkmoth-pollinated) showed strong pollinator-mediated selection on the number of flowers and length of the floral spurs (Sletvold & Ågren 2010; Sletvold et al. 2010). Pollinators also selected for larger flowers in *G. conopsea* and the bee-pollinated *Penstemon digitalis* (Parachnowitsch & Kessler 2010) (Fig. 9.8B). In agreement with the survey of Harder and Johnson (2009), selection gradients for the number of flowers were typically high (Fig. 9.8A), although in the hummingbird-pollinated *Lobelia cardinalis* (Bartkowska & Johnston 2012) and one population of *G. conopsea*, selection in hand-pollinated plants was not significantly reduced relative to open-pollinated plants. In these instances, even when pollination was ensured, a greater number of flowers still had a positive effect on fitness. However, in *L. cardinalis*, some significant correlational selection (γ) gradients (Section 2.2.3.5) for several trait pairs *were* significantly greater in open-pollinated plants relative to hand-pollinated plants (Bartkowska & Johnston 2012). For example, flower number interacted with flowering date such that there was significant pollinator-mediated selection ($\Delta\gamma_{poll} = 0.58$, $p = 0.04$) for plants that flowered early and made many flowers. It is important to recognize that floral traits sometimes function as integrated units (Reynolds et al. 2010; Conner et al. 2014); traits that do not appear to be subject to pollinator-mediated selection (e.g., flower number in *L. cardinalis*, Fig. 9.8A) can still be selected indirectly via correlation with one or more additional traits.

Given the diversity of selection agents that might affect floral traits, path analysis (Section 2.2.3.6) has sometimes been used effectively to quantify direct and indirect effects of both pollinators and floral herbivores (florivores), as well as particular floral traits, on plant fitness (e.g., Conner et al. 1996; Cariveau et al. 2004; Irwin et al.

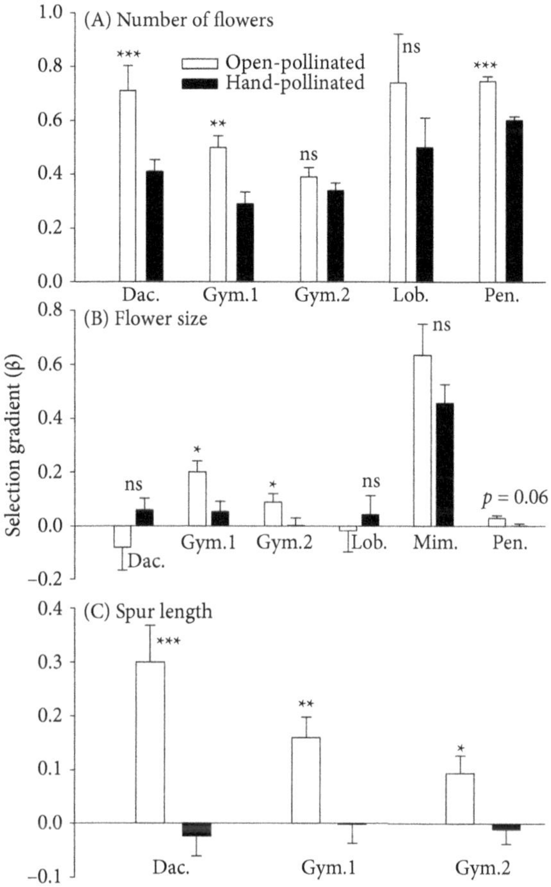

FIGURE 9.8

(A, B) Standardized directional selection gradients (β) ± standard error for number of flowers (A), flower size (B), and spur length (C) in open-pollinated and hand-pollinated plants of herbaceous species. Dac., *Dactylorhiza lapponica* (Sletvold et al. 2010), Gym. 1 and Gym. 2, two populations of *Gymnadenia conopsea* (Sletvold & Ågren 2010), Lob., *Lobelia cardinalis* (Bartkowska & Johnston 2012); Mim., *Mimulus guttatus* (Fishman & Willis 2008); Pen., *Penstemon digitalis* (Parachnowitsch & Kessler 2010). Statistical significance of differences between open-pollinated and hand-pollinated groups are indicated: ns, not significant; $^{*}p \leq 0.05$; $^{**}p < 0.01$; $^{***}p < 0.001$.

2004; Parachnowitsch & Caruso 2008; Gómez et al. 2009b; Bartkowska & Johnston 2012). The push and pull of mutualistic (pollinators) versus antagonistic (florivores) animals as agents of selection on floral traits parallels that found for seed dispersers versus seed predators on fruit and seed traits (Section 9.4). Path analysis affords easy visualization of the relative causal relationship (and significance) of those factors measured by the investigators that potentially affect fitness.

As an example of the utility of path analysis to depict complex plant–animal interactions in a simplified way, we consider the hummingbird-pollinated scarlet gilia (*Ipomopsis aggregata*) and its nectar-robbing bumblebees (Irwin et al. 2004; Irwin 2006). Plants at four sites in Colorado were measured for sugar concentration in the nectar, and both pollinator visitation (estimated by stigma pollen loads) and nectar

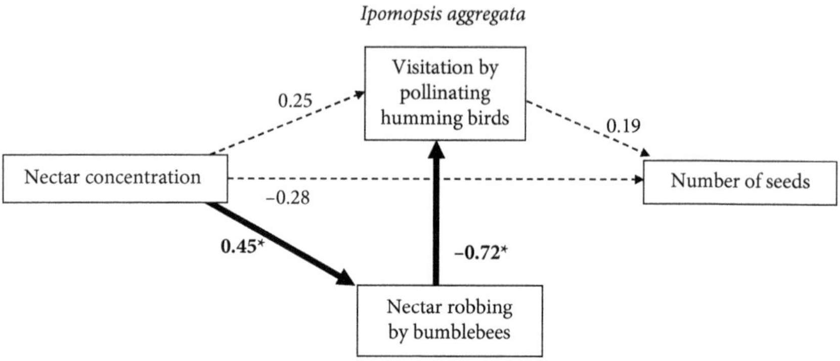

FIGURE 9.9
Path diagram for scarlet gilia and its interactions with hummingbird pollinators and nectar-robbing bumblebees showing standardized coefficients and their statistical significance (*$p < 0.05$). Insignificant paths are denoted by a thin dashed line. Copyright © 2004 by the Ecological Society of America, used with permission. Source: Irwin, R.E., Adler, L.S. & Brody, A.K. 2004. The dual role of floral traits: pollination attraction and plant defense. *Ecology* 85: 1503–1511.

robbing (estimated by counts of floral holes) were assessed twice a week. Four path models were explored; the best-fitting one is depicted in Figure 9.9. Only two paths were statistically significant ($p < 0.05$); nectar concentration increased nectar robbing, which in turn reduced visitation by pollinators. Interestingly, there were no direct effects of nectar concentration on pollinator visits or plant fitness (estimated by seed production). In this system, it may be advantageous for plants to produce dilute nectar because this would reduce nectar robbing and increase pollinator visitation indirectly (Irwin et al. 2004). A subsequent, two-year study showed that greater numbers of flowers increased relative fitness directly, but also increased nectar robbing, which again significantly reduced pollinator visitation (Irwin 2006).

It is evident that pollinators are only one of a number of potential agents of selection on floral traits (Strauss & Whittall 2006). For some species, the strength of selection on floral traits by seed predators can exceed that of pollinators (Cariveau et al. 2004; Parachnowitsch & Caruso 2008), and sometimes the same flower-visiting species acts as both pollinator and seed predator (e.g., Kula et al. 2013). Floral traits in plants probably represent an evolutionary compromise between features that attract pollinators and those that deter flower and seed consumers effectively.

9.3.3 Floral Adaptation

The close match between species-specific floral traits and features of particular pollinator species provides a strong argument for reciprocal selection pressures shaping coevolutionary adaptation of plants and pollinators (Pellmyr 2002; Fenster et al. 2004; Harder & Johnson 2009). The match between floral spur length of species in the columbine genus *Aquilegia* and the tongue lengths of different pollinator species provides an excellent example (Whittall & Hodges 2007).

In studies in which the floral traits of a species have been manipulated experimentally, the original phenotypes have usually been shown to be better adapted (in terms of fitness) than the altered, alternative phenotypes (surveyed by Harder and Johnson [2009]). These negative fitness effects of changes made to the usual phenotypic trait reinforces the contention that floral features represent adaptations that allow successful pollination and seed set. But what about the variation in plant–pollinator relationships *within* species? Is there evidence for variation in selection pressures exerted by pollinators on plants at different sites and do plant populations ever show evidence of floral adaptation to their local group of pollinators?

Studies of populations of several plant species at different sites have, in general, revealed geographic variation in the ecological outcomes of plant–pollinator interactions and in the extent to which pollinator-mediated selection occurs (Thompson & Cunningham 2002; Caruso et al. 2003; Gómez et al. 2009b; Ellis & Anderson 2012). Perhaps not surprising to the evolutionary ecologist, variation in the strength of reciprocal selection in biotic interactions (i.e., a geographic mosaic of selection) is predicted by coevolutionary theory (Thompson 2005). In the herbaceous species *Lithophragma parviflorum* across 12 sites in the northwestern United States, the effects of a co-occurring moth (*Greya politella*) were shown to vary from mutualistic (acting as pollinators) in some habitats to antagonistic (acting as seed predators) in others (Thompson & Cunningham 2002). Geographic variation in other plant–pollinator systems is usually not so extreme. For example, Gómez et al. (2009b) investigated eight populations of *Erysimum mediohispanicum* pollinated by different assemblages of insect species and found that some floral traits (e.g., corolla diameter and length of the floral tube) showed significant directional selection only in specific populations. Again supporting coevolutionary theory, there was a mosaic of selection regimes, with populations experiencing strong selection (coevolutionary hotspots [Thompson 2005]) mixed with other populations (cold spots) for which selection regimes were relatively weak or insignificant (Gómez et al. 2009b).

To test experimentally for possible local adaptation of *Erysimum mediohispanicum* to its pollinators, reciprocal translocation of plants was conducted between four populations: two were in hotspots where pollinators exerted strong selection, and the other two were in cold spots where pollinator-mediated selection was weak (Gómez et al. 2009a). Seeds were collected from all populations, and plants from those seeds were grown in a common garden until they flowered. They were then placed into the four sites, and floral visitation by pollinators was quantified. The attractiveness of flowers to pollinators depended significantly on population origin ($p < 0.0001$). In addition, plants of hotspot populations in their native sites had greater floral visitation than foreign plants from cold-spot populations (Gómez et al. 2009a). However, even in their native sites, plants of cold-spot populations were visited less than those from hotspot populations, indicating a lack of adaptation to their local pollinators.

Evidence of potential reciprocal coadaptation of plants and their pollinators can sometimes be evaluated indirectly by careful measurement of key traits of both the flower and its pollinating species across multiple sites. The pattern of covariation, if any, in the floral and pollinator traits can then be examined across the different populations. Anderson and Johnson (2008) investigated such a relationship across 16 populations of the plant *Zaluzianskya microsiphon* in South Africa, a variable species that has flowers with long corolla tubes that are pollinated by a long-tongued fly (*Prosoeca ganglbaueri*). Across their geographic ranges, there was great variation in corolla length and fly proboscis length, and a highly significant correlation between the two traits ($r^2 = 0.69$, $p < 0.001$). This pattern of covariation is consistent with the idea that reciprocal selection pressures have resulted in coevolution of this plant species and its natural pollinator. Strengthening this contention, the researchers later found that other plant species pollinated by the same fly showed a similar pattern of geographic covariation in flower depth and proboscis length (Anderson & Johnson 2009).

To test the hypothesis experimentally, that floral tube length is an adaptive trait, Anderson and Johnson (2009) conducted reciprocal translocations among two populations of the orchid *Disa nivea*, a species that shows a positive relationship ($r^2 = 0.69$, $p < 0.001$) between flower depth and proboscis length of the fly *Prosoeca ganglbaueri*. The two orchid populations differed greatly in length of the floral spur: 42.1 ± 0.7 mm (mean ± SE) at the Lodge site versus 24.3 ± 0.7 mm at the Rama site (27 km away). Cut inflorescences from both populations were placed into test tubes with water and left at the two sites for a week. The proportion of flowers that had been pollinated and the proportion of flowers with pollinaria removed were recorded as measures of female and male pollination success, respectively (Anderson & Johnson 2009). Plants from the Rama population (long floral spurs) showed significantly greater female (Fig. 9.10A) and male pollination success (Fig. 9.10B) than plants from the Lodge population (short spurs) at *both* sites.

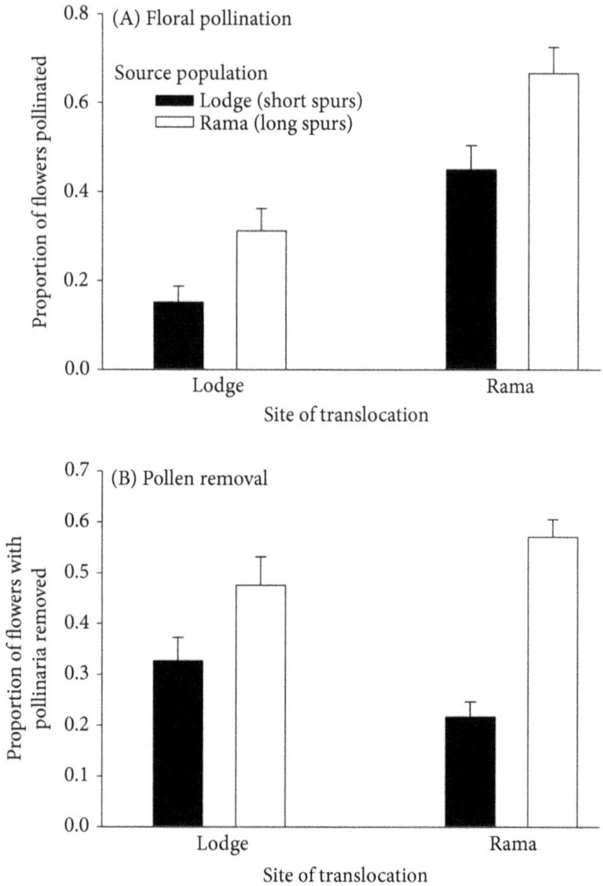

FIGURE 9.10

(A, B) Mean ± standard error of the proportion of flowers pollinated (A) and the proportion of flowers with their pollinaria removed (B) in reciprocally translocated plants of the fly-pollinated orchid *Disa nivea* at two sites (Lodge and Rama) in South Africa. Flowers of plants native to the Lodge site have shorter spurs (24 mm) than those native to the Rama site (42 mm). Data from Anderson and Johnson (2009).

The plants with short floral spurs were especially at a disadvantage relative to plants with long spurs at the Rama site with regard to pollen removal because flies have the longest proboscides at this site. Therefore, pollinator-mediated selection favoring long spurs had likely been important to the evolution of floral form in this orchid species.

9.3.4 Molecular Genetic Tools

The tools of molecular genetics have begun to be applied to the study of plant–pollinator interactions (Clare et al. 2013) and the evolution of floral traits (Clegg & Durbin 2003; Sapir 2009). A key objective in this research is to connect the molecular genetic variation of floral traits to floral phenotypes that affect pollinator-mediated selection and plant fitness, through both male and female components.

Polymorphism in floral color resulting from proteins coded by one or a few gene loci constitutes one of the simplest traits in which links have been established from the gene(s) involved to real-world plant–pollinator relationships (Conner 2006; Sapir 2009). A well-known example is the locus *YELLOW UPPER* (*YUP*) , which controls the deposition of yellow carotenoids in the petals of *Mimulus* flowers (Bradshaw & Schemske 2003). The pink flowers (low carotenoids, mostly anthocyanins) of *Mimulus lewisii* are pollinated by bumblebees whereas the red flowers (high levels of anthocyanins) of *Mimulus cardinalis* are pollinated by hummingbirds. Using near-isogenic lines in which the YUP locus had the allele of one species substituted for the usual allele in the genome of the second species, Bradshaw and Schemske (2003) were able to perform field experiments to show that pollinator preferences were governed by the particular allelic variant present at this single gene locus. Genetic variation at several loci regulating anthocyanin biosynthesis (thereby controlling floral pigmentation) in *Ipomoea*, the morning glory genus, has been shown to be critical to pollinator preference; the genes involved are therefore likely to be targets of pollinator-mediated selection (Clegg & Durbin 2003; Baucom et al. 2011). Floral color in *Raphanus sativus*, controlled by a pair of alleles at each of two loci, appears to be subject to opposing selection mediated both by pollinators and herbivores (Strauss & Whittall 2006).

For continuously varying floral traits controlled by QTLs, genomic tools based on specific DNA markers can provide information on the molecular genetic variation underlying ecologically relevant phenotypes (Herrera & Bazaga 2008, 2009; Sapir 2009; Clare et al. 2013). Genome scans of neutral genetic markers can be followed by outlier tests to identify loci that show greater divergence among populations than expected (Section 5.3) and that may correlate with floral traits. Herrera and Bazaga (2008) screened many AFLP loci and used a neutrality test to identify nine outlier loci in the hawkmoth-pollinated, perennial violet *Viola cazorlensis*. These outlier loci showed a highly significant ($p < 0.0005$) deviation from what would be expected based on selective neutrality, and therefore were likely to have been under selection. But was this pollinator-mediated selection? The authors next examined the correlations of three floral traits with the allele frequencies of the outlier loci. Statistically significant correlations were found between petal, spur, and peduncle length and allele frequency of four loci (r range, 0.50–0.79; p range, <0.01–0.07). Although it was not known what these particular loci (or loci to which they were linked) actually coded for, they apparently have diverged along with floral phenotype in *V. cazorlensis*. Unfortunately, the precise role of pollinators in this "adaptive floral divergence" (Herrera & Bazaga 2008) is unclear.

Much remains to be learned about the ecological genetics and coevolution of plant–pollinator interactions. Many of the key questions that may guide future research have been assembled in a lengthy paper by a dozen pollination ecologists (Mayer et al. 2011). No doubt, both traditional and molecular-based approaches will continue to be used in future studies on the ecology and evolution of pollination systems (Mitchell et al. 2009; Clare et al. 2013).

9.4 FRUIT AND SEED DISPERSAL

The plant–animal interaction of fruit and seed dispersal is another mostly mutualistic one; the dispersing agent receives the food reward provided by the fruit (or the seeds within) whereas the plant presumably receives a fitness benefit when some fraction of its seeds are transported and deposited into a new location. I use "presumably" in the last sentence to emphasize that (1) the fate of ingested seeds is often not known, but some proportion of them are destined to be destroyed during fruit consumption and/or digestion (Janzen 1983; Traveset et al. 2007), and (2) the suitability of a site to which seeds have been dispersed for germination and seedling establishment is often not known, but clearly is likely to be highly variable in space and time (Harper 1977; Rey & Alcántara 2000; Schupp 2007). There is no clear dividing line between dispersal and seed predation by animals (Fig. 9.11), and some predispersal loss of seeds to predation (and also to microbial degradation) is part of the cost of reliable dispersal (Janzen 1971; McKey 1975; Wenny 2000; vander Wall 2001). For dispersal-related adaptations to evolve in plants, the fitness benefits of dispersal must outweigh the costs.

The ecological study of plant dispersal has a venerable history (Ridley 1930; McKey 1975; Howe & Smallwood 1982; van der Pijl 1982 [earlier editions in 1969 and 1972]; Herrera 1985), with early work emphasizing fruit and seed adaptations to different dispersal agents and the distances to which seeds travel. Dispersal ecology has since developed into a complex, integrative discipline (Clobert et al. 2012)

FIGURE 9.11
Seed disperser or seed predator? A song sparrow (*Melospiza melodia*) consumes the caryopses of a beachgrass (*Panicum amarum*) on Long Island, New York. Photo courtesy of S. Mitra.

addressing long-standing questions about dispersal systems from both the plant and animal perspective as two selection agents that reciprocally affect traits important to their coevolution. The edited volumes in Table 9.1 contain diverse sets of articles that review many topics relevant to the evolutionary ecology of dispersal in plants and provide a good overview of the methods used to address them.

9.4.1 Selection Mediated by Fruit Consumers

What types of traits are likely to be subject to selection by animal dispersers and fruit or seed consumers? I compiled some of them in Table 9.3 and provide conventional expectations for the selective importance of fruit and seed characteristics. Some of these, such as the elaiosomes, lipid-rich structures attached to seeds of ant-dispersed plants (Beattie & Hughes 2002), are adaptations for a specific animal group. However, most phenotypic features in Table 9.3 are more general. Most fruit and seed traits vary greatly among vertebrate-dispersed species (e.g., Herrera 1987), and quantitative genetic variation within and among populations provide the raw material necessary for the evolution of traits that influence dispersal (Zera & Brisson 2012).

Features such as the color, odor, size, and nutrient composition of fruits are well known to be important to animal dispersal (Willson & Whelan 1990; Schaefer & Schaefer 2007; Cousens et al. 2008; Rodriguez et al. 2013; Renoult et al. 2014). Phenotypic selection pressures mediated by vertebrate consumers may favor large, brightly colored fruits rich in nutrients if fruit consumption improves seed dispersal and, ultimately, seedling establishment (Herrera et al. 1994; Schupp et al. 2010). However, the same features of fruits that make them attractive to potential dispersers can also make them attractive to predators and microbes that reduce plant fitness (Table 9.3). These latter agents may select for increased levels of toxic secondary compounds in fruits (Cipollini & Levey 1997), such as the capsaicin made by chili peppers, which deters both herbivores and decomposing fungi (Levey et al. 2007).

This push and pull of mutualistic versus antagonistic agents of selection creates a complicated scenario for the evolution of fruit or seed traits. Some fruit features such as nutrient content and the type or levels of defensive secondary compounds may represent evolutionary compromises in response to opposing selection pressures exerted by animal dispersers and organisms that consume fruits (and/or seeds) without dispersing them. Increased levels of energy-rich nutrients such as sugars attract both dispersers and predators; although selection by dispersers might favor high nutrient concentration if vertebrate dispersal of seeds improves plant fitness, strong counter-selection by predators and microbes should favor reductions in fruit nutrient concentration (Fig. 9.12). With regard to secondary compounds, the opposite situation occurs. If such compounds render fruits unattractive or toxic to potential animal dispersers, selection should favor lower levels within fruits. However, significant loss of fitness resulting from predation or microbial degradation could select for greater levels of protective secondary compounds in fruits and seeds (Fig. 9.12).

Because of the myriad biotic and abiotic factors that could affect the evolution of fruit and seed characteristics, it has been difficult to demonstrate pairwise coevolution definitively between a particular vertebrate-dispersed plant species and a particular vertebrate disperser. Relative to other types of plant–animal interactions, evidence for coadaptation resulting from reciprocal selection between plants and their vertebrate dispersers is weak, possibly because of the slow evolutionary rates

Table 9.3 Types of fruit and seed characteristics that can be subject to selection mediated by animal dispersers and/or fruit and seed consumers (e.g., insects, microbes).

Category	*Characteristic*	*Comment ("conventional wisdom")*
General	Fruiting phenology	Timing of fruit maturation selected to coincide with presence of dispersers; lengthy duration can improve chances of dispersal, but also increase chances of attack by predators and microbes
Fruit	Color	Brightly colored fruits selected to attract dispersers
	Size	Width/length selected to match gape size of dispersers
	Odor (volatile compounds)	Specific odors selected to attract specific dispersers
	Nutrients (carbohydrates, lipids, proteins)	Nutrient-rich fruits selected to attract dispersers
	Minerals	General effects on animal-mediated dispersal unknown
	Water content	Relative water content determines nutrient, mineral, and secondary chemical concentration; fleshy fruits may attract dispersers in dry environments
	Secondary compounds (alkaloids, phenolics, and so on)	Selected to provide defense against insect or microbial attack
Seeds	Number per fruit	Variable effects on animal-mediated dispersal; often shows trade-off with mass
	Size and mass	Can affect whether disperser drops or swallows seeds while consuming fruit
	Coat thickness	Can determine whether seeds are viable following passage through animal gut and also affects germination
	Nutrients	Nutrient-rich seeds selected to attract dispersers
	Secondary compounds	Selected to provide defense against seed predators or microbial attack
	Elaiosomes (lipid-rich accessory structures)	Provides food reward to ants that disperse seeds
	Bristles, hairs, hooks, barbs, awns[a]	Density and length can determine the effectiveness of dispersal by adhesion to animal fur

[a]All structures can also apply to fruits.

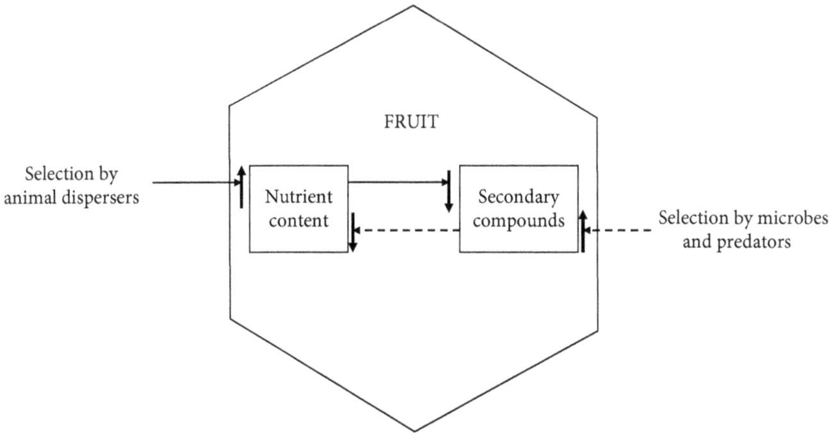

FIGURE 9.12
General scheme for opposing selection pressures exerted by animal dispersers (solid lines) versus microbes and predators (dashed lines) on the levels of energy-rich nutrients and secondary compounds found in vertebrate-dispersed fruits.

of long-lived plants, unpredictability of sites to which seeds may be dispersed, and generally weak selection exerted by any one dispersal agent (Herrera 1985, 2002). Nonetheless, when fitness is assessed as the relative number of seeds dispersed, phenotypic selection analysis (Lande & Arnold 1983) is possible and can shed light on what traits are most critical for successful dispersal.

Two frugivore-mediated selection studies are presented here as examples, both using tree species in the Rosaceae that produce one-seeded fruits (drupes) dispersed by birds. Jordano (1995) examined a population of *Prunus mahaleb* (St. Lucie's cherry) in southeastern Spain and quantified the size of the fruit crop (for each of 60 trees), the proportion of fruits taken by birds, and the level of effective seed dispersal (estimated with ground traps in which regurgitated and defecated seeds could be counted). Relative fitness was defined as the number of dispersed seeds of a tree relative to the population mean number of dispersed seeds. Both the computation of directional selection gradients (β) and a modified path analysis model revealed the overriding importance of fruit crop size (fecundity) to plant fitness ($\beta = 0.89$, $p < 0.001$). Selection on fruit diameter and seed mass was weak and insignificant (Fig. 9.13); however, the dry mass of dispersed seeds was significantly ($p < 0.01$) less than that of seeds in fruits before dispersal in the two years of the study (Jordano 1995). Because seed mass and fruit diameter correlate positively ($r = 0.68$, $p < 0.01$), birds may select for smaller fruits (with lighter weight seeds), although the negative selection gradient for fruit diameter was weak (Fig. 9.13).

The second selection study involved a population of *Crategus monogyna* (a common hawthorn), also growing in Spain (Sobral et al. 2010). Thirty trees were included, and measures were made of fruit crop size and fruit (and seed) size. The assumption was made, as in the Jordano (1995) study, that "dispersed seeds are more likely to have higher fitness than undispersed seeds" (Sobral et al. 2010, p. 1282). Thus, relative fitness was based on the number of seeds dispersed per tree. Although the β value for fruit crop size was relatively low, it was positive and highly significant

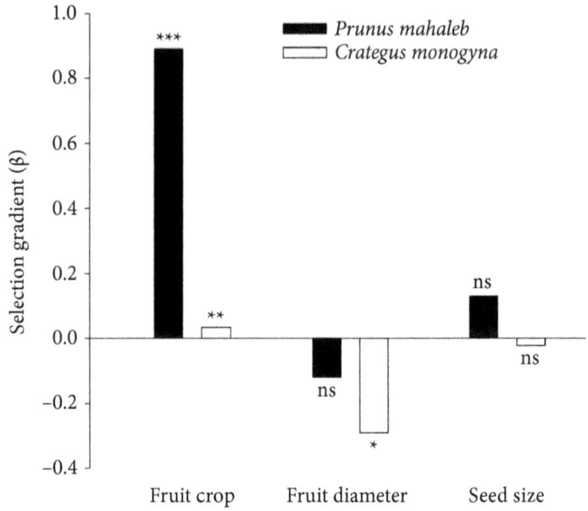

FIGURE 9.13
Standardized directional selection gradients (β) for size of the fruit crop, fruit diameter, and seed size, all based on relative seed dispersal as a fitness metric, in two bird-dispersed species in Spain. Statistical significance of the estimates are indicated: ns, not significant; $^{*}p \leq 0.05$; $^{**}p < 0.01$; $^{***}p < 0.001$. Seed size is dry mass (in *Prunus mahaleb*) or seed diameter (in *Crategus monogyna*). Data for *P. mahaleb* from Jordano (1995) and for *C. monogyna* from Sobral et al. (2010).

(Fig. 9.13). In addition, there was significant selection for smaller fruits, as indicated by the negative β value for fruit diameter (Fig. 9.13). Interestingly, there were significant correlational selection gradients (γ) for some pairs of traits (see Section 2.2.3.5 for a discussion of correlational selection). Fitness surfaces (Brodie et al. 1995) were used to illustrate significant correlational selection for fruit diameter × seed length ($\gamma = -0.073$) and fruit length × seed length ($\gamma = -0.033$). These analyses indicate that the relative success of dispersal from frugivore-mediated selection on this species was greatest for plants having larger fruits with smaller seeds.

Of course, the efficiency with which seeds are removed by animal dispersers is only one component of the overall effectiveness of a particular dispersal system (Schupp et al. 2010). To the fruiting plant, relative fitness is also determined by what happens inside the vertebrate gut and seed fate after discharge by the animal (Janzen 1983; Traveset et al. 2007). Selection may favor a particular fruit trait when it improves dispersal *and* results in the transport of seeds to habitats favorable to seedling establishment and survival (Herrera et al. 1994; de Casas et al. 2012). Bridging the gaps between dispersal by frugivores, deposition onto the soil, germination, and seedling establishment (Herrera et al. 1994; Rey & Alcántara 2000; Wenny 2000; Alcántara & Rey 2003) is challenging, but necessary, if ecologists are to glean a more complete understanding of the relative role of dispersing animals as agents of selection (in comparison with other agents) in plants.

9.4.2 Molecular Genetic Tools

We saw earlier (Section 5.4.1) how the tools of molecular genetics have been used to characterize gene flow via pollen or seeds between plant populations, and to estimate dispersal distances. For example, molecular markers (polymorphic microsatellites) in parental trees and fruit endocarps have been used to distinguish the relative effectiveness of different frugivores in dispersing the seeds of *Prunus mahaleb* to various distances (Jordano et al. 2007). Might the use of molecular markers and the identification of candidate gene loci also enable improved understanding of disperser-mediated selection on fruit and seed traits?

Some aspects of the molecular genetics of traits important to dispersal have been reviewed, but examples are drawn mostly from animal populations (Zera & Brisson 2012) or from plants that are not dispersed by animals, such as *Arabidopsis thaliana* (Hall & Donohue 2012). For example, in a review of the "Genetics of Plant Dispersal" (Hall & Donohue 2012), the molecular loci considered are several genes that control dehiscence of the dry fruits of *A. thaliana* and fruit development (and size) in tomato plants. Even if the gene loci were not known, the success of artificial selection for large fruits in tomato provides a stunning, textbook example (e.g., Herron & Freeman 2014) of how selection imposed on genetically variable populations can change the phenotypic characteristics of fruits in a dramatic way.

Although there is little doubt that phenotypic traits such as fruit size and color are genetically variable within and among populations, for most wild plants the relevant candidate gene loci have not been identified. Thus, adaptive differentiation in molecular genetic loci that correlate with fruit phenotypes important to animal dispersal has not been demonstrated in the way that molecular adaptation to abiotic factors has (see Section 5.3.2). In short, there is clearly much opportunity for future work on the molecular basis of plant adaptation to frugivore-mediated dispersal.

9.5 WRAP-UP

Plant–animal interactions span the range from antagonistic to mutualistic, involving reciprocal selective effects that create the potential for coevolutionary adaptation. Herbivores, seed predators, pollinators, and fruit/seed dispersers can all function as significant agents of natural selection on plant populations and affect the evolution of diverse sets of phenotypic traits. The multiplicity of biotic agents that can affect a plant population at any one time makes it a continuing challenge to delimit the precise contribution of individual animal species to the evolution of morphological, biochemical, and molecular traits that currently function as adaptations to herbivory, pollination, or dispersal.

10 Future Directions

Throughout the 30+ years of my personal research career I have looked at many journals and read countless papers on plant ecology and evolution. Since beginning graduate school, I have sometimes wondered "What more is there to do?" or "What new research avenues remain?" or "What hasn't yet been done?" Yet, every time I begin to think the discipline has become saturated with repetitive investigations of the same basic topics, I peruse the most recent issues of the relevant journals and find there are still many basic questions to address and many hypotheses to test (and retest). The proverbial well with research topics never appears to run dry. Ask any editor of a major journal and he or she will note there is no shortage of manuscripts submitted. As I write, new articles are being published, some of which will shed new light on the topics covered by this book. The authors of these pieces use both new and old approaches to address both new and old questions.

10.1 A FEW PREDICTIONS

What can be said about the future directions that the study of evolutionary ecology might take? Some predictions are so easy to make that they border on being trivial:

- Interdisciplinary teams of biologists with complementary expertise will provide a powerful collective body capable of tackling some of the broader issues in the field that require wide-ranging approaches and techniques.
- Because of the primacy of genetics to research in evolutionary ecology (as pointed out in Chapter 1), the inclusion of molecular tools to explore many key questions regarding population microevolution and adaptation, as well as the connections between genotype and phenotype, is likely to become increasingly important

(Barrett & Hoekstra 2011; Strasburg et al. 2012; Andrew et al. 2013; Des Marais et al. 2013). This should be apparent from this book, in which many examples have been provided on the use of molecular genetics to address issues related to natural selection by both abiotic and biotic agents, and to adaptive differentiation of populations.

- More studies will investigate the effects of global environmental changes on plant microevolution and adaptation using the tools of both quantitative and molecular genetics (Reusch & Wood 2007; Hoffman & Sgrò 2011; Shaw & Etterson 2012). Dormant propagules, or ancestral genotypes of plant species from germplasm collections (Beaton et al. 2011), could be used as "evolutionary time machines" to predict population adaptation to changing environments (Orsini et al. 2013). These studies will explore not only adaptation to climate and drought (Weinig et al. 2014), but also evolutionary processes affected by diverse anthropogenic disturbances ranging from habitat fragmentation to pollution (Briggs 2009; Weller et al. 2013).
- The future will likely see greater application of the basic theoretical concepts and methodology of evolutionary ecology to problems imposed by introduced and invasive species, the conservation of threatened or endangered species, and the use of plants for restoration of degraded habitats (Hufford & Mazer 2003; Bossdorf et al. 2005; Kinnison & Hairston Jr. 2007; Bischoff et al. 2010).

10.2 MORE THAN A FEW QUESTIONS

On the occasion of the 100th anniversary of the British Ecological Society, and to highlight priorities for future research, a group of 34 ecologists identified 100 questions deemed to be important in basic ecology (Sutherland et al. 2013). It is a useful exercise to see which questions mesh with the set of topics included in the domain of plant evolutionary ecology (Chapter 1). It may also be germane to make an attempt to identify potential gaps in our ecological understanding (Agrawal et al. 2007) of specific topics within this domain.

10.2.1 Natural Selection and Adaptation

We have seen throughout this book many examples of how phenotypic selection has been estimated for a variety of traits for plants in different habitats exposed to a diversity of selection agents. Local adaptation and differentiation of populations in relation to specific habitat conditions have also been demonstrated in many species. Some important questions for the future and relevant source information are provided here.

- How can field experiments be better designed to identify the specific agents of selection that cause the evolution of particular phenotypic traits (MacColl 2011)?
- Is there a unified set of approaches and statistical procedures that can be applied to the detection of local adaptation and adaptive genetic divergence among populations (Blanquart et al. 2013)?
- How can artificial selection experiments be used to provide insight into fundamental issues regarding the adaptive responses of populations to different agents of natural selection (Conner 2003)?

- How can traditional ecological techniques such as reciprocal transplants be integrated with molecular approaches for studies of the evolutionary genetics of adaptation (Anderson et al. 2011)?
- How can the tools of ecological genomics best be used to delineate natural selection and adaptation at the molecular level (Hohenlohe et al. 2010; Barrett & Hoekstra 2011; Strasburg et al. 2012; Andrew et al. 2013; Des Marais et al. 2013)?
- What are the best procedures for evaluating the adaptive significance of putative candidate gene loci (especially for nonmodel species) (Andrew et al. 2013)?
- What are the natural scales at which evolutionary divergence occurs in relation to specific habitat factors? In other words, "how local is adaptation" (Sutherland et al. 2013, p. 61; Richardson et al. 2014)?
- What are the ecological and genetic factors that limit directional selection (Antonovics 1976; Bradshaw 1991; Kingsolver & Diamond 2011)?
- How does phenotypic plasticity influence the microevolution and evolutionary trajectories of natural populations (Sutherland et al. 2013)?
- How can we best predict how plant populations will respond evolutionarily to global environmental change (Reusch & Wood 2007; Shaw & Etterson 2012)?

Of course, I recognize these are very broad and general questions not likely to be answered adequately by any single study or project. They are also mostly basic questions without a specific practical application in mind; however, they can be relevant to applied questions when the selection agents are of anthropogenic origin or the species under investigation is economically important or of conservation interest.

10.2.2 Biotic Interactions

A hefty chunk of this book has been devoted to exploring the many types of biotic interactions that involve plants: competition and facilitation (Chapter 7), microbial symbiosis (Chapter 8), and herbivory, plus animal-mediated pollination and seed dispersal (Chapter 9). As natural components of all populations, biotic factors can be expected to function as potentially significant agents of selection, thereby governing population microevolution and adaptation. Information on the functioning of species interactions across multiple scales and diverse environments is needed to address some of the perceived gaps in our understanding of population and community ecology (Agrawal et al. 2007; Urban 2011). Some of the many possible questions one might ask regarding biotic interactions and plants, and relevant sources, are presented here (organized into subtopics).

Plant–Plant Interactions

- How can experiments be better designed and implemented to separate the role of plant–plant interactions versus abiotic factors as selection agents in local adaptation (Ariza & Tielbörger 2011; Thorpe et al. 2011)?
- How can experiments at the molecular level be designed and implemented to identify and characterize the candidate gene loci that may be important to competitive ability?
- Does the genetic identity and/or relatedness of interacting plants in a population influence its evolutionary trajectory by selecting for genotypes with phenotypic

traits that reduce the negative fitness consequences of competition (Mehrhoff & Turkington 1990; Turkington 1996; File et al. 2012)?
- To what extent are evolutionary outcomes of plant competition mediated by microbial symbionts or herbivores in nature (Clay 1990; Louda et al. 1990; Lankau & Klienbenstein 2009)?
- What is the relative importance of allelopathy as a mechanism in the evolution of plant–plant interactions and what are the most effective ways to detect it in nature (Inderjit & Weiner 2001; Inderjit et al. 2011)?
- How important are facilitative interactions in plant communities, is there genetic variation in the traits involved, and in what ways can facilitative relationships evolve (Brooker & Callaway 2009; Bronstein 2009)?

Microbial Symbiosis

- How might microbial symbionts influence the microevolutionary responses of plant populations to global environmental changes (Kivlin et al. 2013)?
- How does microbial symbiosis with plants influence their interactions with consumers and higher trophic levels (Sutherland et al. 2013)?
- What types of experimental procedures will provide the best evidence of local adaptation that has been mediated by specific groups of microbial symbionts (e.g., pathogens, bacterial symbionts, mycorrhizae, leaf endophytes)?
- What particular gene loci and molecular products are under natural selection and therefore critical to the coevolution of specific plant–microbe interactions (Karasov et al. 2014)?

Herbivory

- What combination of resistance traits will be selected for in plant populations exposed continually to a variable and diverse community of herbivores (Agrawal 2011)?
- How and why does the selective value of tolerance to herbivory (relative costs and benefits to the plant) vary among populations and with environmental conditions (Fornoni et al. 2003a; Núñez-Farfán et al. 2007)?
- What are the QTLs controlling resistance and tolerance to herbivory in wild plant species and for which molecular products do they code (Núñez-Farfán et al. 2007; Agrawal 2011)?
- What are the abiotic and biotic factors that contribute to geographic variation in tolerance and how does this affect the coevolution of plants and their herbivores (Fornoni 2011)?

Pollination

- What are the ecological causes of variation in paternal success (i.e., successful pollination and fertilization) and how does this affect the evolution of floral traits and reproductive systems (Mitchell et al. 2009)?
- What is the relative contribution of biotic versus abiotic agents of selection in shaping the evolution of flowers (Mayer et al. 2011)?
- What is the relative importance of mutualists (pollinators) versus antagonists (florivores, nectar robbers, seed predators, and so on) as selection agents on floral traits and mating systems (Mayer et al. 2011)?

- How can genomic tools be incorporated more effectively into studies of plant–pollinator interactions and which specific types of fundamental issues should they be used to investigate (Clare et al. 2013)?
- What are the ecological and evolutionary consequences of global environmental change for plant–pollinator interactions (Mitchell et al. 2009)?

Seed Dispersal

- What is the genetic basis for fruit and seed traits important to animal-mediated dispersal and how does the phenotypic variation in these traits among genotypes translate into variation in the effectiveness of seed dispersal (Herrera 2002; Cousens et al. 2008; Sutherland et al. 2013)?
- How can molecular genetic markers be better used to track seed movement and provide new insights into the evolutionary dynamics of seed dispersal and its effectiveness at different spatial scales (Grivet et al. 2005; Schupp et al. 2010; Hamrick & Trapnell 2011)?
- Do some plant species show evidence of local adaptation of populations to specific groups of biotic seed dispersers (Janzen 1983)?
- What is the relative contribution of animals as seed dispersers versus seed predators to selection for specific fruit and seed traits (Janzen 1983; Hulme & Benkman 2002)?

General

- How and why does selection mediated by biotic agents vary geographically and how do traits that modify interspecies interactions evolve in natural communities (Urban 2011)?
- How can the tools of experimental evolution be better applied to studies of adaptation, coevolutionary interactions, and other areas of evolutionary ecology (MacColl 2011; Kawecki et al. 2012)?
- To what extent and in what ways do ecoevolutionary dynamics at the population level affect processes at the community and ecosystem levels (Pelletier et al. 2009)?
- What are the combined selective effects of multiple antagonistic and mutualistic interactions in natural communities? Furthermore, how do species evolve in relation to multiple enemies and mutualists (Thompson 2002)?

And so it goes. Question after question, most too overextended and far-reaching ever to be answered satisfactorily by any one group of researchers or any one study system. They are easy to formulate, but challenging to address. The questions can only serve as very general guideposts to contemporary issues in plant evolutionary ecology and perhaps help shape the types of studies that should be proposed and the best approaches that can be developed.

References

Aarssen, L.W. 1983. Ecological combining ability and competitive combining ability in plants: toward a general evolutionary theory of coexistence in systems of competition. *Am. Nat.* 122: 707–731.

Aarssen, L.W. 1989. Competitive ability and species coexistence: a "plant's-eye" view. *Oikos* 56: 386–401.

Aarssen, L.W. 2005. On size, fecundity, and fitness in competing plants. In: *Reproductive Allocation in Plants*, ed. E.G. Reekie & F.A. Bazzaz, 215–244. Burlington, MA: Elsevier Academic Press.

Aarssen, L.W. & Keogh, T. 2002. Conundrums of competitive ability in plants: what to measure? *Oikos* 96: 531–542.

Aarssen, L.W. & Turkington, R. 1985. Biotic specialization between neighbouring genotypes in *Lolium perenne* and *Trifolium repens* from a permanent pasture. *J. Ecol.* 73: 605–614.

Abdala-Roberts, L. & Marquis, R.J. 2007. Tests of local adaptation to biotic interactions and soil abiotic conditions in the ant-tended *Chamaecrista fasciculata* (Fabaceae). *Oecologia* 154: 315–326.

Adams, W.T., Griffin, A.R. & Moran, G.F. 1992. Using paternity analysis to measure effective pollen dispersal in plant populations. *Am. Nat.* 140: 762–780.

Agrawal, A.A. 2011. Current trends in the evolutionary ecology of plant defence. *Funct. Ecol.* 25: 420–432.

Agrawal, A.A., Ackerly, D.D., Adler, F. et al. 2007. Filling key gaps in population and community ecology. *Front. Ecol. Environ.* 5: 145–152.

Agrawal, A.A., Erwin, A.C. & Cook, S.C. 2008. Natural selection on and predicted responses of ecophysiological traits of swamp milkweed (*Asclepias incarnata*). *J. Ecol.* 96: 536–542.

Agrawal, A.A., Hastings, A.P., Johnson, M.T.J., Maron, J.L. & Salminen, J.-P. 2012. Insect herbivores drive real-time ecological and evolutionary change in plant populations. *Science* 338: 113–116.

Aitken, S.N., Yeaman, S., Holliday, J.A., Wang, T. & Curtis-McLane, S. 2008. Adaptation, migration or extirpation: climate change outcomes for tree populations. *Evol. Appl.* 1: 95–111.

Alcántara, J.M., Bastida, J.M. & Rey, P.J. 2010. Linking divergent selection on vegetative traits to environmental variation and phenotypic diversification in the Iberian columbines (*Aquilegia*). *J. Evol. Biol.* 23: 1218–1233.

Alcántara, J.M. & Rey, P.J. 2003. Conflicting selection pressures on seed size: evolutionary ecology of fruit size in a bird-dispersed tree, *Olea europaea*. *J. Evol. Biol.* 16: 1168–1176.

Alexander, H. M. 1989. An experimental field study of anther-smut disease of *Silene alba* caused by *Ustilago violacea*: genotypic variation and disease incidence. *Evolution* 43: 835–847.

Alexander, H.M., Thrall, P.H., Antonovics, J., Jarosz, A.M. & Oudemans, P.V. 1996. Population dynamics and genetics of plant disease: a case study of anther-smut disease. *Ecology* 77: 990–996.

Allan, E. & Pannell, J.R. 2009. Rapid divergence in physiological and life-history traits between northern and southern populations of the British introduced neo-species, *Senecio squalidus. Oikos* 118: 1053–1061.

Allard, R.W. & Adams, J. 1969. Population studies in predominantly self-pollinating species: XIII. Intergenotypic competition and population structure in barley and wheat. *Am. Nat.*103: 621–645.

Allard, R.W. & Jain, S.K. 1962. Population studies in predominantly self-pollinating species: II. Analysis of quantitative genetic changes in a bulk-hybrid population of barley. *Evolution* 16: 90–101.

Allen, M.F. 1991. *The Ecology of Mycorrhizae.* Cambridge: Cambridge University Press.

Anacker, B.L. 2014. The nature of serpentine endemism. *Am. J. Bot.* 101: 219–224.

Andalo, C., Goldringer, I. & Godelle, B. 2001. Inter- and intragenotypic competition under elevated carbon dioxide in *Arabidopsis thaliana. Ecology* 82: 157–164.

Anderson, J.T. & Geber, M.A. 2010. Demographic source–sink dynamics restrict local adaptation in Elliot's blueberry (*Vaccinium elliottii*). *Evolution* 64: 370–384.

Anderson, B. & Johnson, S.D. 2008. The geographical mosaic of coevolution in a plant–pollinator mutualism. *Evolution* 62: 220–225.

Anderson, B. & Johnson, S.D. 2009. Geographical covariation and local convergence of flower depth in a guild of fly-pollinated plants. *New Phytol.* 182: 533–540.

Anderson, J.T. & Mitchell-Olds, T. 2011. Ecological genetics and genomics of plant defences: evidence and approaches. *Funct. Ecol.* 25: 312–324.

Anderson, E. & Stebbins, G.L. 1954. Hybridization as an evolutionary stimulus. *Evolution* 8: 378–388.

Anderson, J.T., Willis, J.H. & Mitchell-Olds, T. 2011. Evolutionary genetics of plant adaptation. *Trends Genet* 27: 258–266.

Andersson, S. 1996. Floral variation in *Saxifraga granulata*: phenotypic selection, quantitative genetics and predicted response to selection. *Heredity* 77: 217–223.

Andrew, R.L., Bernatchez, L., Bonin, A. et al. 2013. A road map for molecular ecology. *Mol. Ecol.* 22: 2605–2626.

Angert, A.L & Schemske, D.W. 2005. The evolution of species' distributions: reciprocal transplants across the elevation ranges of *Mimulus cardinalis* and *M. lewisii. Evolution* 59: 1671–1684.

Antonovics, J. 1968. Evolution in closely adjacent plant populations: V. Evolution of self-fertility. *Heredity* 23: 219–238.

Antonovics, J. 1976. The nature of limits to natural selection. *Ann. Missouri Bot. Gard.* 63: 224–247.

Antonovics, J. 2006. Evolution in closely adjacent plant populations: X. Long-term persistence of prereproductive isolation at a mine boundary. *Heredity* 97: 33–37.

Antonovics, J. & Bradshaw, A.D. 1970. Evolution in closely adjacent plant populations: VIII. Clinal patterns at a mine boundary. *Heredity* 25: 349–362.

Antonovics, J., Bradshaw, A.D. & Turner, R.G. 1971. Heavy metal tolerance in plants. *Adv. Ecol. Res.* 71: 1–85.

Antonovics, J. & Levin, D.A. 1980. The ecological and genetic consequences of density-dependent regulation in plants. *Annu. Rev. Ecol. Syst.* 11: 411–452.

Antonovics, J. & Primack, R.B. 1982. Experimental ecological genetics in *Plantago*: VI. The demography of seedling transplants of *P. lanceolata. J. Ecol.* 70: 55–75.

Arif, I.A., Bakir, M.A., Khan, H.A., et al. 2010. A brief review of molecular techniques to assess plant diversity. *Int. J. Mol. Sci.* 11: 2079–2096.

Ariza, C. & Tielbörger, K. 2011. An evolutionary approach to studying the relative importance of plant–plant interactions along environmental gradients. *J. Ecol.* 25: 932–942.

Arnold, M.L. 1997. *Natural Hybridization and Evolution.* Oxford: Oxford University Press.

Arnold, A.E. 2007. Understanding the diversity of foliar endophytic fungi: progress, challenges, and frontiers. *Fungal Biol. Rev.* 21: 51–66.

Ashley, M.V. 2010. Plant parentage, pollination, and dispersal: how DNA microsatellites have altered the landscape. *Crit. Rev. Plant Sci.* 29: 148–161.

Ashman, T.-L. & Majetic, C.J. 2006. Genetic constraints on floral evolution: a review and evaluation of patterns. *Heredity* 96: 343–352.

Ashraf, M., McNeilly, T. & Bradshaw, A.D. 1986. The potential for evolution of salt (NaCl) tolerance in seven grass species. *New Phytol.* 103: 299–309.

Aston, J.L. & Bradshaw, A.D. 1966. Evolution in closely adjacent plant populations: II. *Agrostis stolonifera* in maritime habitats. *Heredity* 21: 649–664.

Augé, R. M. 2001. Water relations, drought and vesicular–arbuscular mycorrhizal symbiosis. *Mycorrhiza* 11: 11–42.

Ayasse, M. & Arroyo, J., eds. 2011. Pollination and plant reproductive biology [editorial for a special issue]. *Plant Biol.* 13: 1–6.

Bachmann, K. 1994. Molecular markers in plant ecology. *New Phytol.* 126: 403–418.

Bacon, C.W., Hinton, D.M. & Glenn, A.E. 2009. Endophytic niche and grass defense. In: *Defensive Mutualism in Microbial Symbiosis*, ed. J.F. White, Jr. & M.S. Torres, 385–404. Boca Raton, FL: CRC Press.

Bais, H.P., Weir, T.L., Perry, L.G., Gilroy, S. & Vivanco, J.M. 2006. The role of root exudates in rhizosphere interactions with plants and other organisms. *Annu. Rev. Plant Biol.* 57: 233–266.

Baker, H.G. 1963. Evolutionary mechanisms in pollination biology. *Science* 139: 877–883.

Baker, A., ed. 2000. *Molecular Methods in Ecology.* Oxford: Blackwell Science.

Baltzer, J.L., Thomas, S.C., Nilus, R. & Burslem, D.F.R.P. 2005. Edaphic specialization in tropical trees: physiological correlates and responses to reciprocal transplantation. *Ecology* 86: 3063–3077.

Barrett, J. 1985. The gene-for-gene hypothesis: parable or paradigm. In: *Ecology and Genetics of Host–Parasite Interactions*, ed. D. Rollinson, 215–225. London: Academic Press.

Barrett, R.D.H. & Hoekstra, H.E. 2011. Molecular spandrels: tests of adaptation at the genetic level. *Nat. Rev. Genet.* 12: 767–780.

Bartkowska, M.P. & Johnston, M.O. 2012. Pollinators cause stronger selection than herbivores on floral traits in *Lobelia cardinalis* (Lobeliaceae). *New Phytol.* 193: 1039–1048.

Barton, N.H. 2001a. The evolutionary consequences of gene flow and local adaptation: future approaches. In: *Dispersal*, ed. J. Clobert, E. Danchin, A.A. Dhondt, & J.D. Nichols, 329–340. New York: Oxford University Press.

Barton, N.H. 2001b. The role of hybridization in evolution. *Mol. Ecol.* 10: 551–568.

Barton, L.L. & Northup, D.E. 2011. *Microbial Ecology.* Hoboken, NJ: Wiley-Blackwell.

Baskin, C.C. & Baskin, J.M. 2014. *Seeds: Ecology, Biogeography, and Evolution of Dormancy and Germination*, 2nd ed. San Diego, CA: Academic Press.

Baucom, R.S., Chang, S.-M., Kniskern, J.M., Rausher, M.D. & Stinchcombe, J.R. 2011. Morning glory as a powerful model in ecological genomics: tracing adaptation through both natural and artificial selection. *Heredity* 107: 377–385.

Baucom, R.S. & Holt, J.S. 2009. Weeds of agricultural importance: bridging the gap between evolutionary ecology and crop and weed science. *New Phytol.* 184: 741–743.

Bazzaz, F.A. 1996. *Plants in Changing Environments.* New York: Cambridge University Press.

Bazzaz, F.A., Jasieński, M., Thomas, S.C. & Wayne, P. 1995. Microevolutionary responses in experimental populations of plants to CO_2-enriched environments: parallel results from two model systems. *Proc. Natl. Acad. Sci. U S A* 92: 8161–8165.

Beaton, L.L., van Zandt, P.A., Esselman, E.J. & Knight, T.M. 2011. Comparison of the herbivore defense and competitive ability of ancestral and modern genotypes of an invasive plant, *Lespedeza cuneata*. *Oikos* 120: 1413–1419.

Beattie, A.J. & Hughes, L. 2002. Ant–plant interactions. In: *Plant–Animal Interactions: An Evolutionary Approach*, ed. C.M. Herrera & O. Pellmyr, 211–235. Malden, MA: Blackwell Publishing.

Beebee, T. & Rowe, G. 2008. *An Introduction to Molecular Ecology*. 2nd ed. Oxford: Oxford University Press.

Begon, M., Townsend, C.R. & Harper, J.L. 2006. *Ecology: From Individuals to Ecosystems*. 4th ed. Malden, MA: Blackwell Publishing.

Beierkuhnlein, C., Thiel, D., Jentsch, A., Willner, E. & Kreyling, J. 2011. Ecotypes of European grass species respond differently to warming and extreme drought. *J. Ecol.* 99: 703–713.

Belesky, D. P., Devine, O.J., Pallas, J.E.,Jr. & Stringer, W.C. 1987. Photosynthetic activity of tall fescue as influenced by fungal endophyte. *Photosynthetica* 21: 82–87.

Belesky, D. P. & Fedders, J.M. 1996. Does endophyte influence regrowth of tall fescue? *Ann. Bot.* 78: 499–505.

Bell, G. 2008. *Selection: The Mechanism of Evolution*. 2nd ed. Oxford: Oxford University Press.

Bell, D.L. & Galloway, L.F. 2008. Population differentiation for plasticity to light in an annual herb: adaptation and cost. *Am. J. Bot.* 95: 59–65.

Benkman, C.W. 1999. The selection mosaic and diversifying coevolution between crossbills and lodgepole pine. *Am. Nat.* 153: S75–S91.

Benkman, C.W., Smith, J.W., Maier, M., Hansen, L. & Talluto, M.V. 2013. Consistency and variation in phenotypic selection exerted by a community of seed predators. *Evolution* 67: 157–169.

Bennington, C.C., Fetcher, N., Vavrek, M.C., Shaver, G.R., Cummings, K.J. & McGraw, J.B. 2012. Home site advantage in two long-lived arctic plant species: results from two 30-year reciprocal transplant studies. *J. Ecol.* 100: 841–851.

Bennington, C.C. & McGraw, J.B. 1995. Natural selection and ecotypic differentiation in *Impatiens pallida*. *Ecol. Monogr.* 65: 303–323.

Berenbaum, M. 2001. Plant–herbivore interactions. In: *Evolutionary Ecology: Concepts and Case Studies*, ed. C.W. Fox, D.A. Roff & D.J. Fairbairn, 303–314. New York: Oxford University Press.

Berenbaum, M., Zangerl, A.R. & Nitao, J.K. 1986. Constraints on chemical coevolution: wild parsnips and the parsnip webworm. *Evolution* 40: 1215–1228.

Bergelson, J., Kreitman, M., Stahl, E.A. & Tian, D. 2001. Evolutionary dynamics of plant *R*-genes. *Science* 292: 2281–2285.

Bernhardsson, C. & Ingvarsson, P.K. 2012. Geographical structure and adaptive population differentiation in herbivore defence genes in European aspen (*Populus tremula* L., Salicaceae). *Mol. Ecol.* 21: 2197–2207.

Bernhardsson, C., Robinson, K.M., Abreu, I.N., Jansson, S., Albrectsen, B.R. & Ingvarsson, P.K. 2013. Geographic structure in metabolome and herbivore community co-occurs with genetic structure in plant defence genes. *Ecol. Lett.* 16: 791–798.

Bert, V., Macnair, M.R., de Laguerie, P., Saumitou-Laprade, P. & Petit, D. 2000. Zinc tolerance and accumulation in metallicolous and nonmetallicolous populations of *Arabidopsis halleri* (Brassicaceae). *New Phytol.* 146: 225–233.

Bertin, C., Yang, X. & Weston, L.A. 2003. The role of root exudates and allelochemicals in the rhizosphere. *Plant Soil* 256: 67–83.

Biere, A. & Verhoeven, K.J.F. 2008. Local adaptation and the consequences of being dislocated from coevolved enemies. *New Phytol.*180: 265–268.

Biernaskie, J.M. 2011. Evidence for competition and cooperation among climbing plants. *Proc. R. Soc. B* 278: 1989–1996.

Billington, H.L., Mortimer, M.A. & McNeilly, T. 1990. Survival and reproduction in adjacent grass populations: effects of clone, time and environment. *J. Ecol.* 78: 1–14.

Billington, H.L. & Pelham, J. 1991. Genetic variation in the date of budburst in Scottish birch populations: implications for climate change. *Funct. Ecol.* 5: 403–409.

Bird, A. 2007. Perceptions of epigenetics. *Nature* 447: 396–398.

Bischoff, A., Crémieux, L., Smilauerova, M., et al. 2006. Detecting local adaptation in widespread grassland species: the importance of scale and local plant community. *J. Ecol.* 94: 1130–1142.

Bischoff, A. & Hurault, B. 2013. Scales and drivers of local adaptation in *Brassica nigra* (Brassicaceae) populations. *Am. J. Bot.* 100: 1162–1170.

Bischoff, A. & Müller-Schärer, H. 2010. Testing population differentiation in plant species: how important are environmental maternal effects? *Oikos* 119: 445–454.

Bischoff, A., Steinger, T. & Müller-Schärer, H. 2010. The importance of plant provenance and genotypic diversity of seed material used for ecological restoration. *Restor. Ecol.* 18: 338–348.

Björkman, O. & Holmgren, P. 1966. Photosynthetic adaptation to light intensity in plants native to shaded and exposed habitats. *Physiol. Plant.* 19: 854–859.

Blank, R.R. & Sforza, R. 2007. Plant–soil relationships of the invasive grass *Taeniatherum caput-medusae*: a reciprocal transplant experiment. *Plant Soil* 298: 7–19.

Blanquart, F., Kaltz, O., Nuismer, S.L. & Gandon, S. 2013. A practical guide to measuring local adaptation. *Ecol. Lett.* 16: 1195–1205.

Blouin, M.S. 2003. DNA-based methods for pedigree reconstruction and kinship analysis in natural populations. *Trends Ecol. Evol.* 18: 503–511.

Böcher, T.W. 1949. Racial divergences in *Prunella vulgaris* in relation to habitat and climate. *New Phytol.* 48: 285–314.

Bone, E. & Farres, A. 2001. Trends and rates of microevolution in plants. *Genetica* 112–113: 165–182.

Bonser, S.P. & Ladd, B. 2011. The evolution of competitive strategies in annual plants. *Plant Ecol.* 212: 1441–1449.

Borevitz, J. 2004. Genomic approaches to identifying quantitative trait loci: lessons from *Arabidopsis thaliana*. In: *Plant Adaptation: Molecular Genetics and Ecology*, ed. Q.C.B. Cronk, J. Whitton, R.H. Ree & I.E.P. Taylor, 53–60. Ottawa, Ontario: National Research Council Research Press.

Born, C., Hardy, O.J., Chevallier, M.-H., Ossari, S., Attéké, C., Wickings, J. & Hossaert-McKey, M. 2008. Small-scale spatial genetic structure in the central African rainforest tree species *Aucoumea klaineana*: a stepwise approach to infer the impact of limited gene dispersal, population history and habitat fragmentation. *Mol. Ecol.* 17: 2041–2050.

Bos, M., Harmens, H. & Vrieling, K. 1986. Gene flow in *Plantago* I: Gene flow and neighbourhood size in *P. lanceolata*. *Heredity* 56: 43–54.

Boshier, D.H, & Billingham, M.R. 2000. Genetic variation and adaptation in tree populations. In: *The Ecological Consequences of Environmental Heterogeneity*, ed. M. J. Hutchings, E. A. John, & A.J.A. Stewart, 267–289. Oxford: Blackwell Science.

Bossdorf, O., Arcuir, D., Richards, C.L. & Pigliucci, M. 2010. Experimental alteration of DNA methylation affects the phenotypic plasticity of ecologically relevant traits in *Arabidopsis thaliana*. *Evol. Ecol.* 24: 541–553.

Bossdorf, O., Auge, H., Lafuma, L., Rogers, W.E., Siemann, E. & Prati, D. 2005 Phenotypic and genetic differentiation between native and introduced plant populations. *Oecologia* 144: 1–11.

Bossdorf, O., Richards, C.L. & Pigliucci, M. 2008. Epigenetics for ecologists. *Ecol. Lett.* 11: 106–115.

Bower A.D. & Aitken, S.N. 2008. Ecological genetics and seed transfer guidelines for *Pinus albicaulis* (Pinaceae). *Am. J. Bot.* 95: 66–76.

Bowler, P.J. 2005. Variation from Darwin to the modern synthesis. In: *Variation: A Central Concept in Biology*, ed. B. Hallgrimsson & B.K. Hall, 9–28. Burlington, MA: Elsevier/Academic Press.

Box, E.O. 1996. Plant functional types and climate at the global scale. *J. Veget. Sci.* 7: 309–320.

Boyce, S. G. 1954. The salt spray community. *Ecol. Monogr.* 24: 29–67.

Boyd, R.S. & Rajakaruna, N. 2013. Heavy metal tolerance. In: *Oxford Bibliographies Online, Ecology Module*, ed. D. Gibson. New York: Oxford University Press. http://www.oxfordbibliographies.com/obo/page/ecology.

Bradshaw, A.D. 1952. Populations of *Agrostis tenuis* resistant to lead and zinc poisoning. *Nature* 169: 1098.

Bradshaw, A.D. 1959. Population differentiation in *Agrostis tenuis* Sibth: II. The incidence and significance of infection by *Epichloe typhina*. *New Phytol.* 58: 310–315.

Bradshaw, A.D. 1965. Evolutionary significance of phenotypic plasticity in plants. *Adv. Genet.* 13: 115–155.

Bradshaw, A.D. 1972. Some of the evolutionary consequences of being a plant. *Evol. Biol.* 5: 25–47.

Bradshaw, A.D. 1984. The importance of evolutionary ideas in ecology—and vice versa. In: *Evolutionary Ecology*, ed. B. Shorrocks, 1–25. Oxford: Blackwell Scientific.

Bradshaw, A.D. 1991. Genostasis and the limits to evolution. *Phil. Trans. R. Soc. Lond. B* 333: 289–305.

Bradshaw, H.D., Jr. 2005. Mutations in *CAX1* produce phenotypes characteristic of plants tolerant to serpentine soils. *New Phytol.* 167: 81–88.

Bradshaw, A.D. & McNeilly, T. 1991. Evolutionary responses to global climate change. *Ann. Bot.* 67: 5–14.

Bradshaw, H.D.,Jr. & Schemske, D.W. 2003. Allele substitution at a flower colour locus produces a pollinator shift in mankeyflowers. *Nature* 426: 176–178.

Brady, K.U., Kruckeberg, A.R. & Bradshaw, H.D., Jr. 2005. Evolutionary ecology of plant adaptation to serpentine soils. *Annu. Rev. Ecol. Evol. Syst.* 36: 243–266.

Brakefield, P.M. 2003. Artificial selection and the development of ecologically relevant phenotypes. *Ecology* 84: 1661–1671.

Brassard, J.T. & Schoen, D.J. 1990. Analysis of phenotypic selection among locations in *Impatiens pallida* and *Impatiens capensis*. *Can. J. Bot.* 68: 1098–1105.

Bratteler, M., Lexer, C. & Widmer, A. 2006. Genetic architecture of traits associated with serpentine adaptation of *Silene vulgaris*. *J. Evol. Biol.* 19: 1149–1156.

Briggs, D. 2009. *Plant Microevolution and Conservation in Human-Influenced Ecosystems*. Cambridge: Cambridge University Press.

Briggs, D. & Walters, S.M. 1997. *Plant Variation and Evolution*. 3rd ed. Cambridge: Cambridge University Press.

Brock, M.T., Maloof, J.N. & Weinig, C. 2010. Genes underlying quantitative variation in ecologically important traits: PIF4 (PHYTOCHROME INTERACTING FACTOR 4) is associated with variation in internode length, flowering time, and fruit set in *Arabidopsis thaliana*. *Mol. Ecol.* 19: 1187–1199.

Brodie, E.D., III, Moore, A.J. & Janzen, F.J. 1995. Visualizing and quantifying natural selection. *Trends Ecol. Evol.* 10: 313–318.

Bronstein, J.L. 2001. Mutualisms. In: *Evolutionary Ecology: Concepts and Case Studies*, ed. C.W. Fox, D.A. Roff & D.J. Fairbairn, 315–330. New York: Oxford University Press.

Bronstein, J.L. 2009. The evolution of facilitation and mutualism. *J. Ecol.* 97: 1160–1170.

Bronstein, J.L. 2012. Mutualisms and symbioses. In: *Oxford Bibliographies Online, Ecology Module*, ed. D. Gibson. New York: Oxford University Press. http://www.oxfordbibliographies.com/obo/page/ecology.

Brooker, R.W. 2006. Plant–plant interactions and environmental change. *New Phytol.* 171: 271–284.

Brooker, R.W. & Callaway, R.M. 2009. Facilitation in the conceptual melting pot. *J. Ecol.* 97: 1117–1120.

Brooker, R.W., Maestre, F.T., Callaway, R.M., et al. 2008. Facilitation in plant communities: the past, the present, and the future. *J. Ecol.* 96: 18–34.

Broquet, T. & Petit, E.J. 2009. Molecular estimation of dispersal for ecology and population genetics. *Annu. Rev. Ecol. Evol. Syst.* 40: 193–216.

Brundrett, M.C. 2002. Coevolution of roots and mycorrhizas of land plants. *New Phytol.* 154: 275–304.

Bucci, G., González-Martínez, S.C., Le Provost, G.et al. 2007. Range-wide phylogeography and gene zones in *Pinus pisaster* Ait revealed by chloroplast microsatellite markers. *Mol. Ecol.* 16: 2137–2153.

Buckler, E.S., IV & Thornsberry, J.M. 2002. Plant molecular diversity and applications to genomics. *Curr. Opin. Plant Biol.* 5: 107–111.

Buehler, D., Poncet, B.N., Holderegger, R., Manel, S., Taberlet, P. & Gugerli, F. 2013. An outlier locus relevant in habitat-mediated selection in an alpine plant across independent regional replicates. *Evol. Ecol.* 27: 285–300.

Burczyk, J., Adams, W.T., Birkes, D.S. & Chybicki, I.J. 2006. Using genetic markers to directly estimate gene flow and reproductive success parameters in plants on the basis of naturally regenerated seedlings. *Genetics* 173: 363–372.

Burdon, J.J. 1987. *Diseases and Plant Population Biology.* Cambridge: Cambridge University Press.

Burdon, J.J. 1991. Fungal pathogens as selective forces in plant populations and communities. *Aust. J. Ecol.* 16: 423–432.

Burdon, J.J. 1997. The evolution of gene-for-gene interactions in natural pathosystems. In: *The Gene-for-Gene Relationship in Plant–Parasite Interactions*, ed. I.R. Crute, E.B. Holub & J.J. Burdon, 245–262. Oxford: CAB International.

Burdon, J.J. & Thrall, P.H. 2001. The demography and genetics of host–pathogen interactions. In: *Integrating Ecology and Evolution in a Spatial Context*, ed. J. Silvertown & J. Antonovics, 197–217. Oxford: Blackwell Science.

Burdon, J.J. & Thrall, P.H. 2009. Coevolution of plants and their pathogens in natural habitats. *Science* 324: 755–756.

Burgess, K.S., Etterson, J.R. & Galloway, L.F. 2007. Artificial selection shifts flowering phenology and other correlated traits in an autotetraploid herb. *Heredity* 99: 641–648.

Burke, T., Seidler, R. & Smith, H. 1992. Editorial. *Mol. Ecol.* 1:1.

Busch, J.W., Neiman, M. & Koslow, J.M. 2004. Evidence for maintenance of sex by pathogens in plants. *Evolution* 58: 2584–2590.

Bustos-Segura, C., Fornoni, J. & Núñez-Farfán, J. 2014. Evolutionary changes in plant tolerance against herbivory through a resurrection experiment. *J. Evol. Biol.* 27: 488–496.

Byars, S.G., Papst, W. & Hoffmann, A.A. 2007. Local adaptation and cogradient selection in the alpine plant *Poa hiemata*, along a narrow altitudinal gradient. *Evolution* 61: 2925–2941.

Byers, D.L. 2005. Evolution in heterogeneous environments and the potential of maintenance of genetic variation in traits of adaptive significance. *Genetica* 123: 107–124.

Cain, M.L., Subler, S., Evans, J.P. & Fortin, M.-J. 1999. Sampling spatial and temporal variation in soil nitrogen availability. *Oecologia* 118: 397–404.

Callahan, H.S. 2005. Using artificial selection to understand plastic plant phenotypes. *Integr. Comp. Biol.* 45: 475–485.

Callaway, R.M. 2007. *Positive Interactions and Interdependence in Plant Communities.* Dordrecht, The Netherlands: Springer.

Callaway, R.M., Ridenour, W.M., Laboski, T., Weir, T. & Vivanco, J.M. 2005. Natural selection for resistance to the allelopathic effects of invasive plants. *J. Ecol.* 93: 576–583.

Callaway, R.M. & Walker, L.R. 1997. Competition and facilitation: a synthetic approach to interactions in plant communities. *Ecology* 78: 1958–1965.

Campbell, D.R. & Dooley, J.L. 1992. The spatial scale of genetic differentiation in a hummingbird-pollinated plant: comparison with models of isolation by distance. *Am. Nat.* 139: 735–748.

Campbell, L.G. & Snow, A.A. 2009. Can feral weeds evolve from cultivated radish (*Raphanus sativus*, Brassicaceae)? *Am. J. Bot.* 96: 498–506.

Capelle, J. & Neema, C. 2005. Local adaptation and population structure at a micro-geographical scale of a fungal parasite on its host plant. *J. Evol. Biol.* 18: 1445–1454.

Carey, K. 1983. Breeding system, genetic variability, and response to selection in *Plectritis* (Valerianaceae). *Evolution* 37: 947–956.

Cariveau, D., Irwin, R.E., Brody, A.K., Garcia-Mayeya, L.S. & von der Ohe, A. 2004. Direct and indirect effects of pollinators and seed predators to selection on plant and floral traits. *Oikos* 104: 15–26.

Carmona, D. & Fornoni, J. 2013. Herbivores can select for mixed defensive strategies in plants. *New Phytol.* 197: 576–585.

Carmona, D., Lajeunesse, M.J. & Johnson, M.T.J. 2011. Plant traits that predict resistance to herbivores. *Funct. Ecol.* 25: 358–367.

Carroll, S.P., Hendry, A.P., Reznick, D.N. & Fox, C.W. 2007. Evolution on ecological timescales. *Functional Ecology* 21: 387–393.

Carter, K.K. 1996. Provenance tests are indicators of growth response to climate change in 10 north temperate tree species. *Can. J. For. Res.* 26: 1089–1095.

Caruso, C.M., Peterson, S.B. & Ridley, C.E. 2003. Natural selection on floral traits of *Lobelia* (Lobeliaceae): spatial and temporal variation. *Am. J. Bot.* 90: 1333–1340.

Chamberlain, S.A., Bronstein, J.L. & Rudgers, J.A. 2014. How context dependent are species interactions? *Ecol. Lett.* 17: 881–890.

Charpentier, A. 2001. Consequences of clonal growth for plant mating. *Evol. Ecol.* 15: 521–530.

Chase, J.M. & Leibold, M.A. 2003. *Ecological Niches: Linking Classical and Contemporary Approaches.* Chicago, IL: University of Chicago Press.

Cheplick, G.P. 1988. Influence of environment and population origin on survivorship and reproduction in reciprocal transplants of amphicarpic peanutgrass (*Amphicarpum purshii*). *Am. J. Bot.* 75: 1048–1056.

Cheplick, G.P. 1992. Sibling competition in plants. *J. Ecol.* 80: 567–575.

Cheplick, G.P. 1993a. Reproductive systems and sibling competition in plants. *Plant Species Biol.* 8: 131–139.

Cheplick, G.P. 1993b. Sibling competition is a consequence of restricted dispersal in an annual cleistogamous grass. *Ecology* 74: 2161–2164.

Cheplick G.P. 1996. Do seed germination patterns in cleistogamous annual grasses reduce the risk of sibling competition? *J. Ecol.* 84: 247–255.

Cheplick, G.P. 1997. Effects of endophytic fungi on the phenotypic plasticity of *Lolium perenne* (Poaceae). *Am. J. Bot.* 84: 34–40.

Cheplick, G.P. 2003. Evolutionary significance of genotypic variation in developmental reaction norms for a perennial grass under competitive stress. *Evol. Ecol.* 17: 175–196.

Cheplick, G.P. 2005. Biomass partitioning and reproductive allocation in the invasive, cleistogamous grass *Microstegium vimineum*: influence of the light environment. *J. Torrey Bot. Soc.* 132: 214–224.

Cheplick, G.P. 2008. Host genotype overrides fungal endophyte infection in influencing tiller and spike production of *Lolium perenne* (Poaceae) in a common garden experiment. *Am. J. Bot.* 95: 1063–1071.

Cheplick, G.P. 2009. Conceptual model for the analysis of plant–endophyte symbiosis in relation to abiotic stress. In: *Defensive Mutualism in Microbial Symbiosis*, ed. J.F. White, Jr. & M.S. Torres, 319–334. Boca Raton, FL: CRC Press.

Cheplick, G.P. 2011. Endosymbiosis and population differentiation in wild and cultivated *Lolium perenne* (Poaceae). *Am. J. Bot.* 98: 829–838.

Cheplick, G. P. 2015. Fungal endophytes. In: *Oxford Bibliographies Online, Ecology Module*, ed. D. Gibson. New York: Oxford University Press. http://www.oxfordbibliographies.com/obo/page/ecology.

Cheplick, G.P. 2014. Population differentiation in the tiller architecture of *Microstegium vimineum* (Poaceae) in relation to habitat. *Plant Species Biol.* doi: 10.1111/1442-1984.12040

Cheplick, G.P. & Chui, T. 2001. Effects of competitive stress on vegetative growth, storage, and regrowth after defoliation in *Phleum pratense. Oikos* 95: 291–299.

Cheplick, G.P. & Demetri, H. 1999. Impact of saltwater spray and sand deposition on the coastal annual *Triplasis purpurea* (Poaceae). *Am. J. Bot.* 86: 703–710.

Cheplick, G.P. & Faeth, S.H. 2009. *Ecology and Evolution of the Grass–Endophyte Symbiosis.* New York: Oxford University Press.

Cheplick, G.P. & Fox, J. 2011. Density-dependent growth and reproduction of *Microstegium vimineum* in contrasting light environments. *J. Torrey Bot. Soc.* 138: 62–72.

Cheplick, G.P. & Gutierrez, C.M. 2000. Clonal growth and storage in relation to competition in genets of the rhizomatous perennial *Amphibromus scabrivalvis. Can. J. Bot.* 78: 537–546.

Cheplick, G.P., Harrichandra, A.P. & Liu, A. 2014. Competitive outcomes depend on host genotype, but not clavicipitaceous fungal endophytes, in *Lolium perenne* (Poaceae). *Am. J. Bot.* 101: 2068–2078.

Cheplick G.P. & Kane, K.H. 2004. Genetic relatedness and competition in *Triplasis purpurea* (Poaceae): resource partitioning or kin selection? *Int. J. Plant Sci.* 165: 623–630.

Cheplick, G.P., Perera, A. & Koulouris, K. 2000. Effect of drought on the growth of *Lolium perenne* genotypes with and without fungal endophytes. *Funct. Ecol.* 14: 657–667.

Cheplick, G.P. & Salvadori, G.M. 1991. Intra- and interclonal competition in the cleistogamous grass *Amphibromus scabrivalvis. Am. J. Bot.* 78: 1494–1502.

Cheplick, G.P. & Sung, L. 1998. Effects of maternal nutrient environment and maturation position on seed heteromorphism, germination, and seedling growth in *Triplasis purpurea* (Poaceae). *Int. J. Plant Sci.* 159: 338–350.

Cheplick G.P. & White, T.P. 2002. Saltwater spray as an agent of natural selection: no evidence of local adaptation within a coastal population of *Triplasis purpurea* (Poaceae). *Am. J. Bot.* 89: 623–631.

Cheptou, P.-O., Lepart, J. & Escarre, J. 2001. Inbreeding depression under intraspecific competition in a highly outcrossing population of *Crepis sancta* (Asteraceae): evidence for frequency-dependent variation. *Am. J. Bot.* 88: 1424–1429.

Chou, C.-H. 2006. Introduction to allelopathy. In: *Allelopathy: A Physiological Process with Ecological Implications*, ed. M.J. Reigosa, N. Pedrol & L. González, 1–9. Dordrecht, The Netherlands: Springer.

Christensen, M.J. & Voisey, C.R. 2007. The biology of the endophyte/grass partnership. In: *Proceedings of the 6th International Symposium on Fungal Endophytes of Grasses*, ed. A.J. Popay & E.R. Thom, 123–133. Christchurch, New Zealand: New Zealand Grassland Association.

Chun, Y.J., Le Corre, V. & Bretagnolle, F. 2011. Adaptive divergence for a fitness-related trait among invasive *Ambrosia artimisiifolia* populations in France. *Mol. Ecol.* 20: 1378–1388.

Chun, Y.J., Nason, J.D. & Moloney, K.A. 2009. Comparison of quantitative and molecular genetic variation of native vs. invasive populations of purple loosestrife (*Lythrum salicaria* L., Lythraceae). *Mol. Ecol.* 18: 3020–3035.

Cipollini, M.L. & Levey, D.J. 1997. Secondary metabolites of fleshy vertebrate-dispersed fruits: adaptive hypotheses and implications for seed dispersal. *Am. Nat.* 150: 346–372.

Clare, E.L., Schiestl, F.P., Leitch, A.R. & Chittka, L. 2013. The promise of genomics in the study of plant–pollinator interactions. *Genome Biol.* 14: 207–217.

Clausen, J. 1951. *Stages in the Evolution of Plant Species.* Ithaca, NY: Cornell University Press.

Clausen, J. & Hiesey, W.M. 1958. Experimental studies on the nature of species: IV. Genetic structure of ecological races. Publication 615. Washington, DC: Carnegie Institute of Washington.

Clausen, J., Keck, D.D. & Hiesey, W.M. 1940. Experimental studies on the nature of species: I. The effects of varied environments on western North American plants. Publication 520. Washington, DC: Carnegie Institute of Washington.

Clausen, J., Keck, D.D. & Hiesey, W.M. 1948. Experimental studies on the nature of species: III. Environmental responses of climatic races of *Achillea.* Publication 581. Washington, DC: Carnegie Institute of Washington.

Clay, K. 1990. The impact of parasitic and mutualistic fungi on competitive interactions among plants. In: *Perspectives on Plant Competition*, ed. J.B. Grace & D. Tilman, 391–412. San Diego, CA: Academic Press.

Clay, K. 1991. Parasitic castration of plants by fungi. *Trends Ecol. Evol.* 6: 162–166.

Clay, K. & Kover, P.X. 1996. The Red Queen Hypothesis and plant–pathogen interactions. *Annu. Rev. Phytopathol.* 34: 29–50.

Clay, K. & Levin, D.A. 1986. Environment-dependent intraspecific competition in *Phlox drummondii. Ecology* 67: 37–45.

Clegg, M.T. 1989. Molecular diversity in plant populations. In: *Plant Population Genetics, Breeding, and Genetic Resources*, ed. A.H.D. Brown, M.T. Clegg, A.L. Kahler & B.S. Weir, 98–115. Sunderland, MA: Sinauer Associates.

Clegg, M.T. & Allard, R.W. 1972. Patterns of genetic differentiation in the slender wild oat species *Avena barbata. Proc. Nat. Acad. Sci. U S A* 69: 1820–1824.

Clegg, M.T. & Durbin, M.L. 2003. Tracing floral adaptations from ecology to molecules. *Nat. Rev. Genet.* 4: 206–215.

Clegg, M.T., Kahler, A.L. & Allard, R.W. 1978a. Estimation of life cycle components of selection in an experimental plant population. *Genetics* 89: 765–792.

Clegg, M.T., Kahler, A.L. & Allard, R.W. 1978b. Genetic demography of plant populations. In: *Ecological Genetics: The Interface*, ed. P.F. Brussard, 173–188. New York: Springer-Verlag.

Clements, F.E. 1905. *Research Methods in Ecology.* Lincoln, NE: University Publishing.

Clements, F.E. 1929. Experimental studies in adaptation and morphogeny. *J. Ecol.* 17: 356–379.

Clements, F.E., Weaver, J.E. & Hanson, H.C. 1929. *Plant Competition: An Analysis of Community Functions.* Vol. 398. Washington, DC: Carnegie Institution of Washington.

Climent, J., Prada, M.A., Calama, R., Chambel, M.R., Sánchez de Ron, D. & Alía, R. 2008. To grow or to seed: ecotypic variation in reproductive allocation and cone production by young female Aleppo pine (*Pinus halepensis*, Pinaceae). *Am. J. Bot.* 95: 833–842.

Clobert, J., Baguette, M., Benton, T.G. & Bullock, J.M., eds. 2012. *Dispersal Ecology and Evolution.* Oxford: Oxford University Press.

Clobert, J., Danchin, E., Dhondt, A.A. & Nichols, J.D., eds. 2001. *Dispersal.* Oxford: Oxford University Press.

Cluster, P.D. & Allard, R.W. 1995. Evolution of ribosomal DNA (rDNA) genetic structure in colonial Californian populations of *Avena barbata. Genetics* 139: 941–954.

Colautti, R.I., Lee, C.-R. & Mitchell-Olds, T. 2012. Origin, fate, and architecture of ecologically relevant genetic variation. *Curr. Opin. Plant Biol.* 15: 199–204.

Colautti, R.I., Maron, J.L. & Barrett, S.C.H. 2008. Common garden comparisons of native and introduced plant populations: latitudinal clines can obscure evolutionary inferences. *Evol. Appl.* 2: 187–199.

Collinge, S.K. & Ray, C., eds. 2006. *Disease Ecology: Community Structure and Pathogen Dynamics.* Oxford: Oxford University Press.

Conner, J.K. 2003. Artificial selection: a powerful tool for ecologists. *Ecology* 84: 1650–1660.

Conner, J.K. 2006. Ecological genetics of floral evolution. In: *Ecology and Evolution of Flowers*, ed. L.D. Harder & S.C.H. Barrett, 260–277. Oxford: Oxford University Press.

Conner, J.K., Cooper, I.A., La Rosa, R.J., Pérez, S.G. & Royer, A.M. 2014. Patterns of phenotypic correlations among morphological traits across plants and animals. *Phil. Trans. R. Soc. B*: 20130246.

Conner, J.K. & Hartl, D.L. 2004. *A Primer of Ecological Genetics.* Sunderland, MA: Sinauer Associates.

Conner, J.K., Rush, S. & Jennetten, P. 1996. Measurements of natural selection on floral traits in wild radish (*Raphanus raphanistrum*): I. Selection through lifetime female fitness. *Evolution* 50: 1127–1136.

Coop, G., Witonsky, D., di Rienzo, A. & Pritchard, J.K. 2010. Using environmental correlations to identify loci underlying local adaptation. *Genetics* 185: 1411–1423.

Cordeiro, M.A., Moriuchi, K.S., Fotinos, T.D., Miller, K.E., Nuzhdin, S.V., von Wettberg, E.J. & Cook, D.R. 2014. Population differentiation for germination and early seedling root growth traits under saline conditions in the annual legume *Medicago truncatula* (Fabaceae). *Am. J. Bot.* 101: 488–498.

Cousens, R., Dytham, C. & Law, R. 2008. *Dispersal in Plants: A Population Perspective.* Oxford: Oxford University Press.

Cousens, R. & Mortimer, M. 1995. *Dynamics of Weed Populations.* Cambridge: Cambridge University Press.

Crémieux, L., Bischoff, A., Smilauerova, M., et al. 2008. Potential contribution of natural enemies to patterns of local adaptation in plants. *New Phytol.* 180: 524–533.

Culley, T.M. 2013. Changing technologies offer new opportunities in the plant sciences. *Appl. Plant Sci.* 1: 1–2.

Culley, T.M. & Klooster, M.R. 2007. The cleistogamous breeding system: a review of its frequency, evolution, and ecology in angiosperms. *Bot. Rev.* 73: 1–30.

Curtis, P.S., Snow, A.A. & Miller, A.S. 1994. Genotype-specific effects of elevated CO_2 on fecundity in wild radish (*Raphanus raphanistrum*). *Oecologia* 97: 100–105.

Dalziel, A.C., Rogers, S.M. & Schulte, P.M. 2009. Linking genotypes to phenotypes and fitness: how mechanistic biology can inform molecular ecology. *Mol. Ecol.* 18: 4997–5017.

Darwin, C. 1859. *The Origin of Species* (2005; repr. *The Darwin Compendium*, New York: Barnes & Noble Publishing).

Darwin, C. 1868. *The Variation of Animals and Plants under Domestication.* 2 Vols. The Project Gutenberg etext (2001 release). www.gutenberg.org/ebooks.

Darwin, C. 1877. *The Different Forms of Flowers on Plants of the Same Species.* London: John Murray.

Darwin, C. 1888. *The Various Contrivances by which Orchids are Fertilized by Insects*, 2nd ed. London: John Murray.

Davelos, A.L., Alexander, H.M. & Slade, N.A. 1996. Ecological genetic interactions between a clonal host plant (*Spartina pectinata*) and associated rust fungi (*Puccinia seymouriana* and *Puccinia sparganioides*). *Oecologia* 105: 205–213.

Davies, M.S. & Snaydon, R.W. 1976. Rapid population differentiation in a mosaic environment: III. Measures of selection pressures. *Heredity* 36: 59–66.

de Casas, R. R., Vargas, P., Pérez-Corona, E., Manrique, E., García-Verdugo, C. & Balaguer, L. 2011. Sun and shade leaves of *Olea europaea* respond differently to plant size, light availability and genetic variation. *Funct. Ecol.* 25: 802–812.

de Casas, R. R., Willis, C.G. & Donohue, K. 2012. Plant dispersal phenotypes: a seed perspective of maternal habitat selection. In: *Dispersal Ecology and Evolution*, ed. J. Clobert, M. Baguette, T.G. Benton & J.M. Bullock, 171–184. Oxford: Oxford University Press.

Dechaine, J.M., Burger, J.C., Chapman, M.A., Seiler, G.J., Brunick, R., Knapp, S.J. & Burke, J.M. 2009. Fitness effects and genetic architecture of plant–herbivore interactions in sunflower crop–wild hybrids. *New Phytol.* 184: 828–841.

Dechamps, C., Norer, N., Mozek, R., Escarré, J., Lefèbvre, C., Gruber, W. & Meerts, P. 2008. Cost of adaptation to a metalliferous environment for *Thlaspi caerulescens*: a field reciprocal transplantation approach. *New Phytol.* 177: 167–177.

de Jong, G. 1994. The fitness of fitness concepts and the description of natural selection. *Q. Rev. Biol.* 69: 3–29.

de Jong, T. & Klinkhamer, P. 2005. *Evolutionary Ecology of Plant Reproductive Strategies.* Cambridge: Cambridge University Press.

de Kort, H., Vandepitte, K. & Honnay, O. 2013. A meta-analysis of the effects of plant traits and geographical scale on the magnitude of adaptive differentiation as measured by the difference between Q_{ST} and F_{ST}. *Evol. Ecol.* 27: 1081–1097.

del Moral, L.F.G., Rharrabti, Y., Villegas, D., & Reyo, C. 2003. Evaluation of grain yield and its components in durum wheat under Mediterranean conditions. *Agron. J.* 95: 266–274.

Delph, L.F. & Herlihy, C.R. 2012. Sexual, fecundity, and viability selection on flower size and number in a sexually dimorphic plant. *Evolution* 66: 1154–1166.

de Meaux, J., Cattan-Toupance, I., Lavigne, C., Langin, T. & Neema, C. 2003. Polymorphism of a complex resistance gene candidate family in wild populations of common bean (*Phaseolus vulgaris*) in Argentina: comparison with phenotypic resistance polymorphism. *Mol. Ecol.* 12: 263–273.

de Meaux, J. & Mitchell-Olds, T. 2003. Evolution of plant resistance at the molecular level: ecological context of species interactions. *Heredity* 91: 345–352.

Dennis, A.J., Schupp, E.W. Green, R.A. & Westcott, D.A., eds. 2007. *Seed Dispersal: Theory and Its Application in a Changing World.* New York: CAB International.

Des Marais, D.L., Hernandez, K.M. & Juenger, T.E. 2013. Genotype-by-environment interaction and plasticity: exploring genomic responses of plants to the abiotic environment. *Annu. Rev. Ecol. Evol. Syst.* 44: 5–29.

Dewey, D.R. & Lu, K.H. 1959. A correlation and path-coefficient analysis of components of created wheatgrass seed production. *Agron. J.* 51: 515–518.

Dirzo, R. & Harper, J.L. 1982. Experimental studies on slug–plant interactions: IV. The performance of cyanogenic and acyanogenic morphs of *Trifolium repens* in the field. *J. Ecol.* 70: 119–138.

Dlugosch, K.M. & Parker, I.M. 2008. Founding events in species invasions: genetic variation, adaptive evolution, and the role of multiple introductions. *Mol. Ecol.* 17: 431–449.

Dodd, R.S., Mayer, W., Nettel, A. & Afzal-Rafii, Z. 2013. Clonal growth and fine-scale genetic structure in tanoak (*Notholithocarpus densiflorus*: Fagaceae). *J. Hered.* 104: 105–114.

Donohue, K. 2003. The influence of neighbor relatedness on multilevel selection in the Great Lakes sea rocket. *Am. Nat.* 162: 77–92.

Donohue, K. 2004. Density-dependent multilevel selection in the Great Lakes sea rocket. *Ecology* 85: 180–191.

Donohue, K., ed. 2014. The epigenetics of adaptation: focusing on epigenetic stability as an evolving trait. *Evolution* 68: 617–683.

Donohue, K., Messiqua, D., Pyle, E.H., Heschel, M.S. & Schmitt, J. 2000a. Evidence of adaptive divergence in plasticity: density- and site-dependent selection on shade-avoidance responses in *Impatiens capensis. Evolution* 54: 1956–1968.

Donohue, K., Pyle, E.H., Messiqua, D., Heschel, M.S. & Schmitt, J. 2000b. Density dependence and population differentiation of genetic architecture in *Impatiens capensis* in natural environments. *Evolution* 54: 1969–1981.

Donohue, K., Pyle, E.H., Messiqua, D., Heschel, M.S. & Schmitt, J. 2001. Adaptive divergence in plasticity in natural populations of *Impatiens capensis* and its consequences for performance in novel habitats. *Evolution* 55: 692–702.

Donovan, L.A., Dudley, S.A., Rosenthal, D.M. & Ludwig, F. 2007. Phenotypic selection on leaf water use efficiency and related ecophysiological traits for natural populations of desert sunflowers. *Oecologia* 152: 13–25.

Donovan, L.A., Ludwig, F, Rosenthal, D.M., Rieseberg, L.H. & Dudley, S.A. 2009. Phenotypic selection on leaf ecophysiological traits in *Helianthus. New Phytol.* 183: 868–879.

Dow, B.D. & Ashley, M.V. 1996. Microsatellite analysis of seed dispersal and parentage of saplings in bur oak, *Quercus macrocarpa. Mol. Ecol.* 5: 615–627.

Dow, B.D. & Ashley, M.V. 1998. High levels of gene flow in bur oak revealed by paternity analysis using microsatellites. *J. Hered.* 89: 62–70.

Dudley, S.A. 1996. Differing selection on plant physiological traits in response to environmental water availability: a test of adaptive hypotheses. *Evolution* 50: 92–102.

Dudley, L.S., Hove, A.A. & Mazer, S.J. 2012. Physiological performance and mating system in *Clarkia* (Onagraceae): does phenotypic selection predict divergence between sister species? *Am. J. Bot.* 99: 488–507.

Dudley, S.A. & Schmitt, J. 1995. Genetic differentiation in morphological responses to simulated foliage shade between populations of *Impatiens capensis* from open and woodland sites. *Funct. Ecol.* 9: 655–666.

Dudley, S.A. & Schmitt, J. 1996. Testing the adaptive plasticity hypothesis: density-dependent selection on manipulated stem length in *Impatiens capensis. Am. Nat.* 147: 445–465.

Duffus, J.H. 2002. "Heavy metals": a meaningless term? *Pure Appl. Chem.* 74: 793–807.

Dutech, C., Sork, V.L., Irwin, A.J., Smouse, P.E. & Davis, F.W. 2005. Gene flow and fine-scale genetic structure in a wind-pollinated tree species, *Quercus lobata* (Fagaceae). *Am. J. Bot.* 92: 252–261.

Ebeling, S.K., Stöcklin, J., Hensen, I. & Auge, H. 2011. Multiple common garden experiments suggest lack of local adaptation in an invasive ornamental plant. *J. Plant Ecol.* 4: 209–220.

Eckert, A.J., Bower, A.D. González-Martínez, S.C., Wegrzyn, J.L., Coop, G. & Neale, D.B. 2010a. Back to nature: ecological genomics of loblolly pine (*Pinus taeda*, Pinaceae). *Mol. Ecol.* 19: 3789–3805.

Eckert, A.J., van Heerwaarden, J., Wegrzyn, J.L., Nelson, C.D., Ross-Ibarra, J., González-Martínez, S.C. & Neale, D.B. 2010b. Patterns of population structure and environmental associations to aridity across the range of loblolly pine (*Pinus taeda* L., Pinaceae). *Genetics* 185: 969–982.

Ehlers, B.K. & Thompson, J. 2004. Do co-occurring plant species adapt to one another? The response of *Bromus erectus* to the presence of different *Thymus vulgaris* chemotypes. *Oecologia* 141: 511–518.

Ehrenreich, I.M. & Purugganan, M.D. 2006. The molecular genetic basis of plant adaptation. *Am. J. Bot.* 93: 953–962.

Ehrlich, P.R. & Raven, P.H. 1964. Butterflies and plants: a study in coevolution. *Evolution* 18: 586–608.

Elle, E. 2004. Floral adaptation and biotic and abiotic selection pressures. In: *Plant Adaptation: Molecular Genetics and Ecology*, ed. Q.C.B. Cronk, J. Whitton, R.H. Ree & I.E.P. Taylor, 111–118. Ottawa, Ontario, Canada: National Research Council Research Press.

Ellis, A.G. & Anderson, B. 2012. Pollinator mediated floral divergence in the absence of pollinator shifts. In: *Evolution of Plant–Pollinator Relationships*, ed. S. Patiny, pp. 237–262. Cambridge: Cambridge University Press.

Ellis, A.G. & Weis, A.E. 2006. Coexistence and differentiation of "flowering stones": the role of local adaptation to soil microenvironment. *J. Ecol.* 94: 322–335.

Ellstrand, N.C. 1992. Gene flow among seed plant populations. *New Forests* 6: 241–256.

Ellstrand, N.C. 2003. Current knowledge of gene flow in plants: implications for transgene flow. *Phil. Trans. R. Soc. Lond. B* 358: 1163–1170.

Ellstrand, N.C. 2014. Is gene flow the most important evolutionary force in plants? *Am. J. Bot.* 101: 737–753.

Ellstrand, N.C., Devlin, B. & Marshall, D.L. 1989. Gene flow by pollen into small populations: data from experimental and natural stands of wild radish. *Proc. Nat. Acad. Sci. U S A* 86: 9044–9047.

Ellstrand, N.C. & Marshall, D.L. 1985. Interpopulation gene flow by pollen in wild radish, *Raphanus sativus*. *Am. Nat.* 126: 606–616.

Endler, J.A. 1986. *Natural Selection in the Wild*. Princeton, NJ: Princeton University Press.

Endler, J.A. 2010. Editorial. *Evol. Ecol.* 24: 955–956.

Ennos, R.A. 1989. Detection and measurement of selection: genetic and ecological approaches. In: *Plant Population Genetics, Breeding, and Genetic Resources*, ed. A.H.D. Brown, M.T. Clegg, A.L. Kahler & B.S. Weir, 200–214. Sunderland, MA: Sinauer Associates.

Ennos, R.A. 2001. Inferences about spatial processes in plant populations from the analysis of molecular markers. In: *Integrating Ecology and Evolution in a Spatial Context*, ed. J. Silvertown & J. Antonovics, 45–71. Oxford: Blackwell Science.

Epperson, B. K. 1989. Spatial patterns of genetic variation within plant populations. In: *Plant Population Genetics, Breeding, and Genetic Resources*, ed. A.H.D. Brown, M.T. Clegg, A.L. Kahler & B.S. Weir, 229–253. Sunderland, MA: Sinauer Associates.

Epperson, B.K. 2007. Plant dispersal, neighbourhood size and isolation by distance. *Mol. Ecol.* 16: 3854–3865.

Erickson, V.J., Mandel, N.L. & Sorensen, F.C. 2004. Landscape patterns of phenotypic variation and population structuring in a selfing grass, *Elymus glaucus* (blue wildrye). *Can. J. Bot.* 82: 1776–1789.

Espeland, E.K. & Rice, K.J. 2007. Facilitation across stress gradients: the importance of local adaptation. *Ecology* 88: 2404–2409.

Estrada, A. & Fleming, T.H., eds. 1986. *Frugivores and Seed Dispersal*. Dordrecht, The Netherlands: Dr. W. Junk.

Etterson, J.R. 2004a. Evolutionary potential of *Chamaecrista fasciculata* in relation to climate change: I. Clinal patterns of selection along an environmental gradient in the Great Plains. *Evolution* 58: 1446–1458.

Etterson, J.R. 2004b. Evolutionary potential of *Chamaecrista fasciculata* in relation to climate change: II. Genetic architecture of three populations reciprocally planted along an environmental gradient in the Great Plains. *Evolution* 58: 1459–1471.

Eveno, E., Collada, C., Guevara, M.A., et al. 2008. Contrasting patterns of selection at *Pinus pinaster* Ait drought stress candidate genes as revealed by genetic differentiation analyses. *Mol. Biol. Evol.* 25: 417–437.

Ewald, P.W. 1987. Transmission modes and evolution of the parasitism–mutualism continuum. *Ann. N.Y. Acad. Sci.* 503: 295–306.

Faeth, S.H. 2009. Asexual fungal symbionts alter reproductive allocation and herbivory over time in their native perennial grass hosts. *Am. Nat.* 173: 554–565.

Faeth, S. H. & Saikkonen, K. 2007. Variability is the nature of the endophyte–grass interaction. In: *Proceedings of the 6th International Symposium on Fungal Endophytes of Grasses*, ed. A. J. Popay & E. R. Thom, 37–48. Christchurch, New Zealand: New Zealand Grassland Association.

Faeth, S. H. & Sullivan, T.J. 2003. Mutualistic asexual endophytes in a native grass are usually parasitic. *Am. Nat.* 161: 310–325.

Fairbairn, D.J. & Reeve, J.P. 2001. Natural selection. In: *Evolutionary Ecology: Concepts and Case Studies*, ed. C.W. Fox, D.A. Roff & D.J. Fairbairn, 29–43. New York: Oxford University Press.

Falconer, D.S. & T.F.C. Mackay. 1996. *Introduction to Quantitative Genetics*. 4th ed. New York: Pearson/Prentice Hall.

Farris, M.A. & Mitton, J.B. 1984. Population density, outcrossing rate and heterozygote superiority in Ponderosa pine. *Evolution* 38: 1151–1154.

Fay, J.C. & Wittkopp, P.J. 2008. Evaluating the role of natural selection in the evolution of gene expression. *Heredity* 100: 191–199.

Fenster, C.B., Armbruster, W.S., Wilson, P., Dudash, M.R. & Thomson, J.D. 2004. Pollination syndromes and floral specialization. *Annu. Rev. Ecol. Evol. Syst.* 35: 375–403.

Fenster, C.B., Vekemans, X. & Hardy, O.J. 2003. Quantifying gene flow from spatial genetic structure data in a metapopulation of *Chamaecrista fasciculata* (Leguminosae). *Evolution* 57: 995–1007.

File, A.L., Murphy, G.P. & Dudley, S.A. 2012. Fitness consequences of plants growing with siblings: reconciling kin selection, niche partitioning and competitive ability. *Proc. R. Soc. B* 279: 209–218.

Fishman, L. & Willis, J.H. 2008. Pollen limitation and natural selection on floral characters in the yellow monkeyflower, *Mimulus guttatus. New Phytol.* 177: 802–810.

Fitter, A.H. 1991. Costs and benefits of mycorrhizas: implications for functioning under natural conditions. *Experentia* 47: 350–355.

Fitter, A., Hodge, A. & Robinson, D. 2000. Plant responses to patchy soils. In: *The Ecological Consequences of Environmental Heterogeneity*, ed. M.J. Hutchings, E.A. John & A.J.A. Stewart, 71–90. Oxford: Blackwell Science.

Fleming, T.H. & Estrada, A., eds. 1993. *Frugivory and Seed Dispersal: Ecological and Evolutionary Aspects*. Dordrecht, The Netherlands: Springer Science + Business Media.

Flor, H.H. 1951. Genes for resistance to rust in Victory flax. *Agron. J.* 43: 527–531.

Flory, S.L., Long, F. & Clay, K. 2011. Invasive *Microstegium* populations consistently outperform native range populations across diverse environments. *Ecology* 92: 2248–2257.

Fornoni, J. 2011. Ecological and evolutionary implications of plant tolerance to herbivory. *Funct. Ecol.* 25: 399–407.

Fornoni, J. & Núñez-Farfán, J. 2000. Evolutionary ecology of *Datura stramonium*: genetic variation and costs for tolerance to defoliation. *Evolution* 54: 789–797.

Fornoni, J., Núñez-Farfán, J. & Valverde, P.L. 2003a. Evolutionary ecology of tolerance to herbivory: advances and perspectives. *Comm. Theor. Biol.* 8: 643–663.

Fornoni, J., Valverde, P.L. & Núñez-Farfán, J. 2003b. Quantitative genetics of plant tolerance and resistance against natural enemies of two populations of *Datura stramonium. Evol. Ecol. Res.* 5: 1049–1065.

Fox, C.W., Roff, D.A. & Fairbairn, D.J., eds. 2001. *Evolutionary Ecology: Concepts and Case Studies*. Oxford: Oxford University Press.

Fraenkel, G.S. 1959. The raison d'être of secondary plant substances. *Science* 129: 1466–1470.

Frank, S.A. 2011. Natural selection: II. Developmental variability and evolutionary rate. *J. Evol. Biol.* 24: 2310–2320.

Franks, S.J., Sim, S. & Weis, A.E. 2007. Rapid evolution of flowering time by an annual plant in response to a climate fluctuation. *Proc. Nat. Acad. Sci. U S A* 104: 1278–1282.

Freeland, J.R., Biss, P., Conrad, K.F. & Silvertown, J. 2010. Selection pressures have caused genome-wide population differentiation of *Anthoxanthum odoratum* despite the potential for high gene flow. *J. Evol. Biol.*23: 776–782.

Freeland, J.R., Kirk, H. & Petersen, S.D. 2011. *Molecular Ecology*. 2nd ed. West Sussex, UK: Wiley-Blackwell.

Frenck, G., van der Linden, L., Mikkelsen, T.N., Brix, H. & Jørgensen, R.B. 2013. Response to multi-generational selection under elevated [CO_2] in two temperature regimes suggests enhanced carbon assimilation and increased reproductive output in *Brassica napus* L. *Ecol. Evol.* 3: 1163–1172.

Friesen, M.L., Porter, S.S., Stark, S.C., von Wettberg, E.J., Sachs, J.L. & Martinez-Romero, E. 2011. Microbially mediated plant functional traits. *Annu. Rev. Ecol. Evol. Syst.* 42: 23–46.

Fritz, R.S. 1990. Effects of genetic and environmental variation on resistance of willow to sawflies. *Oecologia* 82: 325–332.

Fritz, R.S. & Simms, E.L., eds. 1992. *Plant Resistance to Herbivores and Pathogens: Ecology, Evolution and Genetics*. Chicago, IL: University of Chicago Press.

Fuchs, E.J. & Hamrick, J.L. 2010. Spatial genetic structure within size classes of the endangered tropical tree *Guaiacum sanctum* (Zygophyllaceae). *Am. J. Bot.* 97: 1200–1207.

Futuyma, D.J. & Slatkin, M., eds. 1983. *Coevolution*. Sunderland, MA: Sinauer Associates.

Gadgil, M. & Solbrig, O.T. 1972. The concept of *r*- and *K*-selection: evidence from wild flowers and some theoretical considerations. *Am. Nat.* 106: 14–31.

Gaiero, J.R., McCall, C.A., Thompson, K.A., Day, N.J., Best, A.S. & Dunfield, K.E. 2013. Inside the root microbiome: bacterial root endophytes and plant growth promotion. *Am. J. Bot.* 100: 1738–1750.

Galen, C. 1989. Measuring pollinator-mediated selection on morphometric floral traits: bumblebees and the alpine sky pilot, *Polemonium viscosum*. *Evolution* 43: 882–890.

Galen, C., Shore, J.S. & Deyoe, H. 1991. Ecotypic divergence in alpine *Polemonium viscosum*: genetic structure, quantitative variation, and local adaptation. *Evolution* 45: 1218–1228.

Galloway, L.F. & Burgess, K.S. 2012. Artificial selection on flowering time: influence on reproductive phenology across natural light environments. *J. Ecol.* 100: 852–861.

Galloway, L.F. & Fenster, C.B. 2000. Population differentiation in an annual legume: local adaptation. *Evolution* 54: 1173–1181.

Gange, A.C. & Ayres, R.L. 1999. On the relation between arbuscular mycorrhizal colonization and plant "benefit." *Oikos* 87: 615–621.

Garant, D. & Kruuk, L.E.B. 2005. How to use molecular marker data to measure evolutionary parameters in wild populations. *Mol. Ecol.* 14: 1843–1859.

Garrido, E., Andraca-Gómez, G. & Fornoni, J. 2012. Local adaptation: simultaneously considering herbivores and their host plants. *New Phytol.* 193: 445–453.

Gartside, D.W. & McNeilly, T. 1974. The potential for evolution of heavy metal tolerance in plants: II. Copper tolerance in normal populations of different plant species. *Heredity* 32: 335–348.

Geber, M.A. & Dawson, T.E. 1993. Evolutionary responses of plants to global change. In: *Biotic Interactions and Global Change*, ed. P.M. Kareiva, J.G. Kingsolver & R.B. Huey, 179–197. Sunderland, MA: Sinauer Associates.

Geber, M.A. & Griffen, L.R. 2003. Inheritance and natural selection on functional traits. *Int. J. Plant Sci.* 164: S21–S42.

Gepts, P. 2004. Crop domestication as a long-term selection experiment. *Plant Breed. Rev.* 24: 1–44.

Gianoli, E. & Saldaña, A. 2013. Phenotypic selection on leaf functional traits of two congeneric species in a temperate rainforest is consistent with their shade tolerance. *Oecologia* 173: 13–21.

Gibert, A., Volaire, F., Barre, P. & Hazard, L. 2012. A fungal endophyte reinforces population adaptive differentiation in its host grass species. *New Phytol.* 194: 561–571.

Gibson, G. 2002. Microarrays in ecology and evolution: a preview. *Mol. Ecol.* 11: 17–24.

Gibson, D.J. 2015. *Methods in Comparative Plant Population Ecology.* 2nd ed. Oxford: Oxford University Press.

Gibson, D.J., Connolly, J., Hartnett, D.C. & Weidenhammer, J.D. 1999. Designs for greenhouse studies of interactions between plants. *J. Ecol.* 87: 1–16.

Gibson, D.J. & Risser, P.G. 1982. Evidence for the absence of ecotypic development in *Andropogon virginicus* (L.) on metalliferous mine wastes. *New Phytol.* 92: 589–599.

Gilbert, G.S. 2002. Evolutionary ecology of plant diseases in natural ecosystems. *Annu. Rev. Phytopathol.* 40: 13–43.

Gilbert, L.E. & Raven, P.H., eds. 1975. *Coevolution of Animals and Plants.* Austin, TX: University of Texas Press.

Givnish, T. J. 1979. On the adaptive significance of leaf form. In: *Topics in Plant Population Biology*, ed. O.T. Solbrig, S. Jain, G.B. Johnson & P.H. Raven, 375–407. New York: Columbia University Press.

Givnish, T. J. 1987. Comparative studies of leaf form: assessing the relative roles of selective pressures and phylogenetic constraints. *New Phytol.* 106: 131–160.

Goff, L.J. 1982. Symbiosis and parasitism: another viewpoint. *Bioscience* 32: 255–256.

Goldberg, D.E. 1988. Response of *Solidago canadensis* clones to competition. *Oecologia* 77: 357–364.

Goldberg, D.E. 1996. Competitive ability: definitions, contingency and correlated traits. *Phil. Trans. R. Soc. Lond. B* 351: 1377–1385.

Gómez, J.M., Abdelaziz, M., Camacho, J.P.M., Muñoz-Pajares, A.J. & Perfectti, F. 2009a. Local adaptation and maladaptation to pollinators in a generalist geographic mosaic. *Ecol. Lett.*12:672–682.

Gómez, J.M., Perfectti, F., Bosch, J. & Camacho, J.P.M. 2009b. A geographic selection mosaic in a generalized plant–pollinator–herbivore system. *Ecol. Monogr.* 79: 245–263.

Gomulkiewicz, R., Nuismer, S.L. & Thompson, J.N. 2003. Coevolution in variable mutualisms. *Am. Nat.* 162: S80–S93.

González-Martínez, S.C., Gerber, S., Cervera, M.T., Martínez-Zapater, J.M., Gil, L. & Alía, R. 2002. Seed gene flow and fine-scale structure in a Mediterranean pine (*Pinus pinaster* Ait) using nuclear microsatellite markers. *Theor. Appl. Genet.* 104: 1290–1297.

González-Martínez, S.C., Krutovsky, K.V. & Neale, D.B. 2006. Forest–tree population genomics and adaptive evolution. *New Phytol* 170: 227–238.

Gonzalo-Turpin, H. & Hazard, L. 2009. Local adaptation occurs along altitudinal gradient despite the existence of gene flow in the alpine plant species *Festuca eskia. J. Ecol.* 97: 742–751.

Göransson, P., Andersson, S., Falkengren-Grerup, U. 2009. Genetic adaptation to soil acidification: experimental evidence from four grass species. *Evol. Ecol.* 23: 963–978.

Goss, E.M. & Bergelson, J. 2006. Variation in resistance and virulence in the interaction between *Arabidopsis thaliana* and a bacterial pathogen. *Evolution* 60: 1562–1573.

Gossmann, T.I., Song, B.H., Windsor, A.J., et al. 2010. Genome wide analyses reveal little evidence for adaptive evolution in many plant species. *Mol. Biol. Evol.* 27: 1822–1832.

Gotelli, N.J. & Ellison, A.M. 2004. *A Primer of Ecological Statistics.* Sunderland, MA: Sinauer Associates.

Gottlieb, L.D. 1971. Gel electrophoresis: new approach to the study of evolution. *Bioscience* 21: 939–944.

Gottschling, D.E. 2007. Epigenetics: from phenomenon to field. In: *Epigenetics*, ed. C.D. Allis, T. Jenuwein & D. Reinberg, 1–14. Cold Spring Harbor, NY: Cold Spring Harbor Laboratory Press.

Grace, J.B. 1995. On the measurement of plant competition intensity. *Ecology* 76: 305–308.

Grace, J.B. & Tilman, D., eds. 1990. *Perspectives on Plant Competition.* San Diego, CA: Academic Press.

Gram, W.K. & Sork, V.L. 2001. Association between environmental and genetic heterogeneity in forest tree populations. *Ecology* 82: 2012–2021.

Gravois, K.A. & Helms, R.S. 1992. Path analysis of rice yield and yield components as affected by seeding rate. *Agron. J.* 84: 1–4.

Gregor, J.W. 1946. Ecotypic differentiation. *New Phytol.* 45: 254–270.

Gregor, J.W. & Sansome, F.W. 1927. Experiments on the genetics of wild populations: part 1: grasses. *J. Genet.* 17: 349–364.

Gregor, J.W. & Watson, P.J. 1961. Ecotypic differentiation: observations and reflections. *Evolution* 15: 166–173.

Gregory, R.P.G. & Bradshaw, A.D. 1965. Heavy metal tolerance in populations of *Agrostis tenuis* Sibth. and other grasses. *New Phytol.* 64: 131–143.

Greischar, M.A. & Koskella, B. 2007. A synthesis of experimental work on parasite local adaptation. *Ecol. Lett.* 10: 418–434.

Griffin, A.S. & West, S.A. 2002. Kin selection: fact and fiction. *Trends Ecol. Evol.* 17: 15–21.

Griffith, T.M. & Watson, M.A. 2005. Stress avoidance in a common annual: reproductive timing is important for local adaptation and geographic distribution. *J. Evol. Biol.* 18: 1601–1612.

Griffith, T.M. & Watson, M.A. 2006. Is evolution necessary for range expansion? Manipulating reproductive timing of a weedy annual transplanted beyond its range. *Am. Nat.* 167: 153–164.

Griffiths, P.E. & Gray, R.D. 2004. The developmental systems perspective: organism–environment systems as units of development and evolution. In: *Phenotypic Integration: Studying the Ecology and Evolution of Complex Phenotypes*, ed. M. Pigliucci & K. Preston, 409–431. Oxford: Oxford University Press.

Grillo, M.A., Li, C., Hammond, M., Wang, L. & Schemske, D.W. 2013. Genetic architecture of flowering time differentiation between locally adapted populations of *Arabidopsis thaliana. New Phytol.* 197: 1321–1331.

Grivet, D., Sebastiani, F., Alía, R., et al. 2011. Molecular footprints of local adaptation in two Mediterranean conifers. *Mol. Biol. Evol.* 28: 101–116.

Grivet, D., Smouse, P.E. & Sork, V.L. 2005. A novel approach to an old problem: tracking dispersed seeds. *Mol. Ecol.* 14: 3585–3595.

Grøndahl, E. & Ehlers, B.K. 2008. Local adaptation to biotic factors: reciprocal transplants of four species associated with aromatic *Thymus pulegiodes* and *T. serpyllum. J. Ecol.* 96: 981–992.

Grover, C.E., Salmon, A. & Wendel, J.F. 2012. Targeted sequence capture as a powerful tool for evolutionary analysis. *Am. J. Bot.* 99: 312–319.

Gundel, P.E., Garibaldi, L.A., Martínez-Ghera, M.A. & Ghersa, C.M. 2012. Trade-off between seed number and weight: influence of a grass–endophyte symbiosis. *Basic Appl. Ecol.* 13: 32–39.

Gupta, P.K., Sharma, P.K., Balyan, H.S., et al. 2002. Polymorphism at rDNA loci in barley and its relation with climatic variables. *Theor. Appl. Genet.* 104: 473–481.

Gurevitch, J., Taub, D.R., Morton, T.C., Gomez, P.L. & Wang, I.-N. 1996. Competition and genetic background in a rapid-cycling cultivar of *Brassica rapa* (Brassicaceae). *Am. J. Bot.* 83: 932–938.

Haasl, R.J. & Payseur, B.A. 2012. Microsatellites as targets of natural selection. *Mol. Biol. Evol.* 30: 285–298.

Hairston, N.G., Jr., Ellner, S.P., Geber, M.A., Yoshida, Y. & Fox, J.A. 2005. Rapid evolution and the convergence of ecological and evolutionary time. *Ecol. Lett.* 8: 1114–1127.

Hall, M.C., Basten, C.J. & Willis, J.H. 2006. Pleiotropic quantitative trait loci contribute to population divergence in traits associated with life-history variation in *Mimulus guttatus. Genetics* 172: 1829–1844.

Hall, J.C. & Donohue, K. 2012. Genetics of plant dispersal. In: *Dispersal Ecology and Evolution*, ed. J. Clobert, M. Baguette, T.G. Benton & J.M. Bullock, 108–115. Oxford: Oxford University Press.

Hall, M.C. & Willis, J.H. 2006. Divergent selection on flowering time contributes to local adaptation in *Mimulus guttatus* populations. *Evolution* 60: 2466–2477.

Hallgrimsson, B. & Hall, B.K., eds. 2011. *Epigenetics: Linking Genotype and Phenotype in Development and Evolution*. Berkeley, CA: University of California Press.

Hamilton, J.A. & Aitken, S.N. 2013. Genetic and morphological structure of a spruce hybrid (*Picea sitchensis* × *P. glauca*) zone along a climatic gradient. *Am. J. Bot.* 100: 1651–1662.

Hamilton, J.A., Lexer, C. & Aitken, S.N. 2013. Differential introgression reveals candidate genes for selection across a spruce (*Picea sitchensis* × *P. glauca*) hybrid zone. *New Phytol.* 197: 927–938.

Hamrick, J.L. & Allard, R.W. 1972. Microgeographical variation in allozyme frequencies in *Avena barbata. Proc. Nat. Acad. Sci. U S A* 69: 2100–2104.

Hamrick, J.L. & Allard, R.W. 1975. Correlations between quantitative characters and enzyme genotypes in *Avena barbata. Evolution* 29: 438–442.

Hamrick, J.L. & Godt, M.J.W. 1989. Allozyme diversity in plant species. In: *Plant Population/Genetics, Breeding, and Genetic Resources*, ed. A.H.D. Brown, M.T. Clegg, A.L. Kahler & B.S. Weir, 43–63. Sunderland, MA: Sinauer Associates.

Hamrick, J.L. & Godt, M.J.W. 1997. Effects of life history traits on genetic diversity in plant species. In: *Plant Life Histories: Ecology, Phylogeny and Evolution*, ed. J. Silvertown, M. Franco & J.L. Harper, 102–118. Cambridge: Cambridge University Press.

Hamrick, J.L. & Holden, L.R. 1979. Influence of microhabitat heterogeneity on gene frequency distribution and gametic phase disequilibrium in *Avena barbata. Evolution* 33: 521–533.

Hamrick, J.L., Linhart, Y.B. & Mitton, J.B. 1979. Relationships between life history characteristics and electrophoretically detectable genetic variation in plants. *Annu. Rev. Ecol. Syst.* 10: 173–200.

Hamrick, J.L., Murawski, D.A. & Nason, J.D. 1993. The influence of seed dispersal mechanisms on the genetic structure of tropical tree populations. *Vegetatio* 107/108: 281–297.

Hamrick, J.L. & Trapnell, D.W. 2011. Using population genetic analyses to understand seed dispersal patterns. *Acta Oecologica* 37: 641–649.

Hancock, A.M., Brachi, B., Faure, N., et al. 2011. Adaptation to climate across the *Arabidopsis thaliana* genome. *Science* 334: 83–86.

Handel, S.N. 1985. The intrusion of clonal growth patterns on plant breeding systems. *Am. Nat.*125: 367–384.

Hannon, N. & Bradshaw, A.D. 1968. Evolution of salt tolerance in two coexisting species of grass. *Nature* 220: 1342–1343.

Hansen, C.F., Garcia, M.B. & Ehlers, B.K. 2013. Water availability and population origin affect the expression of the tradeoff between reproduction and growth in *Plantago coronopus. J. Evol. Biol.*26: 993–1002.

Harder, L.D. & Barrett, S.C.H., eds. 2006. *Ecology and Evolution of Flowers*. Oxford: Oxford University Press.

Harder, L.D. & Johnson, S.D. 2009. Darwin's beautiful contrivances: evolutionary and functional evidence for floral adaptation. *New Phytol.* 183: 530–545.

Hardy, O.J., Maggia, L., Bandou, E. et al. 2006. Fine-scale genetic structure and gene dispersal inferences in 10 neotropical tree species. *Mol. Ecol.* 15: 559–571.

Hare, J.D. 1980. Variation in fruit size and susceptibility to seed predation among and within populations of the cocklebur, *Xanthium strumarium* L. *Oecologia* 46: 217–222.

Harper, J.L. 1967. A Darwinian approach to plant ecology. *J. Ecol.* 55: 247–270.

Harper, J.L. 1977. *Population Biology of Plants*. New York: Academic Press.

Hartl, D.L. & Clark, A.G. 2006. *Principles of Population Genetics*. 4th ed. Sunderland, MA: Sinauer Associates.

He, Q., Bertness, M.D. & Altieri, A.H. 2013. Global shifts towards positive species interactions with increasing environmental stress. *Ecol. Lett.* 16: 695–706.

Heath, K.D., Burke, P.V. & Stinchcombe, J.R. 2012. Coevolutionary genetic variation in the legume–rhizobium transcriptome. *Mol. Ecol.* 21: 4735–4747.

Heath, K.D. & Tiffin, P. 2007. Context dependence in the coevolution of plant and rhizobial mutualists. *Proc. R. Soc. B* 274: 1905–1912.

Hedrick, P.W. 2011. *Genetics of Populations*. 4th ed. Sudbury, MA: Jones & Bartlett.

Hegason, T. & Fitter, A.H. 2009. Natural selection and the evolutionary ecology of the arbuscular mycorrhizal fungi (Phylum Glomeromycota). *J. Exp. Bot.* 60: 2465–2480.

Helenurm, K. 1998. Outplanting and differential source population success in *Lupinus guadalupensis*. *Conserv. Biol.* 12: 118–127.

Helyar, S.J., Hemmer-Hansen, J., Bekkevold, D., et al. 2011. Application of SNPs for population genetics of nonmodel organisms: new opportunities and challenges. *Mol. Ecol. Res.* 11: 123–136.

Hereford, J. 2009. A quantitative survey of local adaptation and fitness trade-offs. *Am. Nat.* 173: 579–588.

Hereford, J. & Winn, A.A. 2008. Limits to local adaptation in six populations of the annual plant *Diodia teres*. *New Phytol.* 178: 888–896.

Herrera, C.M. 1985. Determinants of plant–animal coevolution: the case of mutualistic dispersal of seeds by vertebrates. *Oikos* 44: 132–141.

Herrera, C.M. 1987. Vertebrate-dispersed plants of the Iberian peninsula: a study of fruit characteristics. *Ecol. Monogr.* 57: 305–331.

Herrera, C.M. 2002. Seed dispersal by vertebrates. In: *Plant–Animal Interactions: An Evolutionary Approach*, ed. C.M. Herrera & O. Pellmyr, 185–208. Malden, MA: Blackwell Publishing.

Herrera, C.M. & Bazaga, P. 2008. Population-genomic approach reveals adaptive floral divergence in discrete populations of a hawk moth-pollinated violet. *Mol. Ecol.* 17: 5378–5390.

Herrera, C.M. & Bazaga, P. 2009. Quantifying the genetic component of phenotypic variation in unpedigreed wild plants: tailoring genomic scan for within-population use. *Mol. Ecol.* 18: 2602–2614.

Herrera, C.M. & Bazaga, P. 2011. Untangling individual variation in natural populations: ecological, genetic and epigenetic correlates of long-term inequality in herbivory. *Mol. Ecol.* 20: 1675–1688.

Herrera, C.M., Jordano. P., López-Soria, L. & Amat, J.A. 1994. Recruitment of a mast-fruiting, bird-dispersed tree: bridging frugivore activity and seedling establishment. *Ecol. Monogr.* 64: 315–344.

Herrera, C.M. & Pellmyr, O., eds. 2002. *Plant–Animal Interactions: An Evolutionary Approach*. Malden, MA: Blackwell Publishing.

Herron, J.C. & Freeman, S. 2014. *Evolutionary Analysis*. 5th ed. New York: Benjamin Cummings.

Heschel, M.S., Donohue, K., Hausmann, N. & Schmitt, J. 2002. Population differentiation and natural selection for water-use efficiency in *Impatiens capsensis* (Balsaminaceae). *Int. J. Plant Sci.* 163: 907–912.

Heslop-Harrison, J. 1964. Forty years of genecology. *Adv. Ecol. Res.* 2: 159–247.

Hester, M.W., Mendelssohn, I.A. & McKee, K.L. 1996. Intraspecific variation in salt tolerance and morphology in the coastal grass *Spartina patens* (Poaceae). *Am. J. Bot.* 83: 1521–1527.

Heywood, J.S. 1991. Spatial analysis of genetic variation in plant populations. *Annu. Rev. Ecol. Syst.* 22: 335–355.

Heywood, J.S. & Levin, D.A. 1984. Genotype–environment interactions in determining fitness in dense, artificial populations of *Phlox drummondii*. *Oecologia* 61: 363–371.

Hirsch, A.M. 2004. Plant–microbe symbiosis: a continuum from commensalism to parasitism. *Symbiosis* 37: 345–363.

Ho, M.R., Stark, S. & Chernyshenko, O. 2012. Graphical representation of structural equation models using path diagrams. In: *Handbook of Structural Equation Modeling*, ed. R.H. Hoyle, 43–55. New York: Guilford Press.

Hodgins, K.A. & Rieseberg, L. 2011. Genetic differentiation in life-history traits of introduced and native common ragweed (*Ambrosia artemisiifolia*) populations. *J. Evol. Biol.* 24: 2731–2749.

Hoeksema, J.D. 2010. Ongoing coevolution in mycorrhizal interactions. *New Phytol.* 187: 286–300.

Hoeksema, J.D., Chaudhary, V.B., Gehring, C.A., et al. 2010. A meta-analysis of context-dependency in plant response to inoculation with mycorrhizal fungi. *Ecol. Lett.* 13: 394–407.

Hoeksema, J.D., Hernandez, J.V., Rogers, D.L., Mendoza, L.L. & Thompson, J.N. 2012. Geographic divergence in a species-rich symbiosis: interactions between Monterey pines and ectomycorrhizal fungi. *Ecology* 93: 2274–2285.

Hoeksema, J.D., Piculell, B.J. & Thompson, J.N. 2009. Within-population genetic variability in mycorrhizal interactions. *Comm. Integr. Biol.* 2: 110–112.

Hoeksema, J.D. & Thompson, J.N. 2007. Geographic structure in a widespread plant–mycorrhizal interaction: pines and false truffles. *J. Evol. Biol.* 20: 1148–1163.

Hoekstra, H.E., Hoekstra, J.M., Berrigan, D., et al. 2001. Strength and tempo of directional selection in the wild. *Proc. Nat. Acad. Sci. U S A* 98: 9157–9160.

Hoelzel, A.R. 2000. Editorial. *Conserv. Genet.* 1: 1–2.

Hoffmann, A.A. & Sgrò, C.M. 2011. Climate change and evolutionary adaptation. *Nature* 470: 479–485.

Hohenlohe, P.A., Phillips, P.C. & Cresko, W.A. 2010. Using population genomics to detect selection in natural populations: key concepts and methodological considerations. *Int. J. Plant Sci.* 171: 1059–1071.

Holeski, L.M. & Kelly, J.K. 2006. Mating system and the evolution of quantitative traits: an experimental study of *Mimulus guttatus*. *Evolution* 60: 711–723.

Holliday, J.A., Ralph, S.G., White, R., Bohlmann, J. & Aitken, S.N. 2008. Global monitoring of autumn gene expression within and among phenotypically divergent populations of Sitka spruce (*Picea sitchensis*). *New Phytol.* 178: 103–122.

Holliday, J.A., Ritland, K. & Aitken, S.N. 2010. Widespread, ecologically relevant genetic markers developed from association mapping of climate-related traits in Sitka spruce (*Picea sitchensis*). *New Phytol.* 188: 501–514.

Holsinger, K.E. & Weir, B.S. 2009. Genetics in geographically structured populations: defining, estimating and interpreting F_{ST}. *Nat. Rev. Genet.* 10: 639–650.

Holzapfel, C. & Mahall, B.E. 1999. Bidirectional facilitation and interference between shrubs and annuals in the Mojave desert. *Ecology* 80: 1747–1761.

Horner, J.D. & Abrahamson, W.G. 1992. Influence of plant genotype and environment on oviposition preference and offspring survival in a gallmaking herbivore. *Oecologia* 90: 323–332.

Hossaert-McKey, M., Valero, M., Magda, D., Jarry, M., Cuguen, J. & Vernet, P. 1996. The evolving genetic history of a population of *Lathyrus sylvestris*: evidence from temporal and spatial genetic structure. *Evolution* 50: 1808–1821.

Hough, J., Williamson, R.J. & Wright, S.I. 2013. Patterns of selection in plant genomes. *Annu. Rev. Ecol. Evol. Syst.* 44: 31–50.

Howe, G.T., Aitken, S.N., Neale, D.B. Jermstad, K.D., Wheeler, N.C. & Chen, T.H.H. 2003. From genotype to phenotype: unraveling the complexities of cold adaptation in forest trees. *Can. J. Bot.* 81: 1247–1266.

Howe, H.F. & Smallwood, J. 1982. Ecology of seed dispersal. *Annu. Rev. Ecol. Syst.* 13: 201–228.

Hufford, K.M. & Mazer, S.J. 2003. Plant ecotypes: genetic differentiation in the age of ecological restoration. *Trends Ecol. Evol.* 18: 147–155.

Hughes, A.R., Stachowicz, J.J. & Williams, S.L. 2009. Morphological and physiological variation among seagrass (*Zostera marina*) genotypes. *Oecologia* 159: 725–733.

Hulme, P.E. & Benkman, C.W. 2002. Granivory. In: *Plant–Animal Interactions: An Evolutionary Approach*, ed. C.M. Herrera & O. Pellmyr, 132–154. Malden, MA: Blackwell Publishing.

Hunter, A.F. & Aarssen, L.W. 1988. Plants helping plants. *Bioscience* 38: 34–40.

Hutchings, M.J. & Wijesinghe, D.K. 2008. Performance of a clonal species in patchy environments: effects of environmental context on yield at local and whole-plant scales. *Evol. Ecol.* 22: 313–324.

Hyatt, L.A. & Evans, A.S. 1998. Is decreased germination fraction associated with risk of sibling competition? *Oikos* 83: 29–35.

Inderjit, Wardle, D.A., Karban, R. & Callaway, R.M. 2011. The ecosystem and evolutionary contexts of allelopathy. *Trends Ecol. Evol.* 26: 655–662.

Inderjit, & Weiner, J. 2001. Plant allelochemical interference or soil chemical ecology? *Persp. Plant Ecol. Evol. Syst.* 4: 3–12.

IPCC. 2013. Summary for policymakers. In: *Climate Change 2013: The Physical Science Basis: Contribution of Working Group I to the Fifth Assessment Report of the Intergovernmental Panel on Climate Change*, ed. T.F. Stocker, D. Qin, G.-K. Plattner, M. Tignor, S.K. Allen, J. Boschung, A. Nauels, Y. Xia, V. Bex & P.M. Midgley. 3–29. Cambridge: Cambridge University Press.

Irwin, R.E. 2006. The consequences of direct versus indirect species interactions to selection on traits: pollination and nectar robbing in *Ipomopsis aggregata*. *Am. Nat.* 167: 315–328.

Irwin, R.E., Adler, L.S. & Brody, A.K. 2004. The dual role of floral traits: pollination attraction and plant defense. *Ecology* 85: 1503–1511.

Isaac, S. 1992. *Fungal–Plant Interactions*. London: Chapman & Hall.

Jain, S.K. & Allard, R.W. 1960. Population studies in predominantly self-pollinated species: I. Evidence for heterozygote advantage in a closed population of barley. *Proc. Natl. Acad. Sci. U S A* 46: 1371–1377.

Jain, S.K. & Bradshaw, A.D. 1966. Evolutionary divergence among adjacent plant populations: I. The evidence and its theoretical analysis. *Heredity* 21: 407–441.

Jain, S.K. & Rai, K.N. 1980. The population biology of *Avena*: VIII. Colonization experiment as a test of the role of natural selection in population divergence. *Am. J. Bot.* 67: 1342–1346.

Jakobsson, A. & Dinnetz, P. 2005. Local adaptation and the effects of isolation and population size: the semelparous perennial *Carolina vulgaris* as a study case. *Evol. Ecol.* 19: 449–466.

James, S.E., Pärtel, M., Wilson, S.D. & Peltzer, D.A. 2003. Temporal heterogeneity of soil moisture in grassland and forest. *J. Ecol.* 91: 234–239.

Janzen, D.H. 1969. Seed-eaters versus seed size, number, toxicity and dispersal. *Evolution* 23: 1–27.

Janzen, D.H. 1971. Seed predation by animals. *Annu. Rev. Ecol. Syst.* 2: 465–492.

Janzen, D.H. 1983. Dispersal of seeds by vertebrate guts. In: *Coevolution*, ed. D.J. Futuyma & M. Slatkin, 232–262. Sunderland, MA: Sinauer Associates.

Janzen, F.J. & Stern, H.S. 1998. Logistic regression for empirical studies of multivariate selection. *Evolution* 52: 1564–1571.

Jaramillo-Correa, J.P., Beaulieu, J. & Bousquet, J. 2001. Contrasting evolutionary forces driving population structure at expressed sequence tag polymorphisms, allozymes and quantitative traits in white spruce. *Mol. Ecol.* 10: 2729–2740.

Ji, B. & Bever, J.D. 2012. Mycorrhizal ecology. In: *Oxford Bibliographies Online, Ecology Module*, ed. D. Gibson. New York: Oxford University Press. http://www.oxfordbibliographies.com/obo/page/ecology.

Ji, B., Gehring, C.A., Wilson, G.W.T., Miller, R.M., Flores-Rentería, L. & Johnson, N.C. 2013. Patterns of diversity and adaptation in Glomeromycota from three prairie grasslands. *Mol. Ecol.* 22: 2573–2587.

Jiang, J., Zhu, M., Ai, X., Xiao, L., Deng, G. & Yi, Z. 2013. Molecular evidence for a natural diploid hybrid between *Miscanthus sinensis* (Poaceae) and *M. sacchariflorus*. *Plant Syst. Evol.* 299: 1367–1377.

Johnson, N.C. 1993. Can fertilization of soil select less mutualistic mycorrhizae? *Ecol. Appl.* 3: 749–757.

Johnson, M.T.J., ed. 2011. Evolutionary ecology of plant defences against herbivores [editorial for a special issue]. *Funct. Ecol.* 25: 305–311.

Johnson, M.T.J., Agrawal, A.A., Maron, J.L. & Salminen, J.-P. 2009. Heritability, covariation and natural selection on 24 traits of common evening primrose (*Oenothera biennis*) from a field experiment. *J. Evol. Biol.* 22: 1295–1307.

Johnson, N.C. & Gehring, C.A. 2007. Mycorrhizas: symbiotic mediators of rhizosphere and ecosystem processes. In: *The Rhizosphere: An Ecological Perspective*, ed. Z.G. Cardon & J.L. Whitlock, 73–100. Burlington, MA: Elsevier.

Johnson, N.C., Graham, J.H., & Smith, F.A. 1997. Functioning of mycorrhizal associations along the mutualism–parasitism continuum. *New Phytol.*135: 575–585.

Johnson, N.C., Wilson, G.W.T., Bowker, M.A., Wilson, J.A. & Miller, R.M. 2010. Resource limitation is a driver of local adaptation in mycorrhizal symbiosis. *Proc. Nat. Acad. Sci. U S A* 107: 2093–2098.

Jolivet, C. & Bernasconi, G. 2007. Molecular and quantitative genetic differentiation in European populations of *Silene latifolia* (Caryophyllaceae). *Ann. Bot.* 100: 119–127.

Jordan, N. 1991. Multivariate analysis of selection in experimental populations derived from hybridization of two ecotypes of the annual plant *Diodia teres* W. (Rubiaceae). *Evolution* 45: 1760–1772.

Jordan, N. 1992. Path analysis of local adaptation in two ecotypes of the annual plant *Diodia teres* Walt. (Rubiaceae). *Am. Nat.* 140: 149–165.

Jordano, P. 1995. Frugivore-mediated selection on fruit and seed size: birds and St. Lucie's cherry, *Prunus mahaleb*. *Ecology* 76: 2627–2639.

Jordano, P., Garcia, C., Godoy, J.A. & Garcia-Castaño, J.L. 2007. Differential contribution of frugivores to complex seed dispersal patterns. *Proc. Nat. Acad. Sci. U S A* 104: 3278–3282.

Joshi, J., Schmid, B., Caldeira, M.C., et al. 2001. Local adaptation enhances performance of common plant species. *Ecol. Lett.* 4: 536–544.

Jowett, D. 1958. Populations of *Agrostis* spp. tolerant of heavy metals. *Nature* 182: 816–817.

Juenger, T.E. 2013. Natural variation and genetic constraints on drought tolerance. *Curr. Opin. Plant Biol.* 16: 274–281.

Juenger, T. & Bergelson, J. 2000. The evolution of compensation to herbivory in scarlet gilia, *Ipomopsis aggregata*: herbivore-imposed natural selection and the quantitative genetics of tolerance. *Evolution* 54: 764–777.

Juenger, T., Lennartsson, T. & Tuomi, J. 2000. The evolution of tolerance to damage in *Gentianella campestris*: natural selection and the quantitative genetics of tolerance. *Evol. Ecol.* 14: 393–419.

Jump, A.S., Hunt, J.M., Martínez-Izquierdo, J.A. & Peñuelas, J. 2006. Natural selection and climate change: temperature-linked spatial and temporal trends in gene frequency in *Fagus sylvatica. Mol. Ecol.* 15: 3469–3480.

Jump, A.S. & Peñuelas, J. 2005. Running to stand still: adaptation and the response of plants to rapid climate change. *Ecol. Lett.* 8: 1010–1020.

Kalisz, S. 1986. Variable selection on the timing of germination in *Collinsia verna* (Scrophulariaceae). *Evolution* 40: 479–491.

Kalisz, S. & Kramer, E.M. 2008. Variation and constraint in plant evolution and development. *Heredity* 100: 171–177.

Kalisz, S., Nason, J.D., Hanzawa, F.M. & Tonsor, S.J. 2001. Spatial population genetic structure in *Trillium grandiflorum*: the roles of dispersal, mating, history, and selection. *Evolution* 55: 1560–1568.

Kalisz, S. & Purugganan, M.D. 2004. Epialleles via DNA methylation: consequences for plant evolution. *Trends Ecol. Evol.* 19: 309–314.

Kalske, A., Muola, A., Laukkanen, L., Mutikainen, P. & Leimu, R. 2012. Variation and constraints of local adaptation of a long-lived plant, its pollinators and specialist herbivores. *J. Ecol.*100: 1359–1372.

Kaltz, O. & Shykoff, J.A. 1998. Local adaptation in host-parasite systems. *Heredity* 81: 361–370.

Karasov, T.L., Horton, M.W. & Bergelson, J. 2014. Genomic variability as a driver of plant–pathogen coevolution? *Curr. Opin. Plant Biol.* 18: 24–30.

Karataglis, S.S. 1980. Gene flow in *Agrostis tenuis* (Poaceae). *Plant Syst. Evol.* 134: 23–31.

Karley, A.J., Hawes, C., Iannetta, P.P.M. & Squire, G.R. 2008. Intraspecific variation in *Capsella bursa-pastoris* in plant quality traits for insect herbivores. *Weed Res.* 48: 147–156.

Karp, A., Seberg, O. & Buiatti, M. 1996. Molecular techniques in the assessment of botanical diversity. *Ann. Bot.* 78: 143–149.

Karron, J.D. & Marshall, D.L. 1993. Effects of environmental variation on fitness of singly and multiply sired progenies of *Raphanus sativus* (Brassicaceae). *Am. J. Bot.* 80: 1407–1412.

Kauth, P.J. & M.E. Kane. 2009. In vitro ecology of *Calopogon tuberosus* var. *tuberosus* (Orchidaceae) seedlings from distant populations: implications for assessing ecotypic differentiation. *J. Torrey Bot. Soc.* 136: 433–444.

Kawecki, T.J. & Ebert, D. 2004. Conceptual issues in local adaptation. *Ecol. Lett.* 7: 1225–1241.

Kawecki, T.J., Lenski, R.E., Ebert, D., Hollis, B., Olivieri, I. & Whitlock, M.C. 2012. Experimental evolution. *Trends Ecol. Evol.* 27: 547–560.

Keddy, P.A. 2001. *Competition*. 2nd ed. Dordrecht, The Netherlands: Kluwer Academic.

Keller, S.R., Levsen, N., Olson, M.S. & Tiffin, P. 2012. Local adaptation in the flowering-time gene network of Balsam poplar, *Populus balsamifera* L. *Mol. Biol. Evol.* 29: 3143–3152.

Kelley, S.E. 1994. Viral pathogens and the advantage of sex in the perennial grass *Anthoxanthum odoratum. Phil. Trans. R. Soc. Lond. B* 346: 295–302.

Kelley, S.E. & Clay, K. 1987. Interspecific competitive interactions and the maintenance of genotypic variation within two perennial grasses. *Evolution* 41: 92–103.

Kelly, J.K. 1996. Kin selection in the annual plant *Impatiens capensis*. *Am. Nat.* 147: 899–918.

Kemp, W.B. 1937. Natural selection within plant species as exemplified in a permanent pasture. *J. Hered.* 28: 329–333.

Kessler, A. & Baldwin, I.T. 2002. Plant responses to insect herbivory: the emerging molecular analysis. *Annu. Rev. Plant Biol.* 53: 299–328.

Khan, M.A., Antonovics, J. & Bradshaw, A.D. 1976. Adaptation to heterogeneous environments: III. The inheritance of response to spacing in flax and linseed (*Linum usitatissimum*). *Aust. J. Agricult. Res.* 27: 649–659.

Kikvidze, Z. & Callaway, R.M. 2009. Ecological facilitation may drive major evolutionary transitions. *Bioscience* 59: 399–404.

Kim, E. & Donohue, K. 2013. Local adaptation and plasticity of *Erysimum capitatum* to altitude: its implications for responses to climate change. *J. Ecol.* 101: 796–805.

Kindell, C.E., Winn, A.A. & Miller, T.E. 1996. The effects of surrounding vegetation and transplant age on the detection of local adaptation in the perennial grass *Aristida stricta*. *J. Ecol.* 84: 745–754.

Kingsolver, J.G. & Diamond, S.E. 2011. Phenotypic selection in natural populations: what limits directional selection? *Am. Nat.* 177: 346–357.

Kingsolver, J.G., Diamond, S.E., Siepielski, A.M. & Carlson, S.M. 2012. Synthetic analyses of phenotypic selection in natural populations: limitations and future directions. *Evol. Ecol.* 26: 1101–1118.

Kingsolver, J.G., Hoekstra, H.E., Hoekstra, J.M., et al. 2001. The strength of phenotypic selection in natural populations. *Am. Nat.* 157: 245–261.

Kingsolver, J.G. & Schemske, D.W. 1991. Path analyses of selection. *Trends Ecol. Evol.* 6: 276–280.

Kinnison, M.T & Hairston, N.G., Jr. 2007. Eco-evolutionary conservation biology: contemporary evolution and the dynamics of persistence. *Funct. Ecol.* 21: 444–454.

Kirk, H. & Freeland, J.R. 2011. Applications and implications of neutral versus non-neutral markers in molecular ecology. *Int. J. Mol. Sci.* 12: 3966–3988.

Kivimäki, M., Kärkkäinen, K., Gaudeul, M., Løe, G. & Ågren, J. 2007. Gene, phenotype and function: *GLABROUS1* and resistance to herbivory in natural populations of *Arabidopsis lyrata*. *Mol. Ecol.* 16: 453–462.

Kivlin, S.N., Emery, S.M. & Rudgers, J.A. 2013. Fungal symbionts alter plant responses to global change. *Am. J. Bot.* 100: 1445–1457.

Klironomos, J.N. 2003. Variation in plant response to native and exotic arbuscular mycorrhizal fungi. *Ecology* 84: 2292–2301.

Knight, C.A., Vogel, H., Kroymann, J., Shumate, A., Witsenboer, H. & Mitchell-Olds, T. 2006. Expression profiling and local adaptation of *Boechera holboellii* populations for water use efficiency across a naturally occurring water stress gradient. *Mol. Ecol.* 15: 1229–1237.

Koch, A.M., Croll, D. & Sanders, I.R. 2006. Genetic variability in a population of arbuscular mycorrhizal fungi causes variation in plant growth. *Ecol. Lett.* 9: 103–110.

Koelewijn, H.P. 2004. Sibling competition, size variation and frequency-dependent outcrossing advantage in *Plantago coronopus*. *Evol. Ecol.* 18: 51–74.

Körner, C. 1991. Some often overlooked plant characteristics as determinants of plant growth: a reconsideration. *Funct. Ecol.* 5: 162–173.

Koslow, J.M. & Clay, K. 2007. The mixed mating system of *Impatiens capensis* and infection by a foliar rust pathogen: patterns of resistance and fitness consequences. *Evolution* 61: 2643–2654.

Kover, P.X. & Schaal, B.A. 2002. Genetic variation for disease resistance and tolerance among *Arabidopsis thaliana* accessions. *Proc. Natl. Acad. Sci. U S A* 99: 11270–11274.

Kromer, M. & Gross, K.L. 1987. Seed mass, genotype, and density effects on growth and yield of *Oenothera biennis* L. *Oecologia* 73: 207–212.

Kroymann, J., Donnerhacke, S., Schnabelrauch, D. & Mitchell-Olds, T. 2003. Evolutionary dynamics of an *Arabidopsis* insect resistance quantitative trait locus. *Proc. Nat. Acad. Sci. U S A* 100: 14587–14592.

Kruckeberg, A.R. 1951. Intraspecific variability in the response of certain native plant species to serpentine soils. *Am. J. Bot.* 38: 408–419.

Kruckeberg, A.R. 1954. The ecology of serpentine soils: III. Plant species in relation to serpentine soils. *Ecology* 35: 267–274.

Kula, A.A.R., Dudash, M.R. & Fenster, C.B. 2013. Choices and consequences of oviposition in a pollinating seed predator, *Hadena ectypa* (Noctuidae), on its host plant, *Silene stellata* (Caryophyllaceae). *Am. J. Bot.* 100: 1148–1154.

Ladiges, P.Y. 1976. Variation in drought resistance to adjacent edaphic populations of *Eucalyptus viminalis* Labill. *Aust. J. Ecol.* 1: 67–76.

Laine, A.-L. 2005. Spatial scale of local adaptation in a plant–pathogen metapopulation. *J. Evol. Biol.* 18: 930–938.

Laine, A.-L. 2006. Evolution of host resistance: looking for coevolutionary hotspots at small spatial scales. *Proc. R. Soc. B* 273: 267–273.

Lal, N. 2010. Molecular mechanisms and genetic basis of heavy metal toxicity and tolerance in plants. In: *Plant Adaptation and Phytoremediation*, ed. M. Ashraf, M. Ozturk, & M.S.A. Ahmed, 35–58. New York: Springer Science.

Lande, R. 2000. Quantitative genetics and phenotypic evolution. In: *Evolutionary Genetics: From Molecules to Morphology*, ed. B.S. Singh & C.B. Krimbas, pp. 335–350. Cambridge, UK: Cambridge University Press.

Lande, R. & S. J. Arnold. 1983. The measurement of correlated characters. *Evolution* 37: 1210–1226.

Langlet, O. 1971. Two hundred years [of] genecology. *Taxon* 20: 653–722.

Lankau, R. 2008. A chemical trait creates a genetic trade-off between intra- and interspecific competitive ability. *Ecology* 89: 1181–1187.

Lankau, R. & Klienbenstein, D.J. 2009. Competition, herbivory and genetics interact to determine the accumulation and fitness consequences of a defense metabolite. *J. Ecol.* 97: 78–88.

Larsen, E. 2005. Developmental origins of variation. In: *Variation: A Central Concept in Biology*, ed. B. Hallgrimsson & B.K. Hall, 113–129. Burlington, MA: Elsevier/Academic Press.

Larson, A. 2009. Adaptation. In: *Princeton Guide to Ecology*, ed. S.A. Levin, 93–100. Princeton, NJ: Princeton University Press.

Latta, R.G. 2009. Testing for local adaptation in *Avena barbata*: a classic example of ecotypic divergence. *Mol. Ecol.* 18: 3781–3791.

Latta, R.G. 2010. Natural selection, variation, adaptation, and evolution: a primer of interrelated concepts. *Int. J. Plant Sci.* 171: 930–944.

Latta, R.G., Linhart, Y.B., Fleck, D. & Elliot, M. 1998. Direct and indirect estimates of seed versus pollen movement within a population of ponderosa pine. *Evolution* 52: 61–67.

Latta, R., Mackenzie, J.L., Vats, A. & Schoen, D.J. 2004. Divergence and variation of quantitative traits between allozyme genotypes of *Avena barbata* from contrasting habitats. *J. Ecol.* 92: 57–71.

Latta, R. & McCain, C. 2009. Path analysis of natural selection via survival and fecundity across contrasting environments in *Avena barbata*. *J. Evol. Biol.* 22: 2458–2469.

Lau, J.A. & Lennon, J.T. 2011. Evolutionary ecology of plant–microbe interactions: soil microbial structure alters selection on plant traits. *New Phytol.* 192: 215–224.

Lau, J.A., Shaw, R.G., Reich, P.B., Shaw, F.H. & Tiffin, P. 2007. Strong ecological but weak evolutionary effects of elevated CO_2 on a recombinant inbred population of *Arabidopsis thaliana*. *New Phytol.* 175: 351–362.

Lau, J.A., Shaw, R.G., Reich, P.B. & Tiffin, P. 2010. Species interactions in a changing environment: elevated CO_2 alters the ecological and potential evolutionary consequences of competition. *Evol. Ecol. Res.* 12: 435–455.

Leclerc, J.C. & Blaise, S. 1991. Photosynthetic and genetic differentiation in sun and shade populations of *Holcus mollis* L. *Vegetatio* 92: 85–93.

Ledig, F.T., Hom, J.L. & Smouse, P.E. 2013. The evolution of the New Jersey Pine Plains. *Am. J. Bot.* 100: 778–791.

Leger, E.A. & Rice, K.J. 2007. Assessing the speed and predictability of local adaptation in invasive California poppies (*Eschscholzia californica*). *J. Evol. Biol.* 20: 1090–1103.

Lehmann, C. & Rebele, F. 2004. Evaluation of heavy metal tolerance in *Calamagrostis epigejos* and *Elymus repens* revealed copper tolerance in a copper smelter population of *C. epigejos*. *Environ. Exp. Bot.* 51: 199–213.

Leimu, R. & Fischer, M. 2008. A meta-analysis of local adaptation in plants. *PLoS One* 3: e4010.

Leinonen, T., O'Hara, R., Cano, J.M. & Merilä, J. 2008. Comparative studies of quantitative trait and neutral marker divergence: a meta-analysis. *J. Evol. Biol.* 21: 1–17.

Lenormand, T. 2002. Gene flow and the limits to natural selection. *Trends Ecol. Evol.* 17: 183–189.

Levey, D.J., Silva, W.R. & Galetti, M., eds. 2002. *Dispersal and Frugivory: Ecology, Evolution and Conservation.* New York: CAB International.

Levey, D.J., Tewksbury, J.J., Izhaki, I., Tsahar, E. & Haak, D.C. 2007. Evolutionary ecology of secondary compounds in ripe fruit: case studies with capsaicin and emodin. In: *Seed Dispersal: Theory and Its Application in a Changing World*, ed. A.J. Dennis, E.W. Schupp, R.A. Green & D.A. Westcott, 37–58. New York: CAB International.

Levin, D.A. 1975. Pest pressure and recombination systems in plants. *Am. Nat.* 109: 437–451.

Levin, D.A. 1981. Dispersal versus gene flow in plants. *Ann. Missouri Bot. Gard.* 68: 233–253.

Levin, D.A. & Brack, E.T. 1995. Natural selection against white petals in *Phlox*. *Evolution* 49: 1017–1022.

Levin, D.A. & Kerster, H.W. 1974. Gene flow in seed plants. *Evol. Biol.* 7: 139–230.

Lewontin, R.C. 1974. *The Genetic Basis of Evolutionary Change.* New York: Columbia University Press.

Lexer, C., Lai, Z. & Rieseberg, L.H. 2004. Candidate gene polymorphisms associated with salt tolerance in wild sunflower hybrids: implications for the origin of *Helianthus paradoxus*, a diploid hybrid species. *New Phytol.* 161: 225–233.

Lexer, C., Randell, R.A. & Rieseberg, L.H. 2003a. Experimental hybridization as a tool for studying selection in the wild. *Ecology* 84: 1688–1699.

Lexer, C., Welch, M.E., Durphy, J.L. & Rieseberg, L.H. 2003b. Natural selection for salt tolerance quantitative trait loci (QTLs) in wild sunflower hybrids: implications for the origin of *Helianthus paradoxus*, a diploid hybrid species. *Mol. Ecol.* 12: 1225–1235.

Liancourt, P., Spence, L.A., Song, D.S., et al. 2013. Plant response to climate change varies with topography, interactions with neighbors, and ecotype. *Ecology* 94: 444–453.

Liancourt, P. & Tielbörger, K. 2009. Competition and a short growing season lead to ecotypic differentiation at the two extremes of the ecological range. *Funct. Ecol.* 23: 397–404.

Liancourt, P. & Tielbörger, K. 2011. Ecotypic differentiation determines the outcome of positive interactions in a dryland annual plant species. *Persp. Plant Ecol. Evol. Syst.* 13: 259–264.

Lindroth, R.L., Roth, S. & Nordheim, E.V. 2001. Genotypic variation in response of quaking aspen (*Populus tremuloides*) to atmospheric CO_2 enrichment. *Oecologia* 126: 371–379.

Linhart, Y.B. 1988. Intrapopulation differentiation in annual plants: III. The contrasting effects of intra- and interspecific competition. *Evolution* 42: 1047–1064.

Linhart, Y.B. & Grant, M.C. 1996. Evolutionary significance of local genetic differentiation in plants. *Annu. Rev. Ecol. Syst.* 27: 237–277.

Livshulz, T., Mead, J.V., Goyder, D.J. & Brannin, M. 2011. Climate niches of milkweed with plesiomorphic traits (Secamonoideae; Apocynaceae) and the milkweed sister group link ancient African climates and floral evolution. *Am. J. Bot.* 98: 1966–1977.

Lloyd, D.G. & Barrett, S.C.H., eds. 1996. *Floral Biology: Studies on Floral Evolution in Animal-Pollinated Plants.* New York: Chapman & Hall.

Loranger, J., Meyer, S.T., Shipley, B., et al. 2012. Predicting invertebrate herbivory from plant traits: evidence from 51 grassland species in experimental monocultures. *Ecology* 93: 2674–2682.

Louda, S.M., Keeler, K.H. & Holt, R.D. Herbivore influences on plant performance and competitive interactions. In: *Perspectives on Plant Competition*, ed. J.B. Grace & D. Tilman, 413–444. San Diego, CA: Academic Press.

Loveless, M.D. & Hamrick, J.L. 1984. Ecological determinants of genetic structure in plant populations. *Ann. Rev. Ecol. Syst.* 15: 65–95.

Lovell, J.T. & Menges, E. 2013. Dominant species responses to drought in seasonal wetlands: evidence from reciprocal transplants across a moisture gradient. *J. Torrey Bot. Soc.* 140: 157–169.

Lovett Doust, L. 1981a. Intraclonal variation and competition in *Ranunculus repens. New Phytol.* 89: 495–502.

Lovett Doust, L. 1981b. Population dynamics and local specialization in a clonal perennial (*Ranunculus repens*): II. The dynamics of leaves, and a reciprocal transplant–replant experiment. *J. Ecol.* 69: 757–768.

Lovett Doust, L., Lovett Doust, J. & Biernacki, M. 1994. American wild celery, *Vallisneria americana*, as a biomonitor of organic contaminants in aquatic ecosystems. *J. Great Lakes Res.* 20: 333–354.

Lowry, D.B. 2012. Ecotypes and the controversy over stages in the formation of new species. *Biol. J. Linn. Soc.* 106: 241–257.

Lowry, D.B., Hall, M.C., Salt, D.E. & Willis, J.H. 2009. Genetic and physiological basis of adaptive salt tolerance divergence between coastal and inland *Mimulus guttatus. New Phytol.* 183: 776–788.

Lowry, D.B., Rockwood, R.C. & Willis, J.H. 2008. Ecological and reproductive isolation of coast and inland races of *Mimulus guttatus. Evolution* 62: 2196–2214.

Luikart, G., Ryman, N., Tallmon, D.A., Schwartz, M.K. & Allendorf, F.W. 2010. Estimation of census and effective population sizes: the increasing usefulness of DNA-based approaches. *Conserv. Genet.* 11: 355–373.

MacColl, A.D.C. 2011. The ecological causes of evolution. *Trends Ecol. Evol.* 26: 514–522.

Macel, M., Lawson, C.S., Mortimer, S.R., et al. 2007. Climate vs. soil factors in local adaptation of two common plant species. *Ecology* 88: 424–433.

Macnair, M.R. 1987. Heavy metal tolerance in plants: a model evolutionary system. *Trends Ecol. Evol.* 2: 354–359.

Macnair, M.R. 1993. The genetics of metal tolerance in vascular plants. *New Phytol.* 124: 541–559.

Maddox, G.D. & Antonovics, J. 1983. Experimental ecological genetics in *Plantago*: a structural equation approach to fitness components in *P. aristata* and *P. patagonica. Ecology* 64: 1092–1099.

Maddox, G.D. & Root, R.B. 1987. Resistance to 16 diverse species of herbivorous insects within a population of goldenrod, *Solidago altissima*: genetic variation and heritability. *Oecologia* 72: 8–14.

Maimoona, A., Naeem, I., Saddiqe, Z. & Jameel, K. 2011. A review on biological, nutraceutical and clinical aspects of French maritime pine bark extract. *J. Ethnopharmacol.* 133: 261–277.

Manzaneda, A.J., Prasad, K.V.S.K., & Mitchell-Olds, T. 2010. Variation and fitness costs for tolerance to different types of herbivore damage in *Boechera stricta* genotypes with contrasting glucosinolate structures. *New Phytol.*188: 464–477.

Marak, H.B., Biere, A. & van Damme, J.M.M. 2000. Direct and correlated responses to selection on iridoid glycosides in *Plantago lanceolata* L. *J. Evol. Biol.* 13: 985–996.

Marañon, T. & Bartolome, J.W. 1993. Reciprocal transplants of herbaceous communities between *Quercus agrifolia* woodland and adjacent grassland. *J. Ecol.* 81: 673–682.

Mariette, S., Chagné, D., Lézier, C., et al. 2001. Genetic diversity within and among *Pinus pinaster* populations: comparison between AFLP and microsatellite markers. *Heredity* 86: 469–479.

Marks S. & Clay, K. 1996. Physiological responses of *Festuca arundinacea* to fungal endophyte infection. *New Phytol.*133: 727–733.

Maron, J.L., Vila, M., Bommarco, R., Elmendorf, S. & Beardsley, P. 2004. Rapid evolution of an invasive plant. *Ecol. Monogr.* 74: 261–280.

Marsden-Jones, E.M. & Turrill, W.B. 1930. Report on the transplant experiments of the British Ecological Society at Potterne, Wilts. *J. Ecol.* 18: 352–378.

Marsden-Jones, E.M. & Turrill, W.B. 1938a. Fifth report of the transplant experiments of the British Ecological Society at Potterne, Wiltshire. *J. Ecol.* 26: 359–379.

Marsden-Jones, E.M. & Turrill, W.B. 1938b. Transplant experiments of the British Ecological Society at Potterne, Wiltshire: summary of results, 1928–1937. *J. Ecol.* 26: 380–389.

Martin, M.M. & Harding, J. 1981. Evidence for the evolution of competition between two species of annual plants. *Evolution* 35: 975–987.

Masclaux, F., Hammond, R.L., Meunier, J., Gouhier-Darimont, C., Keller, L. & Reymond, P. 2010. Competitive ability not kinship affects growth of *Arabidopsis thaliana* accessions. *New Phytol.* 185: 322–331.

Matesanz, S., Gimeno, T.E., de la Cruz, M., Escudero, A. & Valladares, F. 2011. Competition may explain the fine-scale spatial patterns and genetic structure of two co-occurring plant congeners. *J. Ecol.* 99: 838–848.

Mathiasen, P. & Premoli, A.C. 2013. Fine-scale genetic structure of *Nothofagus pumilio* (lenga) at contrasting elevations of the altitudinal gradient. *Genetica* 141: 95–105.

Matsueda, R.L. 2012. Key advances in the history of structural equation modeling. In: *Handbook of Structural Equation Modeling*, ed. R.H. Hoyle, 17–42. New York: Guilford Press.

Mátyás, C. 1996. Climatic adaptation of trees: rediscovering provenance tests. *Euphytica* 92: 45–54.

Maun, M.A. 2009. *The Biology of Coastal Sand Dunes.* New York: Oxford University Press.

Mauricio, R. & Rausher, M.D. 1997. Experimental manipulation of putative selective agents provides evidence for the role of natural enemies in the evolution of plant defense. *Evolution* 51: 1435–1444.

Mauricio, R., Stahl, E.A., Korves, T., Tian. D., Kreitman, M. & Bergelson, J. 2003. Natural selection for polymorphism in the disease resistance gene *Rps2* of *Arabidopsis thaliana. Genetics* 163: 735–746.

Mayer, C., Adler, L., Armbruster, W.S., et al. 2011. Pollination ecology in the 21st century: key questions for future research. *J. Pollin. Ecol.* 3: 8–23.

Mayr, E. 1982. *The Growth of Biological Thought: Diversity, Evolution, and Inheritance.* Cambridge, MA: Harvard University Press.

Mazer, S.J. & Damuth, J. 2001. Evolutionary significance of variation. In: *Evolutionary Ecology: Concepts and Case Studies*, ed. C.W. Fox, D.A. Roff & D.J. Fairbairn, 16–28. New York: Oxford University Press.

Mazer, S.J., Delesalle, V.A. & Neal, P.R. 1999. Responses of floral traits to selection on primary sexual investment in *Spergularia marina*: the battle between the sexes. *Evolution* 53: 717–731.

Mazer, S.J. & LeBuhn, G. 1999. Genetic variation in life-history traits: heritability estimates within and genetic differentiation among populations. In: *Life History Evolution in Plants*, ed. T.O. Vuorisalo & P.K. Mutikainen, 85–171. Dordrecht, The Netherlands: Kluwer Academic.

Mazer, S.J. & Schick, C.T. 1991a. Constancy of population parameters for life history and floral traits in *Raphanus sativus* L: I. Norms of reaction and the nature of genotype by environment interactions. *Heredity* 67: 143–156.

Mazer, S.J. & Schick, C.T. 1991b. Constancy of population parameters for life history and floral traits in *Raphanus sativus* L: II. Effects of planting density on phenotype and heritability estimates. *Evolution* 45: 1888–1907.

McCall, C., Mitchell-Olds, T. & Waller, D.M. 1989. Fitness consequences of outcrossing in *Impatiens capensis*: tests of the frequency-dependent and sib-competition models. *Evolution* 43: 1075–1084.

McCall, A.C., Murphy, S.J., Venner, C. & Brown, M. 2013. Florivores prefer white versus pink petal color morphs in wild radish, *Raphanus sativus*. *Oecologia* 172: 189–195.

McGraw, J.B. 1985. Experimental ecology of *Dryas octopetala* ecotypes: relative response to competitors. *New Phytol.* 100: 233–241.

McGraw, J.B. & Antonovics, J. 1983. Experimental ecology of *Dryas octopetala* ecotypes: I. Ecotypic differentiation and life-cycle stages of selection. *J. Ecol.* 71: 879–897.

McIntyre, P.J. & Strauss, S.Y. 2014. Phenotypic and transgenerational plasticity promote local adaptation to sun and shade environments. *Evol. Ecol.* 28: 229–246.

McKay, J.K., Christian, C.E., Harrison, S. & Rice, K.J. 2005. "How local is local?": a review of practical and conceptual issues in the genetics of restoration. *Restor. Ecol.* 13: 432–440.

McKay, J.K. & Latta, R.G. 2002. Adaptive population divergence: markers, QTL and traits. *Trends Ecol. Evol.* 17: 285–291.

McKey, D. 1975. The ecology of coevolved seed dispersal systems. In: *Coevolution of Animals and Plants*, ed. L.E. Gilbert & P.H. Raven, 159–191. Austin, TX: University of Texas Press.

McKinney, K.K. & Fowler, N.L. 1991. Genetic adaptations to grazing and mowing in the unpalatable grass *Cenchrus incertus*. *Oecologia* 88: 238–242.

McLeod, K.A., Scascitelli, M. & Vellend, M. 2012. Detecting small-scale genotype–environment interactions in apomictic dandelion (*Taraxacum officinale*) populations. *J. Evol. Biol.* 25: 1667–1675.

McMillan, C. 1956. Nature of the plant community: I. Uniform garden and light period studies of five grass taxa in Nebraska. *Ecology* 37: 330–340.

McMillan, C. 1959. The role of ecotypic variation in the distribution of the central grassland of North America. *Ecol. Monogr.* 29: 285–308.

McMillan, C. 1964. Ecotypic differentiation within four North American prairie grasses: I. Morphological variation within transplanted community fractions. *Am. J. Bot.* 51: 1119–1128.

McMillan, C. 1965. Ecotypic differentiation within four North American prairie grasses: II. Behavioral variation within transplanted community fractions. *Am. J. Bot.* 52: 55–65.

McMillan, C. 1967. Phenological variation within seven transplanted grassland community fractions from Texas and New Mexico. *Ecology* 48: 807–813.

McNaughton, S.J. 1984. Grazing lawns: animals in herds, plant form, and coevolution. *Am. Nat.* 124: 863–886.

McNeilly, T. 1968. Evolution in closely adjacent plant populations: III. *Agrostis tenuis* on a small copper mine. *Heredity* 23: 99–108.

McNeilly, T. 1981. Ecotypic differentiation in *Poa annua*: interpopulation differences in response to competition and cutting. *New Phytol.* 88: 539–547.

McNeilly, T. 1984. Ectotypic [*sic*] differentiation in *Poa annua*: within population variation in response to competition and cutting. *New Phytol.* 96: 307–316.

McNeilly, T. & Antonovics, J. 1968. Evolution in closely adjacent plant populations: IV. Barriers to gene flow. *Heredity* 23: 205–218.

McNeilly, T. & Bradshaw, A.D. 1968. Evolutionary processes in populations of copper tolerant *Agrostis tenuis* Sibth. *Evolution* 22: 108–118.

McNeilly, T. & Roose, M.L. 1996. Co-adaptation between neighbours? A case study with *Lolium perenne* genotypes. *Euphytica* 92: 121–128.

McRoberts, N., Finch, R.P., Sinclair, W., et al. 1999. Assessing the ecological significance of molecular diversity data in natural plant populations. *J. Exp. Bot.* 50: 1635–1645.

Meagher, T.R. & Thompson, E. 1987. Analysis of parentage for naturally established seedlings of *Chamaelirium luteum* (Liliaceae). *Ecology* 68: 803–812.

Medel, R., Botto-Mahan, C. & Kalin-Arroyo, M. 2003. Pollinator-mediated selection on the nectar guide phenotype in the Andean monkey flower *Mimulus luteus*. *Ecology* 84: 1721–1732.

Meharg, A.A., Cumbes, Q.J. & Macnair, M.R. 1993. Pre-adaptation of Yorkshire fog, *Holcus lanatus* L. (Poaceae) to arsenate tolerance. *Evolution* 47: 313–316.

Mehrhoff, L.A. & Turkington, R. 1990. Microevolution and site-specific outcomes of competition among pasture plants. *J. Ecol.* 78: 745–756.

Mehrhoff, L.A. & Turkington, R. 1995. Resource use and co-existence: experimental tests from a sequence of different-aged pastures. *Evol. Ecol.* 9: 617–632.

Meindl, G.A., Bain, D.J. & Ashman, T.-L. 2013. Edaphic factors and plant–insect interactions: direct and indirect effects of serpentine soil on florivores and pollinators. *Oecologia* 173: 1355–1366.

Mengoni, A., Barabesi, C., Gonnelli, C., Galardi, F., Gabbrielli, R. & Bazzicalupo, M. 2001. Genetic diversity of heavy metal-tolerant populations in *Silene paradoxa* L. (Caryophyllaceae): a chloroplast microsatellite analysis. *Mol. Ecol.* 10: 1909–1916.

Mercer, K.L., Wyse, D.L. & Shaw, R.G. 2006. Effects of competition on the fitness of wild and crop-wild hybrid sunflower from a diversity of wild populations and crop lines. *Evolution* 60: 2044–2055.

Merilä, J. & Björklund, M. 2004. Phenotypic integration as a constraint and adaptation. In: *Phenotypic Integration: Studying the Ecology and Evolution of Complex Phenotypes*, ed. M. Pigliucci & K. Preston, 107–129. Oxford: Oxford University Press.

Merilä, J. & Crnokrak, P. 2001. Comparison of genetic differentiation at marker loci and quantitative traits. *J. Evol. Biol.* 14: 892–903.

Metcalf, C.J.E. & Pavard, S. 2007. Why evolutionary biologists should be demographers. *Trends Ecol. Evol.* 22: 205–211.

Michalet, R., Xiao, S., Touzard, B., et al. 2011. Phenotypic variation in nurse traits and community feedbacks define an alpine community. *Ecol. Lett.* 14: 433–443.

Miglia, K.J., McArthur, E.D., Moore, W.S., Wang, H., Graham, J.H. & Freeman, D.C. 2005. Nine-year reciprocal transplant experiment in the gardens of the basin and mountain big sagebrush (*Artemisia tridentata*: Asteraceae) hybrid zone of Salt Creek Canyon: the importance of multiple-year tracking of fitness. *Biol. J. Linn. Soc.* 86: 213–225.

Milla, R., del Burgo, A.V., Escudero, A. & Iriondo, J.M. 2012. Kinship rivalry does not trigger specific allocation strategies in *Lupinus angustifolius*. *Ann. Bot.* 110: 165–175.

Milla, R., Forero, D.M., Escudero, A. & Iriondo, J.M. 2009. Growing with siblings: a common ground for cooperation or for fiercer competition among plants? *Proc. R. Soc. B* 276: 2531–2540.

Miller, T.E. 1995. Evolution of *Brassica rapa* L. (Cruciferae) populations in intra- and interspecific competition. *Evolution* 49: 1125–1133.

Miller, R.E. & Fowler, N.L. 1993. Variation in reaction norms among populations of the grass *Bouteloua rigidiseta*. *Evolution* 47: 1446–1455.

Miller, M.R., White, A. & Boots, M. 2006. The evolution of parasites in response to tolerance in their hosts: the good, the bad, and apparent commensalism. *Evolution* 60: 945–956.

Mimura, M. & Aitken, S.N. 2010. Local adaptation at the range peripheries of Sitka spruce. *J. Evol. Biol.* 23: 249–258.

Mitchell, R.J. 2001. Path analysis: pollination. In: *Design and Analysis of Ecological Experiments*, S.M. Scheiner & J. Gurevitch, 217–234. New York: Oxford University Press.

Mitchell, R.J. 2004. Heritability of nectar traits: why do we know so little? *Ecology* 85: 1527–1533.

Mitchell, R.J., Irwin, R.E., Flanagan, R.J. & Karron, J.D. 2009. Ecology and evolution of plant–pollinator interactions. *Ann. Bot.*103: 1355–1363.

Mitchell-Olds, T. & Rutledge, J.J. 1986. Quantitative genetics in natural plant populations: a review of the theory. *Am. Nat.* 127: 379–402.

Mitchell-Olds, T. & Shaw, R.G. 1987. Regression analysis of natural selection: statistical inference and biological interpretation. *Evolution* 41: 1149–1161.

Mizuki, I., Ishida, K., Tani, N. & Tsumura, Y. 2010. Fine-scale spatial structure of genets and sexes in the dioecious plant *Dioscorea japonica*, which disperses by both bulbils and seeds. *Evol. Ecol.* 24: 1399–1415.

Moloney, K.A., Holzapfel, C., Tielbörger, K., Jeltsch, F. & Schurr, F.M. 2009. Rethinking the common garden in invasion research. *Persp. Plant Ecol. Evol. Syst.* 11: 311–320.

Monier, J.-M. 2006. Bacterial assemblages on plant surfaces. In: *Microbial Ecology of Aerial Plant Surfaces*, ed. M.J. Bailey, A.K. Lilley, T.M. Timms-Wilson & P.T.N. Spencer-Phillips, 83–106. Cambridge, MA: CAB International.

Monty, A., LeBeau, J., Meerts, P. & Mahy, G. 2009. An explicit test for the contribution of environmental maternal effects to rapid clinal differentiation in an invasive plant. *J. Evol. Biol.* 22: 917–926.

Monty, A. & Mahy, G. 2009. Clinal differentiation during invasion: *Senecio inaequidens* (Asteraceae) along altitudinal gradients in Europe. *Oecologia* 159: 305–315.

Moon, C. D., Craven, K.D., Leuchtmann, A., Clement, S.L. & Schardl, C.L. 2004. Prevalence of interspecific hybrids amongst asexual fungal endophytes of grasses. *Mol. Ecol.*13: 1455–1467.

Moran, E.V., Willis, J. & Clark, J.S. 2012. Genetic evidence for hybridization in red oaks (*Quercus* Sect. *Lobatae*, Fagaceae). *Am. J. Bot.* 99: 92–100.

Morgan, J.A.W., Bending, G.D. & White, P.J. 2005. Biological costs and benefits to plant–microbe interactions in the rhizosphere. *J. Exp. Bot.* 56: 1729–1739.

Morin, P.J. 2011. *Community Ecology.* 2nd ed. Oxford: Wiley-Blackwell.

Moritz, D.M.L. & Kadereit, J.W. 2001. The genetics of evolutionary change in *Senecio vulgaris* L.: a QTL mapping approach. *Plant Biol.* 3: 544–552.

Morley, F.H.W. 1959. Natural selection and variation in plants. In: *Cold Spring Harbor Symposium on Quantitative Biology. Vol. XXIV: Genetics and Twentieth Century Darwinism*, ed. C. Wooldridge, 47–56. Cold Spring Harbor: Cold Spring Harbor Laboratory.

Morse, L.J., Faeth, S.H. & Day, T.A. 2007. *Neotyphodium* interactions with a wild grass are driven mainly by endophyte haplotype. *Funct. Ecol.* 21: 813–822.

Moyle, L.C., Levine, M., Stanton, M.L. & Wright, J.W. 2012. Hybrid sterility over tens of meters between ecotypes adapted to serpentine and non-serpentine soils. *Evol. Biol.* 39: 207–218.

Müller, T., Strobel, K. & Ulrich, A. 2006. Microorganisms in the phyllosphere of temperate forest ecosystems in a changing environment. In: *Microbial Ecology of Aerial Plant Surfaces*, ed. M.J. Bailey, A.K. Lilley, T.M. Timms-Wilson & P.T.N. Spencer-Phillips, 51–66. Cambridge, MA: CAB International.

Munguía-Rosas, M.A., Ollerton, J., Parra-Tabla, V. & De-Nova, J.A. 2011. Meta-analysis of phenotypic selection on flowering phenology suggests that early flowering plants are favoured. *Ecol. Lett.* 14: 511–521.

Murphy, G.P. & Dudley, S.A. 2009. Kin recognition, competition and cooperation in *Impatiens* (Balsaminaceae). *Am. J. Bot.* 96: 1990–1996.

Murren, C.J., Chang, C.C. & Dudash, M.R. 2009. Patterns of selection of two North American native and nonnative populations of monkeyflower (Phrymaceae). *New Phytol.* 183: 691–701.

Murren, C.J. & Kover, P.X. 2004. QTL mapping: a first step toward an understanding of molecular genetic mechanisms behind phenotypic complexity/integration. In: *Phenotypic Integration: Studying the Ecology and Evolution of Complex Phenotypes*, ed. M. Pigliucci & K. Preston, 195–212. Oxford: Oxford University Press.

Nakamura, R.R. 1980. Plant kin selection. *Evol. Theory* 5: 113–117.

Namroud, M.C., Beaulieu, J., Juge, N., Laroche, J. & Bousquet, J. 2008. Scanning the genome for gene single nucleotide polymorphisms involved in adaptive population differentiation in white spruce. *Mol. Ecol.* 17: 3599–3613.

Neale, D.B. & Savolainen, O. 2004. Association genetics of complex traits in conifers. *Trends Plant Sci.* 9: 325–330.

Nei, M. 1973. Analysis of gene diversity in subdivided populations. *Proc. Nat. Acad. Sci. U S A* 70: 3321–3323.

Nei, M. 1987. *Molecular Evolutionary Genetics*. New York: Columbia University Press.

Neufeld, H.S. & Young, D. R. 2003. Ecophysiology of the herbaceous layer in temperate deciduous forests. In: *The Herbaceous Layer in Forests of Eastern North America*, ed. F.S. Gilliam & M.R. Roberts, 38–90. New York: Oxford University Press.

Neuhauser, C. & Fargione, J.E. 2004. A mutualism–parasitism continuum model and its application to plant–mycorrhizae interactions. *Ecol. Model.* 177: 337–352.

Nevo, E., Beiles, A., Kaplan, D., Golenberg, E.M., Olsvig-Whittaker, L. & Naveh, Z. 1986. Natural selection of allozyme polymorphisms: a microsite test revealing ecological genetic differentiation in wild barley. *Evolution* 40: 13–20.

Nevo, E., Beiles, A. & Krugman, T. 1988. Natural selection of allozyme polymorphisms: a microgeographic climatic differentiation in wild emmer wheat (*Triticum dicoccoides*). *Theor. Appl. Genet.* 75: 529–538.

Nevo, E., Krugman, T., & Beiles, A. 1994. Edaphic natural selection of allozyme polymorphisms in *Aegilops peregrine* at a Galilee microsite in Israel. *Heredity* 72: 109–112.

Nielsen, R. 2005. Molecular signatures of natural selection. *Annu. Rev. Genet.* 39: 197–218.

Novy, A., Flory, S.L. & Hartman, J.M. 2013. Evidence for rapid evolution of phenology in an invasive grass. *J. Evol. Biol.* 26: 443–450.

Nuismer, S.L. & Gandon, S. 2008. Moving beyond common-garden and transplant designs: insight into the causes of local adaptation to species interactions. *Am. Nat.* 171: 658–668.

Núñez-Farfán, J. & Dirzo, R. 1994. Evolutionary ecology of *Datura stramonium* L. in central Mexico: natural selection for resistance to herbivorous insects. *Evolution* 48: 423–436.

Núñez-Farfán, J., Fornoni, J. & Valverde, P.L. 2007. The evolution of resistance and tolerance to herbivores. *Annu. Rev. Ecol. Evol. Syst.* 38: 541–566.

Núñez-Farfán, J. & Schlichting, C.D. 2001. Evolution in changing environments: the "synthetic" work of Clausen, Keck, and Hiesey. *Q. Rev. Biol.* 76: 433–457.

Núñez-Farfán, J. & Schlichting, C.D. 2005. Natural selection in *Potentilla glandulosa* revisited. *Evol. Ecol. Res.* 7: 105–119.

Nussey, D.H., Wilson, A.J. & Brommer, J.E. 2007. The evolutionary ecology of individual phenotypic plasticity in wild populations. *J. Evol. Biol.* 20: 831–844.

Nyberg Berglund, A.B., Dahlgren, S. & Westerbergh, A. 2004. Evidence for parallel evolution and site-specific selection of serpentine tolerance in *Cerastium alpinum* during the colonization of Scandinavia. *New Phytol.* 161: 199–209.

Nybom, H. 2004. Comparison of different nuclear DNA markers for estimating intraspecific genetic diversity in plants. *Mol. Ecol.* 13: 1143–1155.

Nybom, H. & Bartish, I.V. 2000. Effects of life history traits and sampling strategies on genetic diversity estimates obtained with RAPD markers in plants. *Persp. Plant Ecol. Evol. Syst*, 3/2: 93–114.

O'Brien, E.K., Mazanec, R.A. & Krauss, S.L. 2007. Provenance variation of ecologically important traits of forest tress: implications for restoration. *J. Appl. Ecol.* 44: 583–593.

O'Dell, R.E. & Rajakaruna, N. 2011. Intraspecific variation, adaptation and evolution. In: *Serpentine: The Evolution and Ecology of a Model System*, ed. S. Harrison & N. Rajakaruna, 97–137. Berkeley, CA: University of California Press.

Oetjen, K., Ferber, S., Dankert, I. & Reusch, T.B.H. 2010. New evidence for habitat-specific selection to Wadden Sea *Zostera marina* populations revealed by genome scanning using SNP and microsatellite markers. *Mar. Biol.*157: 81–89.

Ollerton, J., Winfree, R. & Tarrant, S. 2011. How many flowering plants are pollinated by animals? *Oikos* 120: 321–326.

O'Neil, P. 1997. Natural selection on genetically correlated phenological characters in *Lythrum salicaria* L. (Lythraceae). *Evolution* 51: 267–274.

Oostermeijer, J.G.B., Luijten, S.H. & den Nijs, J.C.M. 2003. Integrating demographic and genetic approaches in plant conservation. *Biol. Conserv.* 113: 389–398.

Orsini, L., Schwenk, K., de Meester, L., Colbourne, J.K., Pfrender, M.E. & Weider, L.J. 2013. The evolutionary time machine: using dormant propagules to forecast how populations can adapt to changing environments. *Trends Ecol. Evol.* 28: 274–282.

Ortegón-Campos, I., Abdala-Roberts, L., Parra-Tabla, P., Cervera, J.C., Marrufo-Zapata, D. & Herrera, C.M. 2012. Influence of multiple factors on plant local adaptation: soil type and folivore effects in *Ruellia nudiflora* (Acanthaceae). *Evol. Ecol.* 26: 545–558.

Ortegón-Campos, I., Parra-Tabla, P., Abdala-Roberts, L. & Herrera, C.M. 2009. Local adaptation of *Ruellia nudiflora* (Acanthaceae) to biotic counterparts: complex scenarios revealed when two herbivore guilds are considered. *J. Evol. Biol.* 22: 2288–2297.

Ostfeld, R.S., Keesing, F. & Eviner, U.T., eds. 2008. *Infectious Disease Ecology: The Effects of Ecosystems on Disease and of Disease in Ecosystems.* Princeton, NJ: Princeton University Press.

Ouborg, N.J., Piquot, Y. & van Groenendael, J.M. 1999. Population genetics, molecular markers and the study of dispersal in plants. *J. Ecol.* 87: 551–568.

Ouborg, N.J. & Vriezen, W.H. 2007. An ecologist's guide to ecogenomics. *J. Ecol.* 95: 8–16.

Oyarzabal, M., Paruelo, J.M., del Pino, F., Oesterheld, M. & Lauenroth, W.K. 2008. Trait differences between grass species along a climatic gradient in South and North America. *J. Veget. Sci.* 19: 183–192.

Paige, K.N. 1992. Overcompensation in response to mammalian herbivory: from mutualistic to antagonistic interactions. *Ecology* 73: 2076–2085.

Palo, R.T. & Robbins, C.T., eds. 1991. *Plant Defenses against Mammalian Herbivory.* Boca Raton, FL: CRC Press.

Pan, J.J. & Price, J.S. 2001. Fitness and evolution in clonal plants: the impact of clonal growth. *Evol. Ecol.* 15: 583–600.

Pannell, J.R. & Fields, P.D. 2014. Evolution in subdivided plant populations: concepts, recent advances and future directions. *New Phytol.* 201: 417–432.

Paracer, S. & Ahmadjian, V. 2000. *Symbiosis: An Introduction to Biological Associations.* 2nd ed. New York: Oxford University Press.

Parachnowitsch, A.L. & Caruso, C.M. 2008. Predispersal seed herbivores, not pollinators, exert selection on floral traits via female fitness. *Ecology* 89: 1802–1810.

Parachnowitsch, A.L. & Kessler, A. 2010. Pollinators exert natural selection on flower size and floral display in *Penstemon digitalis. New Phytol.* 188: 393–402.

Parker, M.A. 1985. Local population differentiation for compatibility in an annual legume and its host-specific fungal pathogen. *Evolution* 39: 713–723.

Parker, M.A. 1986. Individual variation in pathogen attack and differential reproductive success in the annual legume, *Amphicarpaea bracteata. Oecologia* 69: 253–259.

Parker, M.A. 1987. Pathogen impact on sexual vs. asexual reproductive success in *Arisaema triphyllum. Am. J. Bot.* 74: 1758–1763.

Parker, M.A. 1991. Local genetic differentiation for disease resistance in a selfing annual. *Biol. J. Linn. Soc.* 42: 337–349.

Parker, M.A. 1995. Plant fitness variation caused by different mutualist genotypes. *Ecology* 76: 1525–1535.

Parker, M.A. 1996. Cryptic species within *Amphicarpaea bracteata* (Leguminosae): evidence from isozymes, morphology, and pathogen specificity. *Can. J. Bot.* 74: 1640–1650.

Parker, M.A. 2012. Legumes select symbiosis island sequence variants in *Bradyrhizobium. Mol. Ecol.* 21: 1769–1778.

Parmesan, C. 2006. Ecological and evolutionary responses to recent climate change. *Annu. Rev. Ecol. Evol. Syst.* 17: 637–669.

Patiny, S., ed. 2012. *Evolution of Plant–Pollinator Relationships.* Cambridge: Cambridge University Press.

Paul, J., Vachon, N., Garroway, C.J. & Freeland, J.R. 2010. Molecular data provide strong evidence of natural hybridization between native and introduced lineages of *Phragmites australis* in North America. *Biol. Inv.* 12: 2967–2973.

Pauwels, M., Saumitou-Laprade, P., Holl, A.C., Petit, D. & Bonnin, I. 2005. Multiple origin of metallicolous populations of the pseudometallophyte *Arabidopsis halleri* (Brassicaceae) in central Europe: the cpDNA testimony. *Mol. Ecol.* 14: 4403–4414.

Pelletier, F., Garant, D. & Hendry, A.P. 2009. Introduction: eco-evolutionary dynamics. *Phil. Trans. R. Soc. B.* 364: 1483–1489.

Pellmyr, O. 2002. Pollination by animals. In: *Plant–Animal Interactions: An Evolutionary Approach*, ed. C.M. Herrera & O. Pellmyr, 157–184. Malden, MA: Blackwell Publishing.

Pennings, S.C., Selig, E.R., Houser, L.T. & Bertness, M.D. 2003. Geographic variation in positive and negative interactions among salt marsh plants. *Ecology* 84: 1527–1538.

Peterson, C.A., Fetcher, N., McGraw, J.B. & Bennington, C.C. 2012. Clinal variation in stomatal characteristics of an arctic sedge, *Eriophorum vaginatum* (Cyperaceae). *Am. J. Bot.* 99: 1562–1571.

Petit, R.J., Bahrman, N. & Baradat, PH. 1995. Comparison of genetic differentiation in maritime pine (*Pinus pinaster* Ait) estimated using isozyme, total protein and terpenic loci. *Heredity* 75: 382–389.

Petit, R.J., Duminil, J., Fineschi, S., et al. 2005. Comparative organization of chloroplast, mitochondrial and nuclear diversity in plant populations. *Mol. Ecol.* 14: 689–701.

Petit, C. & Thompson, J.D. 1998. Phenotypic selection and population differentiation in relation to habitat heterogeneity in *Arrhenatherum elatius* (Poaceae). *J. Ecol.* 86: 829–840.

Pfennig, D.W. & Kingsolver, J.G. 2009. Phenotypic selection. In: *Princeton Guide to Ecology*, ed. S.A. Levin, 101–108. Princeton, NJ: Princeton University Press.

Pianka, E.R. 2011. *Evolutionary Ecology*. 7th ed. (ebook repr., based on the 6th ed. from 2000) www.books.google.com/books?id=giFL5bonGhQC&printsec=frontcover&dq=evolutionary+ecology+Pianka&hl=en&sa=X&ei=MbOuVPfIJuHIsQSmiYH4DA&ved=0CB4Q6wEwAA#v=onepage&q=evolutionary%20ecology%20Pianka&f=false

Piculell, B.J., Hoeksema, J.D. & Thompson, J.N. 2008. Interactions of biotic and abiotic environmental factors in an ectomycorrhizal symbiosis, and the potential for selection mosaics. *BMC Biol.* 6: 23. doi: 10.1186/1741-7007-6-23.

Pigliucci, M. 2001. Phenotypic plasticity. In: *Evolutionary Ecology: Concepts and Case Studies*, ed. C.W. Fox, D.A. Roff & D.J. Fairbairn, 58–69. New York: Oxford University Press.

Pigliucci, M. & Kaplan, J. 2006. *Making Sense of Evolution: The Conceptual Foundations of Evolutionary Biology*. Chicago, IL: The University of Chicago Press.

Platenkamp, G.A.J. & Foin, T.C. 1990. Ecological and evolutionary importance of neighbors in the grass *Anthoxanthum odoratum*. *Oecologia* 83: 201–208.

Pluess, A.R. & Stöcklin, J. 2004. Genetic diversity and fitness in *Scabiosa columbaria* in the Swiss Jura in relation to population size. *Conserv. Genet.* 5: 145–156.

Poncet, B.N., Herrmann, D., Gugerli, F., et al. 2010. Tracking genes of ecological relevance using a genome scan in two independent regional population samples of *Arabis alpina*. *Mol. Ecol.* 19: 2896–2907.

Popay, A.J. 2009. Insect herbivory and defensive mutualisms between plants and fungi. In: *Defensive Mutualism in Microbial Symbiosis*, ed. J.F. White, Jr. & M.S. Torres, 347–366. Boca Raton, FL: CRC Press.

Pope, N., Fong, M., Boyd, R.S. & Rajakaruna, N. 2013. The role of elevation and soil chemistry in the distribution and ion accumulation of floral morphs of *Streptanthus polygaloides* Gray (Brassicaeae), a Californian nickel hyperaccumulator. *Plant Ecol. Div.* 7: 421–432.

Potvin, C. & Tousignant, D. 1996. Evolutionary consequences of simulated global change: genetic adaptation or adaptive phenotypic plasticity. *Oecologia* 108: 683–693.

Prasad, K.V.S.K., Song, B.-H., Olson-Manning, C., et al. 2012. A gain-of-function polymorphism controlling complex traits and fitness in nature. *Science* 337: 1081–1084.

Pregitzer, C.C., Bailey, J.K., Hart, S.C. & Schweitzer, J.A. 2010. Soils as agents of selection: feedbacks between plants and soils alter seedling survival and performance. *Evol. Ecol.* 24: 1045–1059.

Price, G.R. 1995. The nature of selection. *J. Theor. Biol.* 175: 389–396.

Price, P.W., Lewinsohn, T.M., Fernandes, G.W. & Benson, W.W., eds. 1991. *Plant–Animal Interactions: Evolutionary Ecology in Tropical and Temperate Regions*. New York: Wiley-Interscience.

Price, S.C., Shumaker, K.M., Kahler, A.L., Allard, R.W. & Hill, J.E. 1984. Estimates of differentiation obtained from enzyme polymorphisms and quantitative characters. *J. Hered.* 75: 141–142.

Primack, R.B. & Kang, H. 1989. Measuring fitness and natural selection in wild plant populations. *Annu. Rev. Ecol. Syst.* 20: 367–396.

Prince, S.D. & Carter, R.N. 1985. The geographical distribution of prickly lettuce (*Lactuca serriola*): III. Its performance in transplant sites beyond its distribution limit in Britain. *J. Ecol.* 73: 49–64.

Proffitt, C.E., Chiasson, R.L., Owens, A.B., Edwards, K.R. & Travis, S.E. 2005. *Spartina alterniflora* genotype influences facilitation and suppression of high marsh species colonizing an early successional salt marsh. *J. Ecol.* 93: 404–416.

Puerma, E. & Aguade, M. 2013. Polymorphism at genes involved in salt tolerance in *Arabidopsis thaliana* (Brassicaceae). *Am. J. Bot.* 100: 384–390.

Pujol, B. & Pannell, J.R. 2008. Reduced responses to selection after species range expansion. *Science* 321: 96.

Pujol, B., Wilson, A.J., Ross, R.I.C. & Pannell, J.R. 2008. Are Q_{ST}–F_{ST} comparisons for natural populations meaningful? *Mol. Ecol.* 17: 4782–4785.

Pulido, F. & Berthold, P. 2004. Microevolutionary responses to climate change. *Adv. Ecol. Res.* 35: 151–183.

Quinn, J.A. 1969. Variability among High Plains populations of *Panicum virgatum*. *Bull. Torrey Bot. Club* 96: 20–41.

Quinn, J.A. 1975. Variability among *Danthonia sericea* (Gramineae) populations in responses to substrate moisture levels. *Am. J. Bot.* 62: 884–891.

Raabová, J., Münzbergová, Z. & Fischer, M. 2007. Ecological rather than geographic or genetic distance affects local adaptation of the rare perennial herb, *Aster amellus*. *Biol. Conserv.* 139: 348–357.

Rae, A.M., Ferris, R., Tallis, M.J. & Taylor, G. 2006. Elucidating genomic regions determining enhanced leaf growth and delayed senescence in elevated CO_2. *Plant Cell Env.* 29: 1730–1741.

Rae, A.M., Tricker, P.J., Bunn, S.M. & Taylor, G. 2007. Adaptation of tree growth to elevated CO_2: quantitative trait loci for biomass in *Populus*. *New Phytol.* 175: 59–69.

Rafferty, N.E. 2013. Pollination ecology. In: *Oxford Bibliographies Online, Ecology Module*, ed. D. Gibson. New York: Oxford University Press. http://www.oxfordbibliographies.com/obo/page/ecology.

Rajakaruna, N. & Bohm, B.A. 1999. The edaphic factor and patterns of variation in *Lasthenia californica* (Asteraceae). *Am. J. Bot.* 86: 1576–1596.

Rajakaruna, N. & Boyd, R.S. 2014. Serpentine soils. In: *Oxford Bibliographies Online, Ecology Module*, ed. D. Gibson. New York: Oxford University Press. http://www.oxfordbibliographies.com/obo/page/ecology.

Rajakaruna, N., Bradfield, G.E., Bohm, B.A. & Whitton, J. 2003. Adaptive differentiation in response to water stress by edaphic races of *Lasthenia californica* (Asteraceae). *Int. J. Plant Sci.* 164: 371–376.

Ramírez-Valiente, J.A., Lorenzo, Z., Soto, A., Valladares, F., Gil, L. & Aranda, I. 2010. Natural selection on cork oak: allele frequency reveals divergent selection in cork oak populations along a temperature cline. *Evol. Ecol.* 24: 1031–1044.

Ramírez-Valiente, J.A., Alia, R. & Aranda, I. 2014. Geographical variation in growth form traits in *Quercus suber* and its relation to population evolutionary history. *Evol. Ecol.* 28: 55–68.

Rapson, G.L. & Wilson, J.B. 1988. Non-adaptation in *Agrostis capillaries* L. (Poaceae). *Funct. Ecol.* 2: 479–490.

Rathcke, B. 1983. Competition and facilitation among plants for pollinators. In: *Pollination Biology*, ed. L. Real, 305–330. Orlando, FL: Academic Press.

Rausher, M.D. 1992. The measurement of selection on quantitative traits: biases due to environmental covariances between traits and fitness. *Evolution* 46: 616–626.

Rausher, M.D. 2001. Co-evolution and plant resistance to natural enemies. *Nature* 411: 857–864.

Rausher, M.D. & Simms, E.L. 1989. The evolution of herbivory in *Ipomoea purpurea*: I. Attempts to detect selection. *Evolution* 43: 563–572.

Real, L.A., ed. 1996. Special feature: disease ecology. *Ecology* 77: 989–1042.

Reed, E.S. 1981. The lawfulness of natural selection. *Am. Nat.* 118: 61–71.

Reed, D.H. & Frankham, R. 2001. How closely correlated are molecular and quantitative measures of genetic variation? A meta-analysis. *Evolution* 55: 1095–1103.

Rehfeldt, G.E., Ying, C.C., Spittlehouse, D.L. & Hamilton, Jr., D.A. 1999. Genetic responses to climate in *Pinus contorta*: niche breadth, climate change, and reforestation. *Ecol. Monogr.* 69: 375–407.

Renoult, J.P., Valido, A., Jordano, P. & Schaefer, H.M. 2014. Adaptation of flower and fruit colours to multiple, distinct mutualists. *New Phytol.* 201: 678–686.

Reusch, T.B.H., Hukriede, W., Stam, W.T. & Olsen, J.L. 1999. Differentiating between clonal growth and limited gene flow using autocorrelation of microsatellites. *Heredity* 83: 120–126.

Reusch, T.B.H. & Wood, T.E. 2007. Molecular ecology of global change. *Mol. Ecol.* 16: 3973–3992.

Rey, P.J. & Alcántara, J.M. 2000. Recruitment dynamics of a fleshy-fruited plant (*Olea europaea*): connecting patterns of seed dispersal to seedling establishment. *J. Ecol.* 88: 622–633.

Reynolds, R.J., Dudash, M.R. & Fenster, C.B. 2010. Multiyear study of multivariate linear and nonlinear phenotypic selection on floral traits of hummingbird-pollinated *Silene virginica*. *Evolution* 64: 358–369.

Reznick, D.N. & Ghalambor, C.K. 2001. The population ecology of contemporary adaptations: what empirical studies reveal about the conditions that promote adaptive evolution. *Genetica* 112–113: 183–198.

Reznick, D.N. & Travis, J. 2001. Adaptation. In: *Evolutionary Ecology: Concepts and Case Studies*, ed. C.W. Fox, D.A. Roff & D.J. Fairbairn, 44–57. New York: Oxford University Press.

Ribeiro, M.M., Mariette, S., Vendramin, G.G., Szmidt, A.E., Plomion, C. & Kremer, A. 2002. Comparison of genetic diversity estimates within and among populations of maritime pine using chloroplast simple-sequence repeat and amplified fragment length polymorphism data. *Mol. Ecol.* 11: 869–877.

Ribeiro, M.M., Plomion, C., Petit, R., Vendramin, G.G. & Szmidt, A.E. 2001. Variation in chloroplast single-sequence repeats in Portuguese maritime pine (*Pinus pinaster* Ait). *Theor. Appl. Genet.* 102: 97–103.

Rice, E.L. 1984. *Allelopathy*. 2nd ed. Orlando, FL: Academic Press.

Rice, K.J. & Knapp, E.E. 2008. Effects of competition and life history stage on the expression of local adaptation in two native bunchgrasses. *Restor. Ecol.* 16: 12–23.

Rice, K.J. & Mack, R.N. 1991. Ecological genetics of *Bromus tectorum*: III. The demography of reciprocally sown populations. *Oecologia* 88: 91–101.

Rice, J. S., Pinkerton, B.W., Stringer, W.C. & Undersander, D.J. 1990. Seed production in tall fescue as affected by fungal endophyte. *Crop Sci.*30: 1303–1305.

Richards, A.J. & Fletcher, A. 2002. The effects of altitude, aspect, grazing and time on the proportion of cyanogenics in neighbouring populations of *Trifolium repens* L. (white clover). *Heredity* 88: 432–436.

Richardson, J.L., Urban, M.C., Bolnick, D.I. & Skelly, D.K. 2014. Microgeographic adaptation and the spatial scale of evolution. *Trends Ecol. Evol.* 29: 165–176.

Ridley, H.N. 1930. *The Dispersal of Plants throughout the World*. Ashford, UK: L. Reeve.

Rieseberg, L.H. & Carney, S.E. 1998. Plant hybridization. *New Phytol.* 140: 599–624.

Rieseberg, L.H., Kim, S.-C., Randell, R.A., Whitney, K.D., Gross, B.L., Lexer, C. & Clay, K. 2007. Hybridization and the colonization of novel habitats by annual sunflowers. *Genetica* 129: 149–165.

Roach, D.A. & Wulff, R.D. 1987. Maternal effects in plants. *Annu. Rev. Ecol. Syst.* 18: 209–235.

Rodríguez, A., Alquézar, B. & Peña, L. 2013. Fruit aromas in mature fleshy fruits as signals of readiness for predation and seed dispersal. *New Phytol.* 197: 36–48.

Rodriguez, R.J., White, J.F., Jr., Arnold, A.E. & Redman, R.S. 2009a. Fungal endophytes: diversity and functional roles. *New Phytol.* 182: 314–330.

Rodriguez, R.J., Woodward, C., Kim, Y.-O. & Redman, R.S. 2009b. Habitat-adapted symbiosis as a defense against abiotic and biotic stresses. In: *Defensive Mutualism in Microbial Symbiosis*, ed. J.F. White, Jr. & M.S. Torres, 335–346. Boca Raton, FL: CRC Press.

Rosenthal, G.A. & Berenbaum, M.R., eds. 1992. *Herbivores: Their Interactions with Secondary Plant Metabolites. Vol. II: Ecological and Evolutionary Processes.* San Diego, CA: Academic Press.

Roughgarden, J. 1979. *Theory of Population Genetics and Evolutionary Ecology: An Introduction.* (1996, repr.). Upper Saddle River, NJ: Prentice-Hall.

Roumet, C., Laurent, G., Canivenc, G. & Roy, J. 2002. Genotypic variation in the response of two perennial grass species to elevated carbon dioxide. *Oecologia* 133: 342–348.

Rout, M.E. & Southworth, D. 2013. The root microbiome influences scales from molecules to ecosystems: the unseen majority. *Am. J. Bot.* 100: 1689–1691.

Roy, B.A., Kirchner, J.W., Christian, C.E. & Rose, L.E. 2000. High disease incidence and apparent disease tolerance in a North American Great Basin plant community. *Evol. Ecol.* 14: 421–438.

Rozema, J., Bijwaard, P., Prast, G. & Bruekman, R. 1985. Ecophysiological adaptations of coastal halophytes from foredunes and salt marshes. *Vegetatio* 62: 499–521.

Rutter, M.T. & Fenster, C.B. 2007. Testing for adaptation to climate in *Arabidopsis thaliana*: a calibrated common garden approach. *Ann. Bot.* 99: 529–536.

Saari, S. & Faeth, S.H. 2012. Hybridization of *Neotyphodium* endophytes enhances competitive ability of the host grass. *New Phytol.* 195: 231–236.

Saari, S., Helander, M., Faeth, S.H. & Saikkonen, K. 2010. The effects of endophytes on seed production and seed predation of tall fescue and meadow fescue. *Microb. Ecol.* 60: 928–934.

Sachs, J.L., Gano, K.A., Hollowell, A.C. & Regus, J.U. 2013. The legume-rhizobium symbiosis. In: *Oxford Bibliographies Online, Ecology Module*, ed. D. Gibson. New York: Oxford University Press. http://www.oxfordbibliographies.com/obo/page/ecology.

Sahli, H.F., Conner, J.K., Shaw, F.H., Howe, S. & Lale, A. 2008. Adaptive differentiation of quantitative traits in the globally distributed weed, wild radish (*Raphanus raphanistrum*). *Genetics* 180: 945–955.

Saikkonen, K., Faeth, S.H., Helander, M. & Sullivan, T.J. 1998. Fungal endophytes: a continuum of interactions with host plants. *Annu. Rev. Ecol. Syst.* 29: 319–343.

Saikkonen, K., Lehtonen, P., Helander, M., Koricheva, J. & Faeth, S.H. 2006. Model systems in ecology: dissecting the endophyte–grass literature. *Trends Plant Sci.* 11: 428–433.

Saikkonen, K., Wäli, P.R. & Helander, M. 2010. Genetic compatibility determines endophyte–grass combinations. *PLoS One* 5: e11395.

Sakai, K. 1961. Competitive ability in plants: its inheritance and some related problems. *Symp. Soc. Exp. Biol.* 15: 245–263.

Saldaña, A., Lusk, C.H., Gonzáles, W.L. & Gianoli, E. 2007. Natural selection on ecophysiological traits of a fern species in a temperate rainforest. *Evol. Ecol.* 21: 651–662.

Salmela, M.J. 2014. Rethinking local adaptation: mind the environment! *For. Ecol. Manage.* 312: 271–281.

Salvador, L., Alía, R., Agúndez, D. & Gill, L. 2000. Genetic variation and migration pathways of maritime pine (*Pinus pinaster* Ait) in the Iberian peninsula. *Theor. Appl. Genet.* 100: 89–95.

Salvaudon, L., Héraudet, V. & Shykoff, J.A. 2005. Parasite–host fitness trade-offs change with parasite identity: genotype-specific interactions in a plant-pathogen system. *Evolution* 59: 2518–2524.

Sambatti, J.B.M. & Rice, K.J. 2006. Local adaptation, patterns of selection, and gene flow in the Californian serpentine sunflower (*Helianthus exilis*). *Evolution* 60: 696–710.

Samis, K.E. & Eckert, C.G. 2009. Ecological correlates of fitness across the northern geographic range limit of a Pacific Coast dune plant. *Ecology* 90: 3051–3061.

Sanders, I.R. 2002. Specificity in the arbuscular mycorrhizal symbiosis. In: *Mycorrhizal Ecology*, ed. M.G.A. van der Heijden & I.R. Sanders, 415–440. Berlin: Springer-Verlag.

Santamaria, L., Figuerola, J., Pilon, J.J., et al. 2003. Plant performance across latitude: the role of plasticity and local adaptation in an aquatic plant. *Ecology* 84: 2454–2461.

Sapir, Y. 2009. Pollination genetics: using molecular genetic [*sic*] underlying floral traits to study pollination ecology in an evolutionary context. *Israel J. Plant Sci.* 57: 141–149.

Sarukhán, J., Martínez-Ramos, M. & Piñero, D. 1984. The analysis of demographic variability at the individual level and its population consequences. In: *Perspectives on Plant Population Ecology*, ed. R. Dirzo & J. Sarukhán, 83–106. Sunderland, MA: Sinauer Associates.

Savolainen, O., Pyhäjärvi, T. & Knürr, T. 2007. Gene flow and local adaptation in trees. *Annu. Rev. Ecol. Evol. Syst.* 38: 595–619.

Scascitelli, M., Whitney, K.D., Randell, R.A., King, M., Buerkle, C.A. & Rieseberg, L.H. 2010. Genome scan of hybridizing sunflowers from Texas (*Helianthus annuus* and *H. debilis*) reveals asymmetric patterns of introgression and small islands of genomic differentiation. *Mol. Ecol.* 19: 521–541.

Schaal, B.A. 1980. Measurement of gene flow in *Lupinus texensis. Nature* 284: 450–451.

Schaefer, H.M. & Schaefer, V. 2007. The evolution of visual fruit signals: concepts and constraints. In: *Seed Dispersal: Theory and Its Application in a Changing World*, ed. A.J. Dennis, E.W. Schupp, R.A. Green & D.A. Westcott, 59–77. New York: CAB International.

Schat, H., Vooijs, R. & Kuiper, E. 1996. Identical major gene loci for heavy metal tolerances that have independently evolved in different local populations and subspecies of *Silene vulgaris. Evolution* 50: 1888–1895.

Schechter, S.P. & Bruns, T.D. 2013. A common garden test of host–symbiont specificity supports a dominant role for soil type in determining AMF assemblage structure in *Collinsia sparsiflora. PLoS One* 8: e55507.

Scheepens, J.F., Frei, E.S. & Stöcklin, J. 2010. Genotypic and environmental variation in specific leaf area in a widespread Alpine plant after transplantation to different altitudes. *Oecologia* 164: 141–150.

Scheepens, J.F. & Stöcklin, J. 2013. Flowering phenology and reproductive fitness along a mountain slope: maladaptive responses to transplantation to a warmer climate in *Campanula thyrsoides. Oecologia* 171: 679–691.

Scheiner, S.M., Mitchell, R.J. & Callahan, H. S. 2000. Using path analysis to measure natural selection. *J. Evol. Biol.* 13: 423–433.

Scheiner, S.M. & Willig, M.R. 2011. A general theory of ecology. In: *The Theory of Ecology*, ed. S.M. Scheiner & M.R. Willig, 3–18. Chicago, IL: University of Chicago Press.

Schemske, D.W. 1984. Population structure and local selection in *Impatiens pallida* (Balsaminaceae), a selfing annual. *Evolution* 38: 817–832.

Schemske, D.W. & Horvitz, C.C. 1988. Plant–animal interactions and fruit production in a neotropical herb: a path analysis. *Ecology* 69: 1128–1137.

Schiestl, F.P. & Johnson, S.D. 2013. Pollinator-mediated evolution of floral signals. *Trends Ecol. Evol.* 28: 307–315.

Schlichting, C.D. & Pigliucci, M. 1998. *Phenotypic Evolution: A Reaction Norm Perspective.* Sunderland, MA: Sinauer Associates, Inc.

Schluter, D. 1988. Estimating the form of natural selection on a quantitative trait. *Evolution* 42: 849–861.

Schmid, B. 1994. Effects of genetic diversity in experimental stands of *Solidago altissima*: evidence for the potential role of pathogens as selective agents in plant populations. *J. Ecol.* 82: 165–175.

Schmidt, K.P. & Levin, D.A. 1985. The comparative demography of reciprocally sown populations of *Phlox drummondii* Hook: I. Survivorships, fecundities, and finite rates of increase. *Evolution* 39: 396–404.

Schmit, J.P. & Mueller, G.M. 2007. An estimate of the lower limit of global fungal diversity. *Biodivers. Conserv.* 16: 99–111.

Schoen, D.J., Stewart, S.C., Lechowicz, M.J. & Bell, G. 1986. Partitioning the transplant site effect in reciprocal transplant experiments with *Impatiens capensis* and *Impatiens pallida. Oecologia* 70: 149–154.

Schoener, T.W. 2011. The newest synthesis: understanding the interplay of evolutionary and ecological dynamics. *Science* 331: 426–429.

Schulz, B. & Boyle, C. 2005. The endophytic continuum. *Mycol. Res.*109: 661–686.

Schupp, E.W. 2007. The suitability of a site for seed dispersal is context-dependent. In: *Seed Dispersal: Theory and Its Application in a Changing World*, ed. A.J. Dennis, E.W. Schupp, R.A. Green & D.A. Westcott, 445–462. New York: CAB International.

Schupp, E.W., Jordano, P. & Gómez, J.M. 2010. Seed dispersal effectiveness revisited: a conceptual review. *New Phytol.* 188: 333–353.

Schweitzer, J.A., Bailey, J.K., Fischer, D.G., LeRoy, C.J., Lonsdorf, E.V., Whitham, T.G. & Hart, S.C. 2008. Plant–soil–microorganism interactions: heritable relationship between plant genotype and associated soil microorganisms. *Ecology* 89: 773–781.

Schwenk, K. & Wagner, G.P. 2004. The relativism of constraints on phenotypic evolution. In: *Phenotypic Integration: Studying the Ecology and Evolution of Complex Phenotypes*, ed. M. Pigliucci & K. Preston, 390–408. Oxford: Oxford University Press.

Scott, J.W., Meyer, S.E., Merrill, K.R. & Anderson, V.J. 2010. Local population differentiation in *Bromus tectorum* L. in relation to habitat-specific selection regimes. *Evol. Ecol.* 24: 1061–1080.

Seckbach, J. & Dubinsky, Z., eds. 2011. *All Flesh Is Grass: Plant–Animal Interrelationships.* Dordrecht, The Netherlands: Springer Science + Business Media.

Seliskar, D.M. 1985. Effects of reciprocal transplanting between extremes of plant zones on morphometric plasticity of five plant species in an Oregon salt marsh. *Can. J. Bot.* 63: 2254–2262.

Service, P. 1984. Genotypic interactions in an aphid–host plant relationship: *Uroleucon rudbeckiae* and *Rudbeckia laciniata. Oecologia* 61: 271–276.

Shaw, R.G. 1986. Response to density in a wild population of the perennial herb *Salvia lyrata*: variation among families. *Evolution* 40: 492–505.

Shaw, A.J., ed. 1990. *Heavy Metal Tolerance in Plants: Evolutionary Aspects.* Boca Raton, FL: CRC Press.

Shaw, R.G. & Etterson, J.R. 2012. Rapid climate change and the rate of adaptation: insight from experimental quantitative genetics. *New Phytol.* 195: 752–765.

Shaw, R.G. & Platenkamp, G.A. 1993. Quantitative genetics of response to competitors in *Nemophila menziesii*: a greenhouse study. *Evolution* 47: 801–812.

Shaw, R.G., Platenkamp, G.A., Shaw, F.H. & Podolsky, R.H. 1995. Quantitative genetics of response to competitors in *Nemophila menziesii*: a field experiment. *Genetics* 139: 397–406.

Shefferson, R.P. & Roach, D.A. 2012. The triple helix of *Plantago lanceolata*: genetics and environment interact to determine population dynamics. *Ecology* 93: 793–802.

Sherrard, M. E. & Maherali, H. 2012. Local adaptation across a fertility gradient is influenced by soil biota in the invasive grass, *Bromus inermis. Evol. Ecol.* 26: 529–544.

Sherwin, W.B. & Moritz, C. 2000. Managing and monitoring genetic erosion. In: *Genetics, Demography and Viability of Fragmented Populations*, ed. A.G. Young & G.M. Clarke, 11–34. Cambridge: Cambridge University Press.

Shipley, B. 2000. *Cause and Correlation in Biology: A User's Guide to Path Analysis, Structural Equations and Causal Inference.* Cambridge: Cambridge University Press.

Shonle, I. & Bergelson, J. 2000. Evolutionary ecology of the tropane alkaloids of *Datura stramonium* L. (Solanaceae). *Evolution* 54: 778–788.

Sicard, D., Pennings, P.S., Grandclément, C., Acosta, J., Kaltz, O. & Shykoff, J.A. 2007. Specialization and local adaptation of a fungal parasite on two host plant species as revealed by two fitness traits. *Evolution* 61: 27–41.

Siepielski, A.M., DiBattista, J.D. & Carlson, S.M. 2009. It's about time: the temporal dynamics of phenotypic selection in the wild. *Ecol. Lett.* 12: 1261–1276.

Siepielski, A.M., Gotanda, K.M., Morrissey, M.B., Diamond, S.E., DiBattista, J.D. & Carlson, S.M. 2013. The spatial patterns of directional phenotypic selection. *Ecol. Let.* 16: 1382–1392.

Silvertown, J. 2001. Plants stand still, but their genes don't: non-trivial consequences of the obvious. In: *Integrating Ecology and Evolution in a Spatial Context*, ed. J. Silvertown & J. Antonovics, 3–20. Oxford: Blackwell Science.

Silvertown, J. & Charlesworth, D. 2001. *Introduction to Plant Population Biology*. Oxford: Blackwell Science.

Silvertown, J., Poulton, P., Johnston, E., Edwards, G., Heard, M. & Biss, P.M. 2006. The Park Grass Experiment 1856–2006: its contribution to ecology. *J. Ecol.* 94: 801–814.

Silvertown, J., Servaes, C., Biss, P. & Macleod, D. 2005. Reinforcement of reproductive isolation between adjacent populations in the Park Grass Experiment. *Heredity* 95: 198–205.

Simms, E.L. 1996. The evolutionary genetics of plant–pathogen systems. *Bioscience* 46: 136–145.

Simms, E.L. & Rausher, M.D. 1989. The evolution of resistance to herbivory in *Ipomoea purpurea*: II. Natural selection by insects and costs of resistance. *Evolution* 43: 573–585.

Simms, E.L. & Rausher, M.D. 1992. Uses of quantitative genetics for studying the evolution of plant resistance. In: *Plant Resistance to Herbivores and Pathogens: Ecology, Evolution and Genetics*, ed. R.S. Fritz & E.L. Simms, pp. 42–68. Chicago, IL: University of Chicago Press.

Siol, M., Wright, S.I. & Barrett, S.C. 2010. The population genomics of plant adaptation. *New Phytol.* 188: 313–332.

Slate, J. 2005. Quantitative trait locus mapping in natural populations: progress, caveats and future directions. *Mol. Ecol.* 14: 363–379.

Sletvold, N. & Ågren, J. 2010. Pollinator-mediated selection on floral display and spur length in the orchid *Gymnadenia conopsea*. *Int. J. Plant Sci.* 171: 999–1009.

Sletvold, N., Grindeland, J.M. & Ågren, J. 2010. Pollinator-mediated selection on floral display, spur length and flowering phenology in the deceptive orchid *Dactylorhiza lapponica*. *New Phytol.*188: 385–392.

Smith, C.C. 1975. The coevolution of plants and seed predators. In: *Coevolution of Animals and Plants*, ed. L.E. Gilbert & P.H. Raven, 53–77. Austin, TX: University of Texas Press.

Smith, H., Burke, T. & Seidler, R. 1993. Editorial. *Mol. Ecol.* 2: 1–2.

Smith, K.P. & Goodman, R.M. 1999. Host variation for interactions with beneficial plant-associated microbes. *Annu. Rev. Phytopathol.* 37: 473–491.

Smith, S.E. & Read, D.J. 1997. *Mycorrhizal Symbiosis*. 2nd ed. San Diego, CA: Academic Press.

Smith, D.S., Schweitzer, J.A., Turk, P., et al. 2012. Soil-mediated local adaptation alters seedling survival and performance. *Plant Soil* 352: 243–251.

Smith, T.M. & Smith, R.L. 2012. *Elements of Ecology*. 8th ed. Boston, MA: Pearson/Benjamin Cummings.

Snaydon, R.W. 1970. Rapid population differentiation in a mosaic environment: I. The response of *Anthoxanthum odoratum* populations to soils. *Evolution* 24: 257–269.

Snaydon, R.W. & Davies, M.S. 1976. Rapid population differentiation in a mosaic environment: IV. Populations of *Anthoxanthum odoratum* at sharp boundaries. *Heredity* 37: 9–25.

Snaydon, R.W. & Davies, M.S. 1982. Rapid divergence of plant populations in response to recent changes in soil conditions. *Evolution* 36: 289–297.

Snow, A.A., Moran-Palma, P., Rieseberg, LH., Wszelaki, A. & Seiler, G.J. 1998. Fecundity, phenology, and seed dormancy of F_1 wild-crop hybrids in sunflower (*Helianthus annuus*, Asteraceae). *Am. J. Bot.* 85: 794–801.

Snow, A.A., Uthus, K.L. & Culley, T.M. 2001. Fitness of hybrids between weedy and cultivated radish: implications for weed evolution. *Ecol. Appl.* 11: 934–943.

Sobral, M., Larrinaga, A.R. & Guitián, J. 2010. Do seed-dispersing birds exert selection on optimal trait combinations? Correlated phenotypic selection on the fruit and seed size of hawthorn (*Crategus monogyna*). *Evol. Ecol.* 24: 1277–1290.

Solbrig, O.T. 1980. Demography and natural selection. In: *Demography and Evolution in Plant Populations*, ed. O.T. Solbrig, 1–20. Berkeley, CA: University of California Press.

Solbrig, O.T. & Simpson, B.B. 1974. Components of regulation of a population of dandelions in Michigan. *J. Ecol.* 62: 473–486.

Solbrig, O.T. & Simpson, B.B. 1977. A garden experiment of competition between biotypes of the common dandelion (*Taraxacum officinale*). *J. Ecol.* 65: 427–430.

Sorensen, A.E. 1986. Seed dispersal by adhesion. *Annu. Rev. Ecol. Syst.* 17: 443–463.

Sork, V.L., Stowe, K.A. & Hochwender, C. 1993. Evidence for local adaptation in closely adjacent subpopulations of northern red oak (*Quercus rubra* L.) expressed as resistance to leaf herbivores. *Am. Nat.* 142: 928–936.

Spoerke, J.M., Wilkinson, H.H. & Parker, M.A. 1996. Nonrandom genotypic associations in a legume-*Bradyrhizobium* mutualism. *Evolution* 50: 146–154.

Springer, Y.P. 2007. Clinal resistance structure and pathogen local adaptation in a serpentine flax–flax rust interaction. *Evolution* 61: 1812–1822.

Springer, C.J., Orozco, R.A., Kelly, J.K. & Ward, J.K. 2008. Elevated CO_2 influences the expression of floral-initiation genes in *Arabidopsis thaliana*. *New Phytol.* 178: 63–67.

Springer, C.J. & Ward, J.K. 2007. Flowering time and elevated atmospheric CO_2. *New Phytol.* 176: 243–255.

Stanton, M.L. & Galen, C. 1997. Life on the edge: adaptation versus environmentally mediated gene flow in the snow buttercup, *Ranunculus adoneus*. *Am. Nat.* 150: 143–178.

Stanton, M.L., Roy, B.A. & Thiede, D.A. 2000. Evolution in stressful environments: I. Phenotypic variability, phenotypic selection, and response to selection in five distinct environmental stresses. *Evolution* 54: 93–111.

Stanton, M.L., Thiede, D.A. & Roy, B.A. 2004. Consequences of intraspecific competition and environmental variation for selection in the mustard *Sinapsis arvensis*: contrasting ecological and evolutionary perspectives. *Am. Nat.* 164: 736–752.

Stearns, S.C. 1992. *The Evolution of Life Histories*. Oxford: Oxford University Press.

Stebbins, G.L., Jr. 1950. *Variation and Evolution in Plants*. New York: Columbia University Press.

Steinger, T., Haldimann, P., Leiss, K.A. &Müller-Schärer, H. 2002. Does natural selection promote population divergence? A comparative analysis of population structure using amplified fragment length polymorphism markers and quantitative traits. *Mol. Ecol.* 11: 2583–2590.

Steinger, T., Stephan, A. & Schmid, B. 2007. Predicting adaptive evolution under elevated atmospheric CO_2 in the perennial grass *Bromus erectus*. *Global Change Biol.* 13: 1028–1039.

Stern, J.T., Jr. 1970. The meaning of "adaptation" and its relation to the phenomenon of natural selection. *Evol. Biol.* 4: 39–66.

Stilwell, K.L., Wilbur, H.M., Werth, C.R. & Taylor, D.R. 2003. Heterozygote advantage in the American chestnut, *Castanea dentata* (Fagaceae). *Am. J. Bot.* 90: 207–213.

Stinchcombe, J.R. & Hoekstra, H.E. 2008. Combining population genomics and quantitative genetics: finding the genes underlying ecologically important traits. *Heredity* 100: 158–170.

Stinchcombe, J.R., Weinig, C., Ungerer, M., et al. 2004. A latitudinal cline in flowering time in *Arabidopsis thaliana* modulated by the flowering time gene FRIGIDA. *Proc. Nat. Acad. Sci. U S A* 101: 4712–4717.

Storz, J.F. 2005. Using genome scans of DNA polymorphism to infer adaptive population divergence. *Mol. Ecol.* 14: 671–688.

Stowe, K.A. 1998. Experimental evolution of resistance in *Brassica rapa*: correlated response of tolerance in lines selected for glucosinolate content. *Evolution* 52: 703–712.

Stowe, K.A., Marquis, R.J., Hochwender, C.G. & Simms, E.L. 2000. The evolutionary ecology of tolerance to consumer damage. *Annu. Rev. Ecol. Syst.* 31: 565–595.

Strasburg, J.L., Sherman, N.A., Wright, K.M., Moyle, L.C., Willis, J.H. & Rieseberg, L.H. 2012. What can patterns of differentiation across plant genomes tell us about adaptation and speciation? *Phil. Trans. R. Soc. B* 367: 364–373.

Stratton, D.A. 1992a. Life cycle components of selection in *Erigeron annuus*: I. Phenotypic selection. *Evolution* 46: 92–106.

Stratton, D.A. 1992b. Life cycle components of selection in *Erigeron annuus*: II. Genetic variation. *Evolution* 46: 107–120.

Stratton, D.A. 1994. Genotype-by-environment interactions for fitness of *Erigeron annuus* show fine-scale selective heterogeneity. *Evolution* 48: 1607–1618.

Strauss, S.Y. 1990. The role of plant genotype, environment and gender in resistance to a specialist chrysomelid herbivore. *Oecologia* 84: 111–116.

Strauss, S.Y. & Agrawal, A.A. 1999. The ecology and evolution of plant tolerance to herbivory. *Trends Ecol. Evol.* 14: 179–185.

Strauss, S.Y. & Irwin, R.E. 2004. Ecological and evolutionary consequences of multispecies plant–animal interactions. *Annu. Rev. Ecol. Evol. Syst.* 35: 435–466.

Strauss, S.Y. & Whittall, J.B. 2006. Non-pollinator agents of selection on floral traits. In: *Ecology and Evolution of Flowers*, ed. L.D. Harder & S.C.H. Barrett, 120–138. New York: Oxford University Press.

Strauss, S.Y. & Zangerl, A.R. 2002. Plant–insect interactions in terrestrial ecosystems. In: *Plant–Animal Interactions: An Evolutionary Approach*, ed. C.M. Herrera & O. Pellmyr, 77–106. Malden, MA: Blackwell Publishing.

Stuber, C.W. 1989. Molecular markers in the manipulation of quantitative characters. In: *Plant Population Genetics, Breeding, and Genetic Resources*, ed. A.H.D. Brown, M.T. Clegg, A.L. Kahler & B.S. Weir, 334–350. Sunderland, MA: Sinauer Associates.

Stuefer, J.F., Anten, N.P.R., de Kroon, H., et al. 2009. Genotypic selection shapes patterns of within-species diversity in experimental plant populations. *J. Ecol.* 97: 1020–1027.

Sullivan, T.J. & Faeth, S.H. 2008. Local adaptation in *Festuca arizonica* infected by hybrid and nonhybrid *Neotyphodium* endophytes. *Microb. Ecol.*55: 697–704.

Sultan, S.E. 1987. Evolutionary implications of phenotypic plasticity in plants. *Evol. Biol.* 21: 127–178.

Sun, M. 1999. Cleistogamy in *Scutellaria indica* (Labiatae): effective mating system and population genetic structure. *Mol. Ecol.* 8: 1285–1295.

Sun, M. & Wong, K.C. 2001. Genetic structure of three orchid species with contrasting breeding systems using RAPD and allozyme markers. *Am. J. Bot.* 88: 2180–2188.

Sutherland, W.J., Freckleton, R.P., Godfray, H.C.J., et al. 2013. Identification of 100 fundamental ecological questions. *J. Ecol.* 101: 58–67.

Sykes, M.T. & Wilson, J.B. 1988. An experimental investigation into the response of some New Zealand sand dune species to salt spray. *Ann. Bot.* 62: 159–166.

Tack, A.J.M., Horns, F. & Laine, A.-L. 2014. The impact of spatial scale and habitat configuration on patterns of trait variation and local adaptation in a wild plant parasite. *Evolution* 68: 176–189.

Taiz, L. & Zeiger, E. 2010. *Plant Physiology.* 5th ed. Sunderland, MA: Sinauer Associates.

Talluto, M.V. & Benkman, C.W. 2013. Landscape-scale eco-evolutionary dynamics: selection by seed predators and fire determine a major reproductive strategy. *Ecology* 94: 1307–1316.

Tan, K.H. 2009. *Environmental Soil Science.* 3rd ed. New York: CRC Press.

Taylor, D.R. & Aarssen, L.W. 1990. Complex competitive relationships among genotypes of three perennial grasses: implications for species coexistence. *Am. Nat.* 136: 305–327.

Taylor, G., Street, N.R., Tricker, P.J., et al. 2005. The transcriptome of *Populus* in elevated CO_2. *New Phytol.* 167: 143–154.

Temeles, E.J., Rah, Y.J., Andicoechea, J., et al. 2013. Pollinator-mediated selection in a specialized hummingbird–*Heliconia* system in the eastern Caribbean. *J. Evol. Biol.* 26: 347–356.

Thomas, S.C. & Bazzaz, F.A. 1993. The genetic component in plant size hierarchies: norms of reaction to density in a *Polygonum* species. *Ecol. Monogr.* 63: 231–249.

Thompson, J.N. 2001. The geographic dynamics of coevolution. In: *Evolutionary Ecology: Concepts and Case Studies*, ed. C.W. Fox, D.A. Roff & D.J. Fairbairn, 331–343. New York: Oxford University Press.

Thompson, J.N. 2002. Plant–animal interactions: future directions. In: *Plant–Animal Interactions: An Evolutionary Approach*, ed. C.M. Herrera & O. Pellmyr, 236–247. Malden, MA: Blackwell Publishing.

Thompson, J.N. 2005. *The Geographic Mosaic of Coevolution.* Chicago, IL: University of Chicago Press.

Thompson, J.N. 2009. The coevolving web of life. *Am. Nat.* 173: 125–140.

Thompson, J.N. & Cunningham, B.M. 2002. Geographic structure and dynamics of coevolutionary selection. *Nature* 417: 735–738.

Thorpe, A.S., Aschehoug, E.T., Atwater, D.Z. & Callaway, R.M. 2011. Interactions among plants and evolution. *J. Ecol.* 99: 729–740.

Thrall, P.H., Burdon, J.J. & Bever, J.D. 2002. Local adaptation in the *Linum marginale–Melampsora lini* host–pathogen interaction. *Evolution* 56: 1340–1351.

Thrall, P.H., Hochberg, M.E., Burdon, J.J. & Bever, J.D. 2007. Coevolution of symbiotic mutualists and parasites in a community context. *Trends Ecol. Evol.* 22: 120–126.

Tilman, D. 1988. *Plant Strategies and the Dynamics and Structure of Plant Communities.* Princeton, NJ: Princeton University Press.

Tonsor, S.J. 1989. Relatedness and intraspecific competition in *Plantago lanceolata*. *Am. Nat.* 134: 897–906.

Torres, E., Iriondo, J.M., Escudero, A. & Pérez, C. 2003. Analysis of within-population spatial genetic structure in *Antirrhinum microphyllum* (Scrophulariaceae). *Am. J. Bot.* 90: 1688–1695.

Torres, M.S., White, J.W., Jr., Zhang, X., Hinton, D.M. & Bacon, C.W. 2012. Endophyte-mediated adjustments in host morphology and physiology and effects on host fitness traits in grasses. *Fungal Ecol.* 5: 322–330.

Trapnell, D.W., Hamrick, J.L., Parker, K.C., Braungart, K.W. & Glenn, T.C. 2011. Evaluating the utility of microsatellites for investigations of autopolyploid taxa. *J. Hered.* 102: 473–478.

Travers, S.E., Tang, Z., Caragea, D., et al. 2010. Variation in gene expression of *Andropogon gerardii* in response to altered environmental conditions associated with climate change. *J. Ecol.* 98: 374–383.

Traveset, A., Robertson, A.W. & Rodríquez-Pérez, J. 2007. A review on the role of endozoochory in seed germination. In: *Seed Dispersal: Theory and Its Application in a*

Changing World, ed. A.J. Dennis, E.W. Schupp, R.A. Green & D.A. Westcott, 78–103. New York: CAB International.

Travis, J. & Reznick, D.N. 1998. Experimental approaches to the study of evolution. In: *Experimental Ecology: Issues and Perspectives*, ed. W.J. Resetarits, Jr. & J. Bernardo, 437–459. New York: Oxford University Press.

Turesson, G. 1922. The genotypical response of the plant species to the habitat. *Hereditas* 3: 211–350.

Turesson, G. 1925. The plant species in relation to habitat and climate: contributions to the knowledge of genecological units. *Hereditas* 6: 147–236.

Turesson, G. 1930. The selective effect of climate upon the plant species. *Hereditas* 14: 99–152.

Turkington, R. 1989. The growth, distribution and neighbour relationships of *Trifolium repens* in a permanent pasture: V. The coevolution of competitors. *J. Ecol.* 77: 717–733.

Turkington, R. 1996. Intergenotypic competition in plant mixtures. *Euphytica* 92: 105–119.

Turkington, R. & Aarssen, L.W. 1984. Local-scale differentiation as a result of competitive interactions. In: *Perspectives on Plant Population Ecology*, ed. R. Dirzo & J. Sarukhán, 107–127. Sunderland, MA: Sinauer Associates, Inc.

Turkington, R. & Harper, J.L. 1979. The growth, distribution and neighbour relationships of *Trifolium repens* in a permanent pasture: IV. Fine-scale biotic differentiation. *J. Ecol.* 67: 245–254.

Turkington, R. & Mehrhoff, L.A. 1990. The role of competition in structuring pasture communities. In: *Perspectives on Plant Competition*, ed. J.B. Grace & D. Tilman, 307–340. San Diego, CA: Academic Press.

Turner, T.L., Bourne, E.C., von Wettberg, E.J. Hu, T.T. & Nuzhdin, S.V. 2010. Population resequencing reveals local adaptation of *Arabidopsis lyrata* to serpentine soils. *Nat. Genet.* 42: 260–263.

Underwood, A.J. 1997. *Experiments in Ecology*. Cambridge: Cambridge University Press.

Underwood, N. & Halpern, S.L. 2012. Insect herbivores, density dependence, and the performance of the perennial herb *Solanum carolinense*. *Ecology* 93: 1026–1035.

Urban, M.C. 2011. The evolution of species interactions across natural landscapes. *Ecol. Lett.* 14: 723–732.

van der Pijl, L. 1982. *Principles of Dispersal in Higher Plants*. 3rd ed. Berlin: Springer-Verlag.

vander Wall, S.B. 2001. The evolutionary ecology of nut dispersal. *Bot. Rev.* 67: 74–117.

van Dijk, H. & Hautekèete, N. 2007. Long day plants and the response to global warming: rapid evolutionary change in day length sensitivity is possible in wild beet. *J. Evol. Biol.* 20: 349–357.

van Kleunen, M. & Fischer, M. 2003. Effects of four generations of density-dependent selection on life history traits and their plasticity in a clonally propagated plant. *J. Evol. Biol.* 16: 474–484.

van Kleunen, M., Fischer, M. & Schmid, B. 2005. Three generations under low versus high neighborhood density affect the life history of a clonal plant through differential selection and genetic drift. *Oikos* 108: 573–581.

van Mölken, T. & Stuefer, J.F. 2011. The potential of plant viruses to promote genotypic diversity via genotype × environment interactions. *Ann. Bot.* 107: 1391–1397.

van Tienderen, P.H. & van der Toorn, J. 1991a. Genetic differentiation between populations of *Plantago lanceolata*: I. Local adaptation in three contrasting habitats. *J. Ecol.* 79: 27–42.

van Tienderen, P.H. & van der Toorn, J. 1991b. Genetic differentiation between populations of *Plantago lanceolata*: II. Phenotypic selection in a transplant experiment in three contrasting habitats. *J. Ecol.* 79: 43–59.

van Zijll de Jong, E., Dobrowolski, M.P., Bannan, N.R., et al. 2008. Global genetic diversity of the perennial ryegrass fungal endophyte *Neotyphodium lolii*. *Crop Sci.* 48: 1487–1501.

Vavrek, M.C. 1998. Within-population genetic diversity of *Taraxacum officinale* (Asteraceae): differential genotype response and effect on interspecific competition. *Am. J. Bot.* 85: 947–954.

Vekemans, X. & Hardy, O.J. 2004. New insights from fine-scale spatial genetic structure analyses in plant populations. *Mol. Ecol.* 13: 921–935.

Vekemans, X. & Lefebvre, C. 1997. On the evolution of heavy-metal tolerant populations in *Armeria maritima*: evidence from allozyme variation and reproductive barriers. *J. Evol. Biol.* 10: 175–191.

Vendramin, G.G., Anzidei, M., Madaghiele, A. & Bucci, G. 1998. Distribution of genetic diversity in *Pinus pinaster* Ait as revealed by chloroplast microsatellites. *Theor. Appl. Genet.* 97: 456–463.

Verhoeven, K.J.F., Jansen, J.J., van Dijk, P.J. & Biere, A. 2010. Stress-induced DNA methylation changes and their heritability in asexual dandelions. *New Phytol.* 185: 1108–1118.

Verhoeven, K.J.F., Poorter, H., Nevo, E. & Biere, A. 2008. Habitat-specific natural selection at a flowering-time QTL is a main driver of local adaptation in two wild barley populations. *Mol. Ecol.* 17: 3416–3424.

Verhoeven, K.J.F., Vanhala, T.K., Biere, A., Nevo, E. & van Damme, J.M.M. 2004. The genetic basis of adaptive population differentiation: a quantitative trait locus analysis of fitness traits in two wild barley populations from contrasting habitats. *Evolution* 58: 270–283.

Violle, C., Navas, M.-L., Vile, D., et al. 2007. Let the concept of trait be functional! *Oikos* 116: 882–892.

Volk, M. & Körner, C. 2001. Genotype × elevated CO_2 interaction and allocation in calcareous grassland species. *New Phytol.* 151: 637–645.

von Wettberg, E.J. & Wright, J. 2011. Genomic approaches to understanding adaptation. In: *Serpentine: The Evolution and Ecology of a Model System*, ed. S. Harrison & N. Rajakaruna, 138–153. Berkeley, CA: University of California Press.

Vos, P., Hogers, R., Bleeker, M. et al. 1995. AFLP: a new technique for DNA fingerprinting. *Nucl. Acids Res.* 23: 4407–4414.

Wade, M.J. & Kalisz, S. 1990. The causes of natural selection. *Evolution* 44: 1947–1955.

Walley, K.A., Khan, M.S.I. & Bradshaw, A.D. 1974. The potential for evolution of heavy metal tolerance in plants: I. Copper and zinc tolerance in *Agrostis tenuis*. *Heredity* 32: 309–319.

Walters, D.R. 2011. *Plant Defense: Warding off Attack by Pathogens, Herbivores, and Parasitic Plants.* Oxford: Blackwell Publishing.

Wang, H., McArthur, E.D., Sanderson, S.C., Graham, J.H. & Freeman, D.C. 1997. Narrow hybrid zone between two subspecies of big sagebrush (*Artemisia tridentate*: Asteraceae): IV. Reciprocal transplant experiments. *Evolution* 51: 95–102.

Wang, X.Y. & Redmann, R.E. 1996. Adaptation to salinity in *Hordeum jubatum* L. populations studied using reciprocal transplants. *Vegetatio* 123: 65–71.

Wang, P., Stieglitz, T., Zhou, D.W. & Cahill, J.F.,Jr. 2010. Are competitive effect and response two sides of the same coin, or fundamentally different? *Funct. Ecol.* 24: 196–207.

Waples, R.S. & Waples, R.K. 2011. Inbreeding effective population size and parentage analysis without parents. *Mol. Ecol. Res.* 11: 162–171.

Ward, J.K., Antonovics, J., Thomas, R.B. & Strain, B.R. 2000. Is atmospheric CO_2 a selective agent on model C_3 annuals? *Oecologia* 123: 330–341.

Ward, J.K. & Kelly, J.K. 2004. Scaling up evolutionary responses to elevated CO_2: lessons from *Arabidopsis*. *Ecol. Lett.* 7: 427–440.

Ward, J.K. & Strain, B.R. 1997. Effects of low and elevated CO_2 partial pressure on growth and reproduction of *Arabidopsis thaliana* from different elevations. *Plant Cell Env.* 20: 254–260.

Warwick, S.I. & Briggs, D. 1978. The genecology of lawn weeds: I. Population differentiation in *Poa annua* L. in a mosaic environment of bowling green lawns and flower beds. *New Phytol.* 81: 711–723.

Warwick, S.I. & Briggs, D. 1980. The genecology of lawn weeds: V. The adaptive significance of different growth habit in lawn and roadside populations of *Plantago major* L. *New Phytol.* 85: 289–300.

Waser, N.M. & Ollerton, J., eds. 2006. *Plant–Pollinator Interactions: From Specialization to Generalization.* Chicago, IL: University of Chicago Press.

Waser, N.M. & Price, M.V. 1981. Pollinator choice and stabilizing selection for flower color in *Delphinium nelsonii. Evolution* 35: 376–390.

Waser, N.M. & Williams, C.F. 2001. Inbreeding and outbreeding. In: *Evolutionary Ecology: Concepts and Case Studies*, ed. C.W. Fox, D.A. Roff & D.J. Fairbairn, 84–97. New York: Oxford University Press.

Watling, J.R. & Press, M.C. 2000. Light heterogeneity in tropical rain forests: photosynthetic responses and their ecological consequences. In: *The Ecological Consequences of Environmental Heterogeneity*, ed. M.J. Hutchings, E.A. John & A.J.A. Stewart, 131–153. Oxford: Blackwell Science.

Weidenhamer, J.D. 2006. Distinguishing allelopathy from resource competition: the role of density. In: *Allelopathy: A Physiological Process with Ecological Implications*, ed. M.J. Reigosa, N. Pedrol & L. González, 85–103. Dordrecht, The Netherlands: Springer.

Weigel, D. 2012. Natural variation in *Arabidopsis*: from molecular genetics to ecological genomics. *Plant Physiol.* 158: 2–22.

Weigensberg, I. & Roff, D.A. 1996. Natural heritabilities: can they be reliably estimated in the laboratory? *Evolution* 50: 2149–2157.

Weiner, J. 1988. The influence of competition on plant reproduction. In: *Plant Reproductive Ecology*, ed. J. Lovett Doust & L. Lovett Doust, 228–245. New York: Oxford University Press.

Weiner, J., Martinez, S., Müller-Schärer, H., Stoll, P. & Schmid, B. 1997. How important are environmental maternal effects in plants? A study with *Centaurea maculosa. J. Ecol.* 85: 133–142.

Weiner, J. & Thomas, S.C. 1986. Size variability and competition in plant monocultures. *Oikos* 47: 211–222.

Weinig, C., Ewers, B.E. & Welch, S.M. 2014. Ecological genomics and process modeling of local adaptation to climate. *Curr. Opin. Plant Biol.* 18: 66–72.

Weinig, C., Stinchcombe, J.R. & Schmitt, J. 2003a. Evolutionary genetics of resistance and tolerance to natural herbivory in *Arabidopsis thaliana. Evolution* 57: 1270–1280.

Weinig, C., Stinchcombe, J.R. & Schmitt, J. 2003b. QTL architecture of resistance and tolerance traits in *Arabidopsis thaliana* in natural environments. *Mol. Ecol.* 12: 1153–1163.

Weller, S.G., Suding, K. & Sakai, A.K. 2013. Botany and a changing world: introduction to the special issue on global biological change. *Am. J. Bot.* 100: 1229–1233.

Wenny, D.G. 2000. Seed dispersal, seed predation, and seedling recruitment of a neotropical montane tree. *Ecol. Monogr.* 70: 331–351.

West, C.P. 2007. Plant influences on endophyte expression. In: *Proceedings of the 6th International Symposium on Fungal Endophytes of Grasses*, ed. A. J. Popay & E. R. Thom, 117–121. Christchurch, New Zealand: New Zealand Grassland Association.

West-Eberhard, M.J. 2003. *Developmental Plasticity and Evolution.* New York: Oxford University Press.

Whitehead, A. & Crawford, D.L. 2006. Variation within and among species in gene expression: raw material for evolution. *Mol. Ecol.* 15: 1197–1211.

Whitlock, M.C. 2008. Evolutionary inference from Q_{ST}. *Mol. Ecol.* 17: 1885–1896.

Whitlock, M.C. & McCauley, D.E. 1999. Indirect measures of gene flow and migration: $F_{ST} \neq 1/(4Nm + 1)$. *Heredity* 82: 117–125.

Whittall, J.B. & Hodges, S.A. 2007. Pollinator shifts drive increasingly long nectar spurs in columbine flowers. *Nature* 447: 706–709.

Wieneke, S., Prati, D., Brandl, R., Stöcklin, J. & Auge, H. 2004. Genetic variation in *Sanguisorba minor* after 6 years *in situ* selection under elevated CO_2. *Global Change Biol.* 10: 1389–1401.

Wilkins, D.A. 1957. A technique for the measurement of lead tolerance in plants. *Nature* 180: 37–38.

Wilkins, D.A. 1978. The measurement of tolerance to edaphic factors by means of root growth. *New Phytol.*80: 623–633.

Wilkinson, H.H., Spoerke, J.M. & Parker, M.A. 1996. Divergence in symbiotic compatibility in a legume–*Bradyrhizobium* mutualism. *Evolution* 50: 1470–1477.

Willi, Y., van Buskirk, J., Schmid, B. & Fischer, M. 2006. Genetic isolation of fragmented populations is exacerbated by drift and selection. *J. Evol. Biol.* 20: 534–542.

Williams, C.F. 1994. Genetic consequences of seed dispersal in three sympatric forest herbs: II. Microspatial genetic structure within populations. *Evolution* 48: 1959–1972.

Williams, J.L., Auge, H. & Maron, J.L. 2008. Different gardens, different results: native and introduced populations exhibit contrasting phenotypes across common gardens. *Oecologia* 157: 239–248.

Williams, J.G.K., Kubelik, A.R., Livak, K.J., Rafalski, J.A. & Tingey, S.V. 1990. DNA polymorphisms amplified by arbitrary primers are useful as genetic markers. *Nucl. Acids Res.* 18: 6531–6535.

Willis, C.G., Brock, M.T. & Weinig, C. 2010. Genetic variation in tolerance of competition and neighbor suppression in *Arabidopsis thaliana. J. Evol. Biol.* 23: 1412–1424.

Willmore, K.E. & Hallgrimsson, B. 2005. Within individual variation: developmental noise versus developmental stability. In: *Variation: A Central Concept in Biology*, ed. B. Hallgrimsson & B.K. Hall, 191–218. Burlington, MA: Elsevier/Academic Press.

Willson, M.F., Thomas, P.A., Hoppes, W.G., Katusic-Malmborg, P.L., Goldman, D.A. & Bothwell, J.L. 1987. Sibling competition in plants: an experimental study. *Am. Nat.* 129: 304–311.

Willson, M.F. & Whelan, C.J. 1990. The evolution of fruit color in fleshy-fruited plants. *Am. Nat.* 136: 790–809.

Wilson, S.D. 2000. Heterogeneity, diversity and scale in plant communities. In: *The Ecological Consequences of Environmental Heterogeneity*, ed. M.J. Hutchings, E.A. John & A.J.A. Stewart, 53–69. Oxford: Blackwell Science.

Wise, M.J. 2007. Evolutionary ecology of resistance to herbivory: an investigation of potential genetic constraints in the multiple-herbivore community of *Solanum carolinense. New Phytol.* 175: 773–784.

Wise, M.J. & Rausher, M.D. 2013. Evolution of resistance to a multiple-herbivore community: genetic correlations, diffuse coevolution, and constraints on the plant's response to selection. *Evolution* 67: 1767–1779.

Wolf, J.B., Allen, C.E. & Frankino, W.A. 2004. Multivariate phenotypic evolution in developmental hyperspace. In: *Phenotypic Integration: Studying the Ecology and Evolution of Complex Phenotypes*, ed. M. Pigliucci & K. Preston, 366–389. Oxford: Oxford University Press.

Woodward, F.I. 1987. *Climate and Plant Distribution.* New York: Cambridge University Press.

Wright, S. 1921. Correlation and causation. *J. Agricult. Res.*10: 557–585.

Wright, S. 1934. The method of path coefficients. *Ann. Math. Stat.* 5: 161–215.

Wright, S. 1943. Isolation by distance. *Genetics* 28: 114–138.

Wright, S. 1951. The genetical structure of populations. *Ann. Eugenics* 15: 323–354.

Wright, J.W. 2007. Local adaptation to serpentine soils in *Pinus ponderosa. Plant and Soil* 293: 209–217.

Wright, S.I. & Gaut, B.S. 2005. Molecular population genetics and the search for adaptive evolution in plants. *Mol. Biol. Evol.* 22: 506–519.

Wright, J.W. & Stanton, M.L. 2011. Local adaptation in heterogeneous landscapes: reciprocal transplant experiments and beyond. In: *Serpentine: The Evolution and Ecology of a Model System*, ed. S. Harrison & N. Rajakaruna, 155–179. Berkeley, CA: University of California Press.

Wright, J.W., Stanton, M.L. & Scherson, R. 2006. Local adaptation to serpentine and non-serpentine soils in *Collinsia sparsiflora. Evol. Ecol. Res.* 8: 1–21.

Wu, L., Bradshaw, A.D. & Thurman, D.A. 1975. The potential for evolution of heavy metal tolerance in plants: III. The rapid evolution of copper tolerance in *Agrostis stolonifera. Heredity* 34: 165–187.

Wulff, R.D. 1995. Environmental maternal effects on seed quality and germination. In: *Seed Development and Germination*, ed. J. Kigel & G. Galili, 491–505. New York: Marcel Dekker.

Yam, T.W., Arditti, J. & Cameron, K.M. 2009. "The orchids have been a splendid sport": an alternative look at Charles Darwin's contribution to orchid biology. *Am. J. Bot.* 96: 2128–2154.

Yost, J.M., Barry, T., Kay, K.M. & Rajakaruna, N. 2012. Edaphic adaptation maintains the coexistence of two cryptic species on serpentine soils. *Am. J. Bot.* 99: 890–897.

Young, C.A., Aiken, G.E., McCulley, R.L., Strickland, J.R. & Schardl, C.L., eds. 2012. *Epichloae, Endophytes of Cool Season Grasses: Implications, Utilization and Biology.* Ardmore, OK: The Samuel Roberts Noble Foundation.

Zane, L., Bargelloni, L. & Patarnello, T. 2002. Strategies for microsatellite isolation: a review. *Mol. Ecol.* 11: 1–16.

Zera, A.J. & Brisson, J.A. 2012. Quantitative, physiological, and molecular genetics of dispersal/migration. In: *Dispersal Ecology and Evolution*, ed. J. Clobert, M. Baguette, T.G. Benton & J.M. Bullock, 63–82. Oxford: Oxford University Press.

Zhao, X., Liu, W. & Zhou, M. 2013. Lack of local adaptation of invasive croton weed (*Ageratina adenophora*) in different climatic areas of Yunnan Province, China. *J. Plant Ecol.* 6: 316–322.

Zietkiewicz, E., Rafalski, A. & Labuda, D. 1994. Genome fingerprinting by simple sequence repeats (SSR)-anchored PCR amplification. *Genomics* 20: 176–183.

Zimmermann, J.K. & Whigham, D.F. 1992. Ecological functions of carbohydrates stored in corms of *Tipularia discolor* (Orchidaceae). *Funct. Ecol.* 6: 575–581.

Züst, T., Heichinger, C., Grossniklaus, U., Harrington, R., Kliebenstein, D.J. & Turnbull, L.A. 2012. Natural enemies drive geographic variation in plant defenses. *Science* 338: 116–119.

Index

www.ingramcontent.com/pod-product-compliance
Ingram Content Group UK Ltd.
Pitfield, Milton Keynes, MK11 3LW, UK
UKHW020848150926
475595UK00003B/74

9 780199 988327